Chris Carter

FAC... A LEVEL

OXFORD
UNIVERSITY PRESS

OXFORD
UNIVERSITY PRESS

Great Clarendon Street, Oxford OX2 6DP

Oxford University Press is a department of the University of Oxford.
It furthers the University's objective of excellence in research, scholarship, and education by publishing worldwide in

Oxford New York

Athens Auckland Bangkok Bogotá Buenos Aires Calcutta Cape Town Chennai Dar es Salaam Delhi Florence Hong Kong Istanbul Karachi Kuala Lumpur Madrid Melbourne Mexico City Mumbai Nairobi Paris São Paulo Singapore Taipei Tokyo Toronto Warsaw

with associated companies in Berlin Ibadan

Oxford is a registered trade mark of Oxford University Press in the UK and in certain other countries

First published 2001

British Library Cataloguing in Publication Data

Data available

ISBN 0 19 914768 X

Typeset by Magnet Harlequin

Printed in Great Britain

Author's acknowledgements

I would like to thank the following for their help in producing this book: P. Baily, J. Highmore, S. Lukito, and Songpol Chuenkhum.

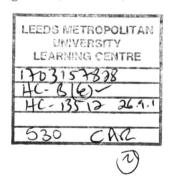

INTRODUCTION

With the change of A levels into the new AS/A2 system, the A-level specifications (syllabuses) are changing. I was asked to write a book to respond to these changes.

I asked my students what they wanted from a physics text book, and their replies are summed up in the following quotations:

> 'A book that you can read easily; something that uses straightforward words, explains things simply, and does not have pages of stuff in it that you do not need to know.'
>
> 'A book that shows you how to answer exam questions; something that gives you a lot of practice at questions, gives you space to do them, and has detailed answers so that you can check your work properly.'

I asked my fellow teachers what they would like in such a book, and their replies are best summarized by the following:

> 'A book that students find easy to understand, that is not too technical for them to read.'
>
> 'A book that has lots of examination-style examples, with well explained answers.'

So here it is. You will find this book different from most other physics books because it is designed to respond to the above requests. It concentrates upon explaining, as simply as possible, what is really important in a particular topic, and gives you a chance to practise applying the material to problem-solving situations. Half of the book is devoted to questions, which cover the range of different types and styles of question that are likely to come up in exams. Space is given in the book to do the questions and is laid out in a similar way to exam papers. Very detailed step-by-step answers are included, showing each stage of a calculation or explanation.

CHRIS CARTER

Chris Carter is currently a director of studies at d'Overbroeck's, an independent school in Oxford, and formerly head of physics at Newbury college of further education.

CONTENTS

Why study physics?		vi
What this book does		vi
How to use this book		vi
How to work and revise		vi
Exam technique		vii
The new A-level system		viii
Key skills		viii
Exam board specifications and modules		ix
Formulae to learn		x
1	Motion	2
2	More motion	6
3	Forces	10
4	Types of force	14
5	Energy	18
6	Forces in action	22
7	Conservation of momentum	26
8	Circular motion and statics	30
9	Electricity	34
10	Electric circuits	38
11	Waves	42
12	Wave phenomena	46
13	Wave–particle duality	50
14	Energy levels	54
15	Atomic and nuclear	58
16	Particle physics I	62
17	Particle physics II	66
18	Electrostatic and gravitational fields	70
19	Applied field theory	74
20	Capacitance	78
21	Magnetic field theory	82
22	Applied magnetism	86
23	Electromagnetic induction	90
24	Simple harmonic motion (SHM)	94
25	Thermal physics	98
26	Thermodynamics	102
27	Materials science and base units	106
28	Practical work	110
Answers		114
Index		

Why study physics?

Physics is the most fundamental of all the sciences, and is the basis of many scientific and technological ideas. The subject looks at how things work and the principles and laws that predict their behaviour. This ranges from the large scale of the Earth and Universe to the very small scale of atoms and subatomic particles. If you enjoy taking things apart, have an inquisitive mind, and like to know what is going on, you will find this subject interesting.

If you wish to embark upon a career in some form of engineering, physics is an essential subject to study at A level. It is also a particularly useful choice for students who want to study medicine, physiotherapy, or veterinary science. It is often chosen in combination with the other sciences, which it complements well. Otherwise it is simply a good standard A level or AS level to include in any combination of subjects.

It is generally regarded as a fairly demanding subject, but this should not put you off. If you liked studying science at GCSE, obtained a good result, and are comfortable with basic algebra, you will be able to cope with it.

What this book does

This book takes each of the main topics in physics in turn, explaining the important key aspects in simple, accessible language, without irrelevant detail. It then gives you an opportunity to apply this knowledge to standard physics problems, based upon AS and A2 examination questions, and finally provides you with detailed answers, explaining each step so you can check your work properly.

How to use this book

This book can be used as a supplement to a standard A-level textbook, or as a free-standing A-level course book in its own right.

The best way to use this book is as a course companion to your A-level studies. It enables you to have a simple, well explained summary of the key facts by your side, and the opportunity to practise applying them whilst you are doing your normal work. It can then act as a revision aid just before the exams.

Alternatively this book can be used simply as a revision guide.

How to work and revise

Most people accept that it is only sensible to work continuously for relatively short periods of time, thirty minutes to one hour, and then have a good break before starting again. This way all the time you are working should be effective and productive. This pattern of working applies at any stage, but is particularly appropriate when revising, as there is a lot of material to be reviewed and this takes a long time.

The best way to revise is to go through your notes and make summaries of all the important facts, information, and formulae that you need to learn. Revision guides and books like this can help you with this process, but you are more likely to remember things if you write them down yourself. You should use the factual recall questions in this book to help test yourself. Do not just read through your notes once: you need to review a topic at least two or three times before you can be assured of remembering it. There is a list on page x of the standard equations that must be learnt.

Exam technique

The single most common mistake that students make in examinations is to not read the question properly. It is always worth reading any question at least twice before starting to write your answer. Make sure you do what the question asks. For example, if a question says 'Explain, using a diagram' then make sure you draw a diagram. This may seem obvious, but having marked many exam papers I can tell you that a huge number of people make this sort of trivial mistake.

When solving problems, always write down any relevant equations and show each important stage of any calculation, because if you make a mathematical mistake you will still get marks for the original equation and method. If you cannot do one part of a calculation, but you can do a later part, show what working you would have used, because if the method is right you will still get credit for it. Other common mistakes are not putting appropriate units to an answer, or not writing an answer to the appropriate number of significant figures (see unit 2).

Most exam questions show the marks for each section on the exam paper; use this as an indicator of the degree of complexity or detail required in your answer.

There are certain important key words used in exam questions: here is a list of them, with explanations of their meanings.

State – a simple answer is required, with no explanation.

Determine – work it out by calculation, usually by putting measured values into an equation.

Sketch – draw a simple, clear diagram with key things labelled (e.g. forces).

Define – give a straightforward definition in words or in the form of an equation with defined symbols.

Estimate – values must be stated roughly and to an appropriate order of magnitude, either from common knowledge or justified from given information, and then other values must be calculated from them.

Calculate – obviously, just work out a numerical answer.

List – simply write out the required items.

Explain what is meant by – normally requires a definition with a brief explanation.

Explain – say how and why something happens.

Deduce – work the answer out from the information provided (rather than recalling information).

Discuss – state the evidence supporting the idea and the evidence against it, and say which is the stronger.

Describe – frequently refers to an experiment and the effects it produces; use words and diagrams to answer the question.

Show – normally associated with manipulating equations to produce a new equation.

Suggest – there may be more than one answer which is valid, so just write down what you think is a plausible one.

What would be observed – usually associated with an experiment: describe what happens in the experiment, particularly the results.

The new A-level system

The main objectives of the new A-level system are to broaden the range of subjects that a student studies in the sixth form, whilst maintaining the old A-level standards.

The new system is designed so that a typical student will study four or five ASs in their first year of study, and then select three from these to continue in their second year of study, called A2. This will then give them three full A levels. The system is designed to be flexible, so that it is possible to do any number of ASs and A2s, but an A2 can only be done if the relevant AS has been completed. In reality, most students' choices will be limited by whatever systems and options their school or college can offer.

Universities vary in their approach to the new system, but the general pattern is that they will primarily base their offers upon the three A2 results or predictions.

Each of the three main examination boards is offering two specifications (syllabuses) in physics. Each AS component contains three modules and each A2 component contains three modules. Most of the content of each specification is the same, because it has been specified by the government, but the content of each module varies from specification to specification. This means that whatever exam board or specification a student follows, most of the content is the same; it is just that they will do it in a different order.

The table on page ix shows which sections of this book correspond to which modules for each specification.

Key skills

As part of the package of the new AS/A2 system the government has also introduced the idea of Key Skills into A levels. These have been a part of GNVQs for some time, but are new to A level, and as yet are not compulsory. In order to make sure that students have basic skills in numeracy, information technology, and communication, all students doing A levels will be encouraged to take a level 3 Key Skills qualification. This is done by accumulating written or other evidence that a student has the necessary skills. This is then assessed, and together with a brief exam, determines whether a student has met the requirements to be awarded the qualification. A school or college may organize its courses to provide opportunities for most of the Key Skills to be met during the normal teaching of AS and A2 subjects.

Exam board specifications and modules

Each specification consists of six modules: three for AS and three for A2, which together make up the full A level. Most of the content of each specification is the same, in order to comply with guidelines laid down by the government, but what goes into each module varies considerably. This table enables you to quickly identify the relevant sections for your specification's modules. If you want to find a particular topic, use the index or the contents page.

	Mod 1	Mod 2	Mod 3	Mod 4	Mod 5	Mod 6
OCR A	1–5, 27	9, 10, 13, 14, 20, 21	11, 12, 28	5–8, 15, 18–26	15–17	1–28
OCR B	9–11, 27	1–5, 12–14	28	6–8, 18–20, 24–26, 28	15, 18, 19, 21–23, 28	1–27
Edexcel A	1–4, 6, 7, 15	9, 10, 25, 26	16, 17, 27, 28	8, 11–14, 24	18–23	15–20, 24
Edexcel B (Salters)	1–6, 9–12, 14, 24	9–13, 27	28	5–8, 15–17, 20–23	15, 18, 19, 24–25, 27	1–28
AQA A	11–17	1–7, 25	9, 10, 27, 28	11, 12, 15, 18–23	15, 28	1–27
AQA B	1–6, 9, 10	11, 12, 14–17	28	5–8, 13, 14, 19, 24–26	15, 17–23	28
WJEC	1–4, 11, 12, 27	9, 10, 14, 15	28	5–8, 20, 24–26	5, 18–23	1–28
NIB	1–4, 8, 27	11–14	28	5–8, 18, 19, 24–26	15, 20–23	16, 17, 28

FORMULAE TO LEARN

All students studying A-level physics must know the following formulae:

(i) the relationship between speed, distance, and time:

$$speed = \frac{distance}{time\ taken} \qquad v = \frac{\Delta s}{\Delta t}$$

(ii) the relationship between force, mass, and acceleration:

$$force = mass \times acceleration \qquad F = ma$$

(iii) the relationship between acceleration, velocity, and time:

$$acceleration = \frac{change\ in\ velocity}{time\ taken} \qquad a = \frac{\Delta v}{\Delta t}$$

(iv) the relationship between density, mass, and volume:

$$density = \frac{mass}{volume} \qquad \rho = \frac{m}{V}$$

(v) the concept of momentum:

$$momentum = mass \times velocity \qquad p = mv$$

(vi) the relationships between force, distance, work, power, and time:

$$work\ done = force \times distance\ moved\ in\ direction\ of\ force \qquad W \text{ or } \Delta W = Fs$$

$$power = \frac{energy\ transferred}{time\ taken} = \frac{work\ done}{time\ taken} \qquad P = \frac{\Delta W}{\Delta t}$$

(vii) the relationships between mass, weight, potential energy, and kinetic energy:

$$weight = mass \times gravitational\ field\ strength \qquad W = mg$$

$$kinetic\ energy = \tfrac{1}{2} \times mass \times (speed)^2 \qquad KE = \tfrac{1}{2}mv^2$$

(viii) the relationship between an applied force, the area over which it acts, and the resulting pressure:

$$pressure = \frac{force}{area} \qquad P = \frac{F}{A}$$

(ix) the ideal gas equation:

$$pressure \times volume = number\ of\ moles \times molar\ gas\ constant \times absolute\ temperature$$

$$pV = nRT$$

(x) the relationships between charge, current, potential difference, resistance, and electrical power:

$$charge = current \times time \qquad \Delta Q = I\Delta t$$

$$potential\ difference = current \times resistance \qquad V = IR$$

$$electric\ power = potential\ difference \times current \qquad P = IV$$

(xi) the relationship between potential difference, energy, and charge:

$$\textit{potential difference} = \frac{\textit{energy transferred}}{\textit{charge}} \qquad V = \frac{E}{Q}$$

potential energy = mass × gravitational field strength × change in height

$$PE = mg\Delta h$$

(xii) the relationship between resistance and resistivity:

$$\textit{resistance} = \frac{\textit{resistivity} \times \textit{length}}{\textit{cross-sectional area}} \qquad R = \frac{\rho l}{A}$$

(xiii) the relationship between energy and charge flow in a circuit:

$$\textit{energy} = \textit{current} \times \textit{potential difference} \times \textit{time} \qquad E = IVt$$

(xiv) the relationship between speed, frequency, and wavelength:

$$\textit{wave speed} = \textit{frequency} \times \textit{wavelength} \qquad v = f\lambda$$

(xv) the relationship between centripetal force, mass, speed, and radius:

$$\textit{centripetal force} = \frac{\textit{mass} \times (\textit{velocity})^2}{\textit{radius}} \qquad F = \frac{mv^2}{r}$$

(xvi) the relationship between capacitance, charge, and potential difference:

$$\textit{capacitance} = \frac{\textit{charge}}{\textit{potential difference}} \qquad C = \frac{Q}{V}$$

(xvii) the relationship between the potential difference across the coils in a transformer and the number of turns of coil:

$$\frac{\textit{potential difference across coil 1}}{\textit{potential difference across coil 2}} = \frac{\textit{number of turns in coil 1}}{\textit{number of turns in coil 2}}$$

$$\frac{V_1}{V_2} = \frac{N_1}{N_2}$$

Unit 1

MOTION

- When you walk down a road you are moving. Your surroundings appear stationary, but they are moving too. The Earth is spinning, it is also in orbit around the Sun, our Solar System is rotating about the centre of our galaxy, and our galaxy is moving through space.

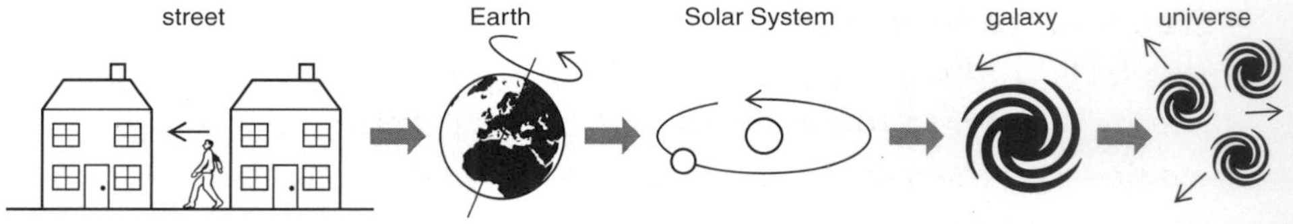

Actual velocity of a man walking down a road.

It is possible to work out the resulting motion, but it is easier to use the idea of **relative velocity**.

Relative velocity is the motion of objects in relation to each other.

'A person walking down the road at a velocity of 4 m/s' really means 'a person walking at a velocity of 4 m/s relative to a stationary road'.

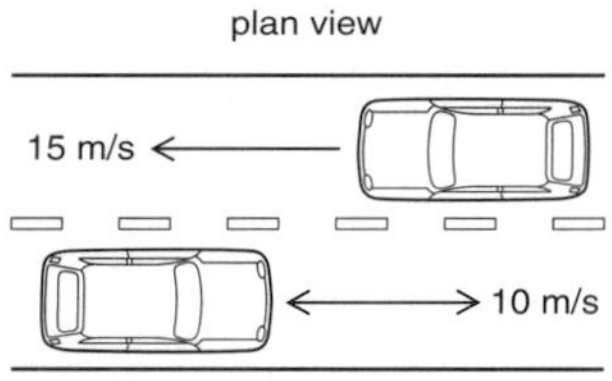

Relative velocity of two cars.

Example *A car is moving with a velocity of 15 m/s. Another car is moving in the opposite direction with a velocity of 10 m/s. What is the velocity of the first car relative to the second car?*

Start by drawing a diagram (see left). Treat the second car as if it is stationary. The velocity of the first car relative to the second is 15 – –10 = **25 m/s**.

A sparrow's mass is the same whether it is stationary or flying.

mass = volume × density
$m = V \times \rho$

- Take a sparrow. Its mass is clearly the same whatever the direction it is flying in, so direction is not important to mass. When it is flying it is always moving in a particular direction, so direction is important to its velocity.

  Direction is therefore important to some quantities, and not others. If it is important to a quantity, the quantity is called a **vector**; if not it is called a **scalar**.

  A **vector** is any quantity that has both size and direction, e.g. velocity.

  A **scalar** is any quantity that has size only, e.g. mass, power, and energy.

- **Mass** is a measure of the **inertia** of an object. It is a scalar, measured in kilograms (kg).

  Inertia is the reluctance of an object to change its state of motion.

- One morning you wake up and go for a bike ride. You leave your house (A) and go directly to a video shop (B), and then on your way home you stop by a friend's house (C). A, B, and C are all on the same straight road.

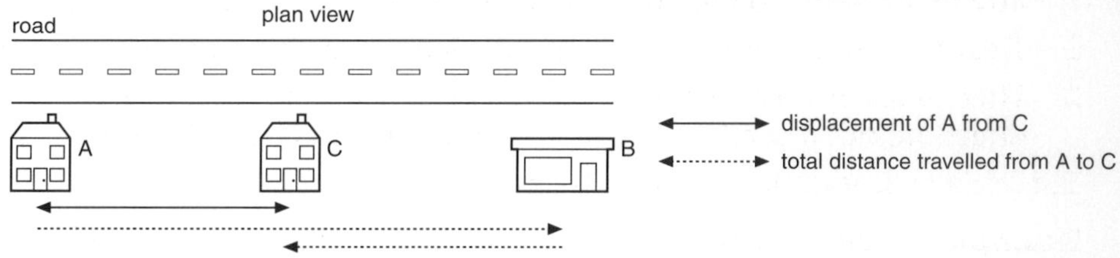

Distance and displacement on a bike ride.

The total length of the journey is AB + BC, and is called the **distance** travelled.

Distance is how far you have actually travelled when moving between two points where the direction of motion may be changing. It is a scalar quantity measured in metres (m).

The difference in position in a straight line between your house and your friend's house is AB – BC and is called your final **displacement**.

Displacement is the length of the straight-line path between two points. It is a vector quantity measured in metres (m).

- Back on the bike ride. If you leave late and the video shop is about to shut, you have to cycle quickly to get there on time. You then cycle slowly back to your friend's house. It is necessary to know how quickly things are moving and have a number that represents it. **Speed** and **velocity** do this. The direction of motion is important for velocity, but not for speed.

  Speed is the rate of change of distance.

  It is a scalar and is measured in m/s. Finding the 'rate of change' means dividing the change in a quantity by the time taken for the change.

  Velocity, v, is the rate of change of displacement, s (see right).

  It is a vector, and is also measured in m/s. Δ means 'change in'.

$$\textbf{speed} = \frac{\textbf{change in distance}}{\textbf{time}}$$

$$\textbf{velocity} = \frac{\textbf{change in displacement}}{\textbf{time}}$$

$$v = \frac{\Delta s}{\Delta t}$$

  Example *An object is moving around in a circle with a constant speed of 2 m/s (see right). What is the change in speed and velocity between A and B?*

  Change in speed = **0 m/s** (because the direction is not important).

  Change in velocity = 2 – –2 = **4 m/s** (because direction does matter).

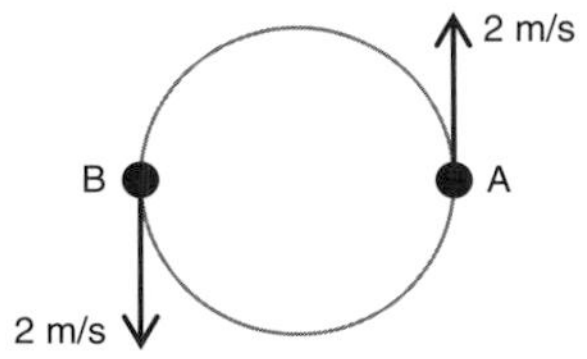

- Returning to the bike ride one last time. You leave your friend's house a couple of hours later, and head home. Your house is down a hill so you go faster and faster as you descend. Your velocity is increasing. It is useful to be able to measure how quickly your velocity is increasing, and we call this **acceleration**.

  Acceleration, a, is the rate of change of velocity. It is a vector, and is measured in m/s^2. An object falling towards the Earth near its surface will always have the same acceleration, whatever its mass. This is known as the **acceleration due to gravity**, g, and is equal to 9.81 m/s^2.

$$\textbf{acceleration} = \frac{\textbf{change in velocity}}{\textbf{time}}$$

$$a = \frac{\Delta v}{\Delta t}$$

- To be able to do calculations where acceleration is zero or constant we use the **equations of motion**. These equations have been derived from several ideas in mechanics. The various quantities used in them are displacement, s, initial velocity, u, final velocity, v, time, t, and acceleration, a.

equations of motion:

$v = u + at$

$s = ut + \frac{1}{2}at^2$

$v^2 = u^2 + 2as$

$s = \frac{1}{2}(v + u)/t$

  Example *A stone is thrown upwards with an initial velocity of 20 m/s. What is the maximum height risen, and how long does it take to reach this height? (Take acceleration due to gravity as 9.8 m/s^2.)*

  Take g as a, the constant acceleration acting downwards. Draw a diagram (see right). List the information given, and pick the appropriate equation.

$u = 20$ m/s	$s = ?$	$v^2 = u^2 + 2as$	$v = u + at$
$v = 0$ m/s	$t = ?$	$0 = 20^2 - (2 \times 9.8s)$	$0 = 20 - 9.8t$
$a = -9.8$ m/s^2		$s = 400/19.6 =$ **20 m**	$t = 20/9.8 =$ **2.0 s**

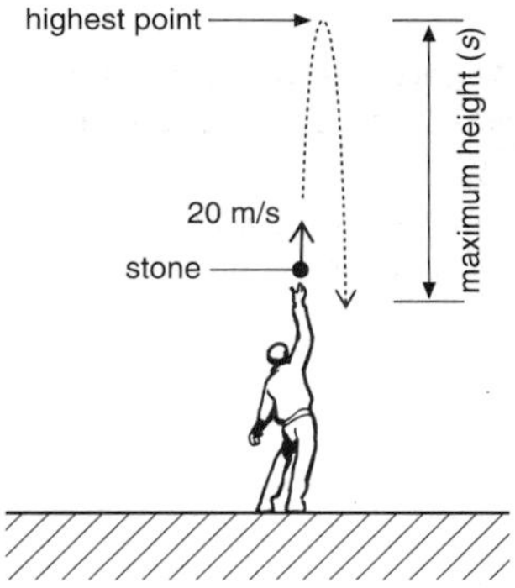

- For any motion in two dimensions, there are two things to remember.
  1. Treat the vertical direction as independent of the horizontal direction.
  2. There is always acceleration due to gravity acting in the vertical direction, but if friction is ignored there is no acceleration horizontally.

  This is sometimes called **projectile motion**.

  Example *A ball is kicked horizontally off the edge of a cliff, 300 m high, at 10 m/s. Find the time taken for it to fall to the ground, and the horizontal distance travelled from the cliff base. (g = 9.8 m/s^2.)*

  Start by drawing a diagram (see right). List the information in the vertical and horizontal directions, and then pick the appropriate equations.

  Vertical motion
  Horizontal motion

$u = 0$ m/s	$v^2 = u^2 + 2as$	$u = 10$ m/s	$s = ut + \frac{1}{2}at^2$
$s = 300$ m	$v^2 = 0 + (2 \times 9.8 \times 300)$	$t = 8$ s	$s = 10 \times 8$
$a = 9.8$ m/s^2	$v = 6000^{1/2} = 78$	$v = u$	**s = 80 m**
$t = ?$	$v = u + at$	$a = 0$ m/s^2	
$v = ?$	$78 = 0 + 9.8t$, **t = 8.0 s**	$s = ?$	

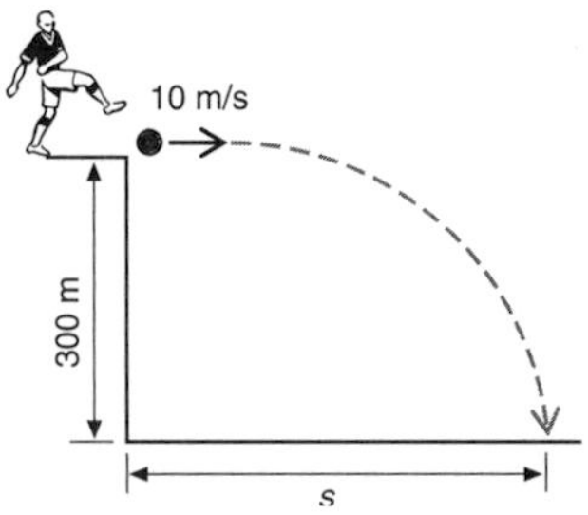

Unit 1

TESTS

RECALL TEST

1 What is meant by 'relative velocity'?

(2)

2 What is the difference between a scalar and a vector quantity?

(2)

3 Why is mass a scalar quantity?

(2)

4 Define 'displacement'.

(2)

5 Define 'distance'.

(2)

6 What is the difference between speed and velocity?

(2)

7 What two things apply to projectile motion?

(2)

8 Define 'acceleration'.

(2)

9 What is meant by the 'rate of change' of a quantity?

(2)

10 When a rock is thrown off the edge of a cliff horizontally, why is there no acceleration in the horizontal direction, if the air resistance is negligible?

(2)

(Total 20 marks)

CONCEPT TEST

Take acceleration due to gravity $g = 9.8\,m/s^2$

1 A car moves 1.0 km in 5.0 minutes in a straight line. What is its average velocity in m/s? It then accelerates, and reaches a new velocity of 30 m/s in another 8.0 seconds. What is its acceleration?

(4)

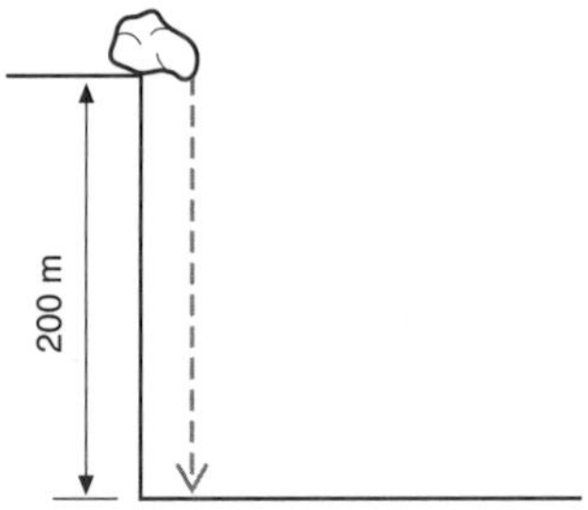

2 A stone falls from the top of a 200 m high cliff, and falls straight downwards (see left). Neglecting air resistance, how long does it take to reach the bottom, and what is the velocity just before it hits the ground?

(4)

3 An aircraft is flying at 200 m/s. It accelerates for 10 seconds to a final velocity of 280 m/s. What is its acceleration, and how far does it travel in this time?

(4)

4 A pebble is dropped down a well, and hits the bottom 4.0 s later. How deep is the well?

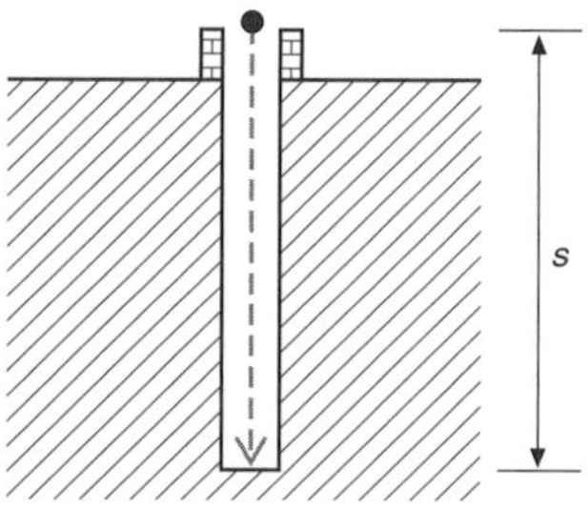

(2)

5 A lorry accelerates uniformly from 34.0 km/h to 60.0 km/h in 5.0 s. Find the acceleration and the distance travelled during this acceleration.

(4)

6 A tennis ball travelling at 10.0 m/s is hit by a tennis racket, and returns in the opposite direction with a velocity of 15.0 m/s (see right). If the duration of the impact is 0.200 seconds, what is the acceleration acting on the ball?

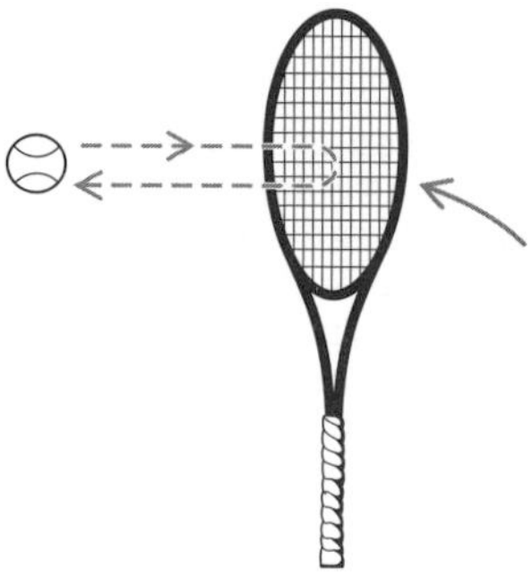

(4)

7 What height does a bullet fired straight upwards with a velocity of 200 m/s reach? How long does it take to return?

(4)

8 A hose pipe ejects water with a velocity of 4 m/s. If the pipe is horizontal, and held 1 m above the ground, how far does the water travel horizontally? (See right.)

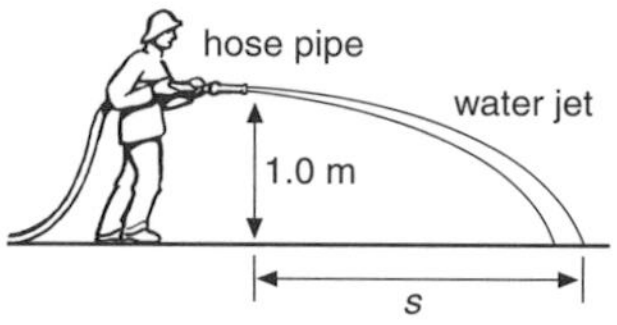

(6)

9 A tennis ball is hit over a net as shown in the figure on the right. What are the initial components of its velocity in the vertical and horizontal directions?

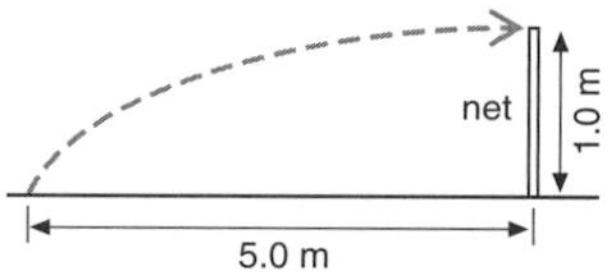

(6)

10 An electron is accelerated from rest to 0.9 times the speed of light in 4.0 s. How far does it travel in this time? (Take the speed of light, c, as 3.0×10^8 m/s.)

(4)

11 Calculate the muzzle velocity of a bullet as it leaves the barrel of a rifle 1.2 m long if it accelerates at 20 000 m/s^2. How far will the bullet drop if it hits a target 200 m away? (See the figure on the right.)

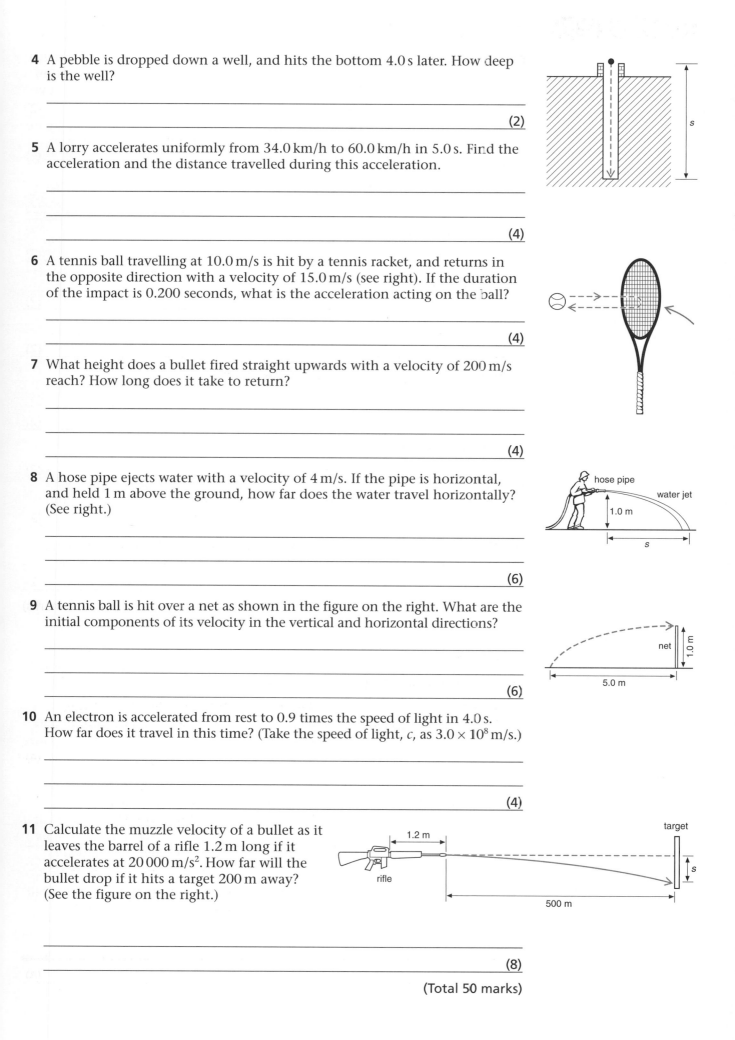

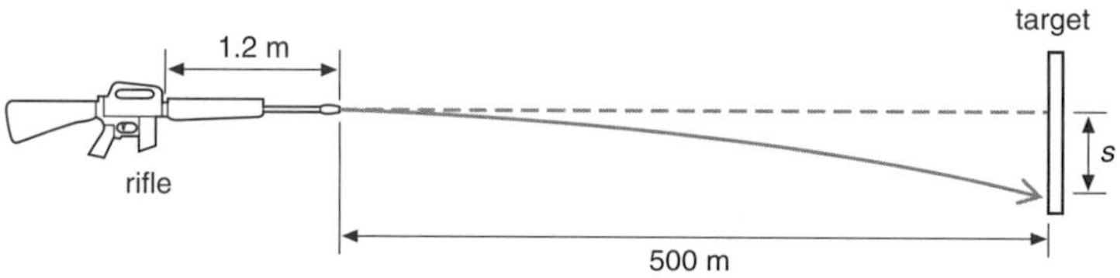

(8)

(Total 50 marks)

Unit 2 MORE MOTION

- We have already looked at moving objects and the equations used to predict their behaviour. It is also very important to be able to represent the motion of an object in a graphical form, and this is what we will be looking at now.

  There are three types of graph used to represent the motion of moving objects: displacement–time, velocity–time, and acceleration–time. The figure shows the shapes you can expect to see, and what they represent for each type of graph is explained below.

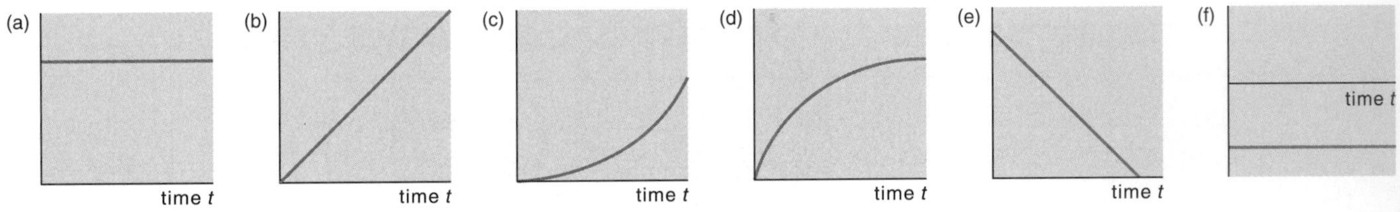

Displacement–time
(a) stationary
(b) constant velocity
(c) acceleration
(d) deceleration
(e) constant negative velocity
(f) stationary

Velocity–time
(a) constant velocity
(b) constant acceleration
(c) increasing acceleration
(d) decreasing acceleration
(e) constant deceleration
(f) constant negative velocity

Acceleration–time
(a) constant acceleration
(b) uniformly increasing acceleration
(c) acceleration increasing at an increasing rate
(d) acceleration increasing at a decreasing rate
(e) uniformly decreasing acceleration
(f) constant deceleration

'Negative velocity' means velocity in the opposite direction to positive velocity.

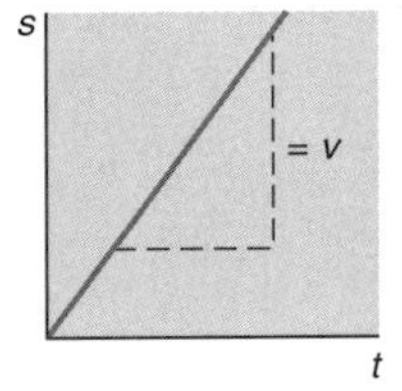

- You should be able to see that from the definition of velocity (velocity = displacement/time, $v = \Delta s/\Delta t$) the slope or gradient of a displacement–time graph represents the velocity (see left).

  The gradient of a displacement–time graph is the velocity.

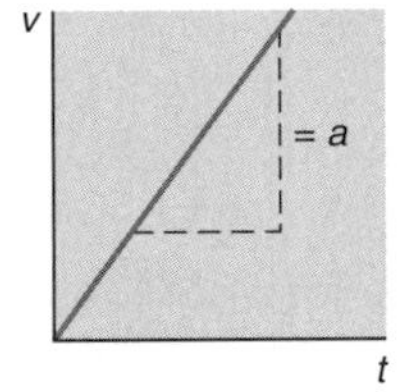

- You should also be able to see that from the definition of acceleration (acceleration = change in velocity/time, $a = \Delta v/t$) the slope or gradient of a velocity–time graph represents the acceleration (see left).

  The gradient of a velocity–time graph is the acceleration.

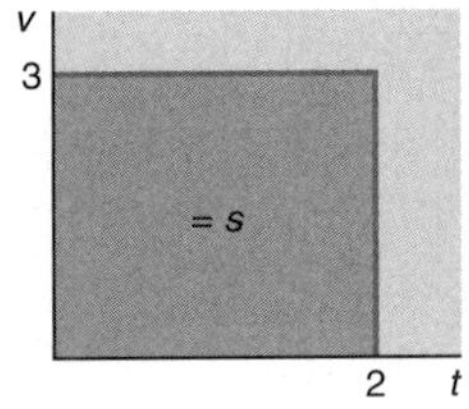

- If an object is moving with a uniform velocity of 3 m/s for 2 s, from the definition of velocity ($v = s/t$), displacement $s = v \times t$, so $s = 3 \times 2 = 6$ m. If you plot a graph of this motion you will see that the area of the shaded region (in the figure left) is the same.

  The area under a velocity–time graph is the displacement.

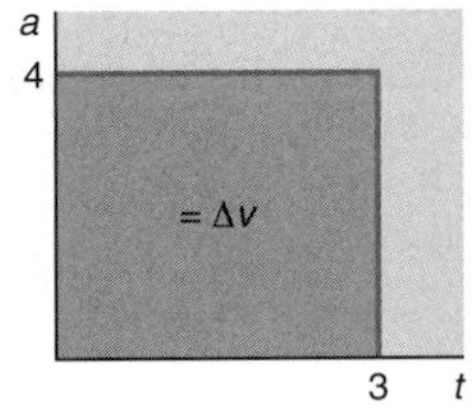

- If an object accelerates uniformly at 4 m/s^2 for 3 s, from the definition of acceleration ($a = \Delta v/\Delta t$) change in velocity is $\Delta v = a \times t$, so $\Delta v = 4 \times 3 = 12$ m/s. If you plot a graph of this motion you will see that the area of the shaded region (left) will give you the same answer.

  The area under an acceleration–time graph is the change in velocity.

  Example *A car accelerates from rest at 2 m/s^2 for 3 s, then continues for 4 s at a constant velocity. It then decelerates to rest in 2 s. Draw a velocity–time graph for this motion. Determine the maximum velocity, the deceleration, and the total distance travelled.*

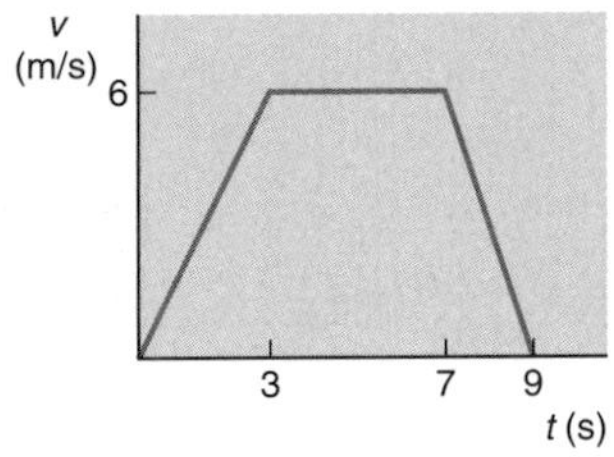

$v = u + at$
$v = 0 + (2 \times 3)$
$\mathbf{v = 6\ m/s}$

$a = \dfrac{\Delta v}{\Delta t} = \dfrac{0 - 6}{2} = \dfrac{-6}{2}$
$\mathbf{a = -3\ m/s^2}$

Area under graph
$= (\tfrac{1}{2} \times 3 \times 6) + (4 \times 6) + (\tfrac{1}{2} \times 2 \times 6)$
Area = 9 + 24 + 6 = 39 m
Total distance travelled = 39 m

Example *A rocket is launched from the surface of the Earth. It accelerates uniformly, directly upwards for 10 minutes. The motors are then switched off. The rocket continues to move upwards for a further 3 minutes, until it reaches its maximum height. It then falls to the ground in a further 5 minutes. Draw displacement–time, velocity–time, and acceleration–time graphs for the motion. Indicate on the velocity–time graph how the total distance travelled and the final displacement are determined. Indicate on the acceleration–time graph how the final velocity just before it hits the ground is determined.*

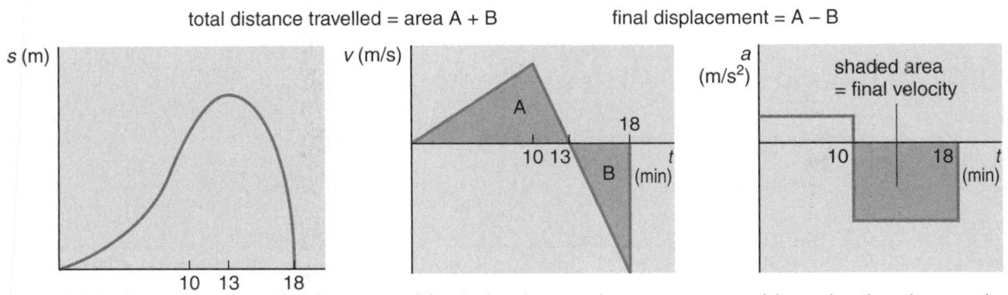

(a) displacement –time graph (b) velocity–time graph (c) acceleration–time graph

Example *Sketch the displacement–time, velocity–time, and acceleration–time graphs for a bouncing ball, for three bounces. Take moving downwards as positive and the release point as zero displacement.*

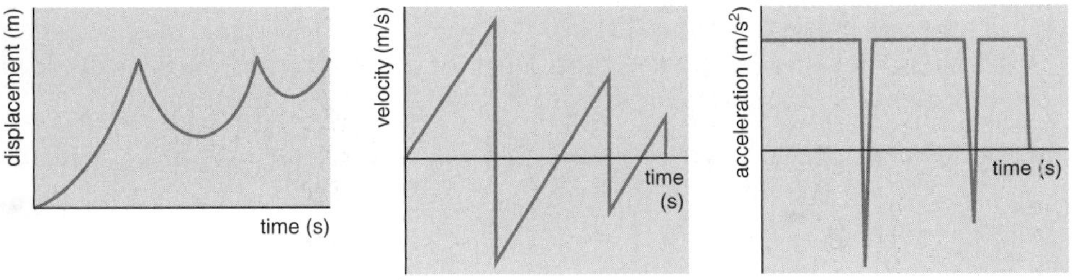

- Graphs appear throughout physics, so it is useful to look at the graphs we have been dealing with to determine some general rules that govern all graphs.

  If you look at the equation for velocity ($v = \Delta s/\Delta t$), a graph of s against t gave us v from its gradient, and a graph of v against t gave Δs from the area under it. In the same way, if you look at the equation for acceleration ($a = \Delta v/\Delta t$), a graph of v against t gave us a from its gradient, and a graph of a against t gave Δv from the area under it. There is an obvious pattern here so we can say, for any equation of the form $p = q/t$:

  In a graph of q against t the gradient is always equal to p.
  In a graph of p against t the area under the graph is always equal to q if p is uniform.

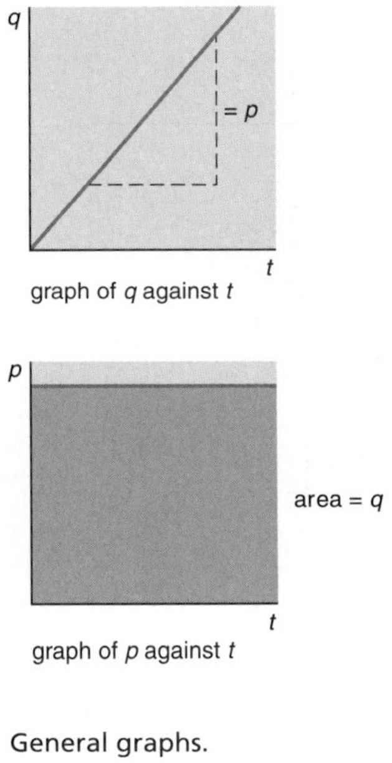

General graphs.

- In this and the last section you have been making numerical calculations, and there is a general rule that applies to all calculations: in any calculation the answer can only be as accurate as the original information.

  In any calculation the initial values are given to a certain number of significant figures (1.2 m/s is 2 sig. figs). The number of sig. figs is the number of digits present. The answer should be to the same number of sig. figs. In dealing with calculations involving several numbers it is the value with the smallest number of sig. figs that sets the limit. For example:

  $$\frac{3.524 \times 6.4}{2.01} = 11.221, \text{ but it must be to 2 sig. figs so} = 11$$

  If in doubt just write your answer to 2 or 3 sig. figs.

- All answers to questions need appropriate units stated after the value. There are two forms in which they may appear: solidus (e.g. m/s) and index (e.g. m s^{-1}). You can use either form, but you need to be familiar with both.

Unit 2

TESTS

RECALL TEST

1 What is the gradient of a velocity–time graph?

____ (2)

2 What is the gradient of a displacement–time graph?

____ (2)

3 What is the area under an acceleration–time graph?

____ (2)

4 What is the area under a velocity–time graph?

____ (2)

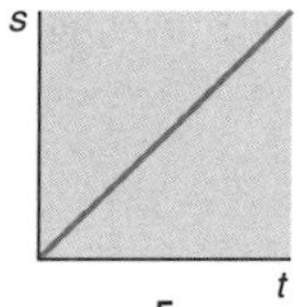

5

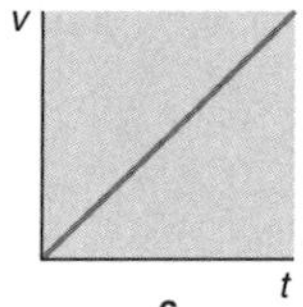

6

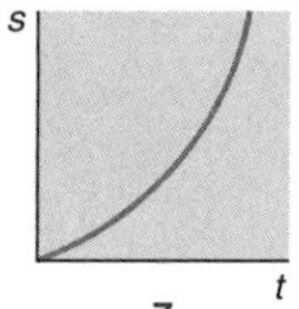

7

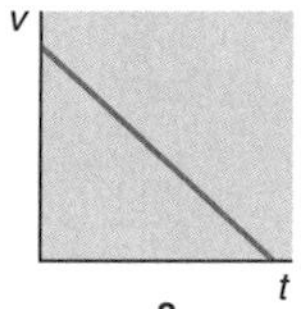

8

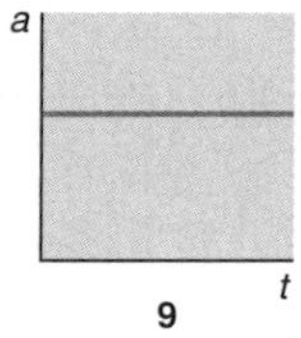

9

In each of the situations shown left describe the motion of the object:

5 ____ (2)

6 ____ (2)

7 ____ (2)

8 ____ (2)

9 ____ (2)

10 If you were given the equation 'the height of a person equals the acceleration due to gravity divided by the number of dental fillings in his mouth', how would you determine the acceleration due to gravity graphically?

____ (2)

(Total 20 marks)

CONCEPT TEST

Take $g = 9.8\ \text{m/s}^2$

1 Draw a velocity–time graph and a displacement–time graph for a coffee mug knocked out of the upstairs window of a house on the axes. Take displacement and motion downwards as positive.

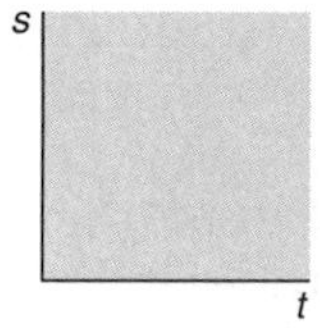

a displacement–time

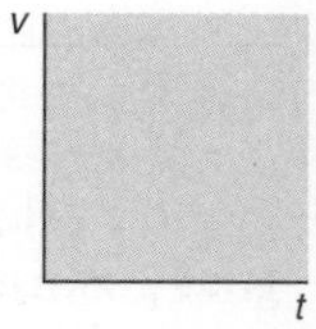

b velocity–time

(4)

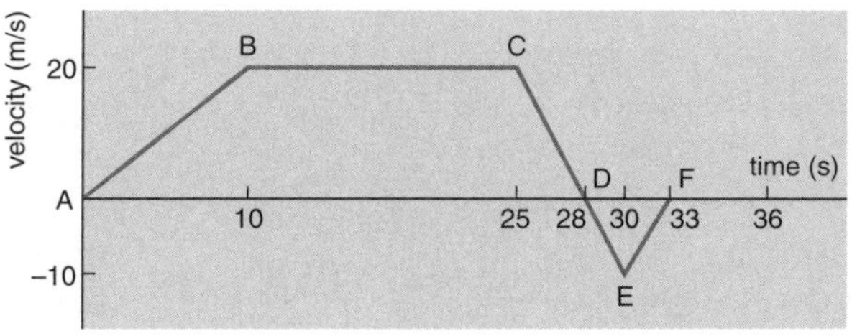

2 Describe the motion of the object in the graph shown left.

____

____

____

____ (5)

3 Draw a displacement–time graph and a velocity–time graph for an apple falling out of a tree and bouncing twice. Take displacement and motion downwards as positive.

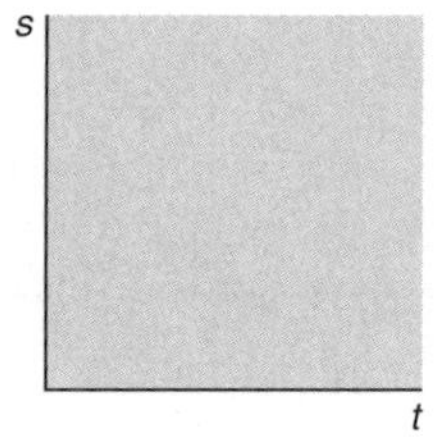

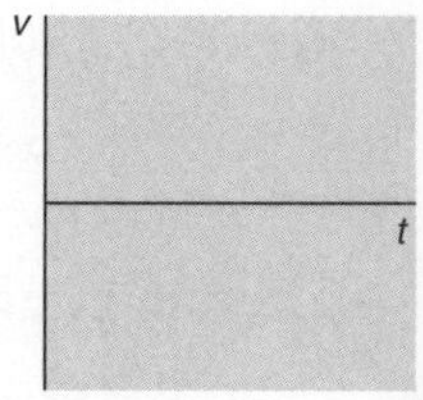

(6)

4 A piece of metal falls off an aircraft which is flying horizontally 30 000 feet above the ground. It reaches terminal velocity (see unit 3) and eventually hits the ground. Draw velocity–time and acceleration–time graphs for the motion in the vertical direction.

(4)

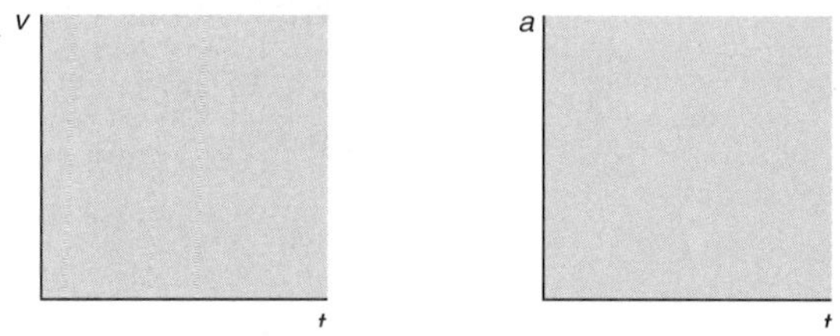

5 The graph on the right is for a moving object. What is the total distance travelled and the final displacement? Draw displacement–time and acceleration–time graphs for the same motion on the axes shown.

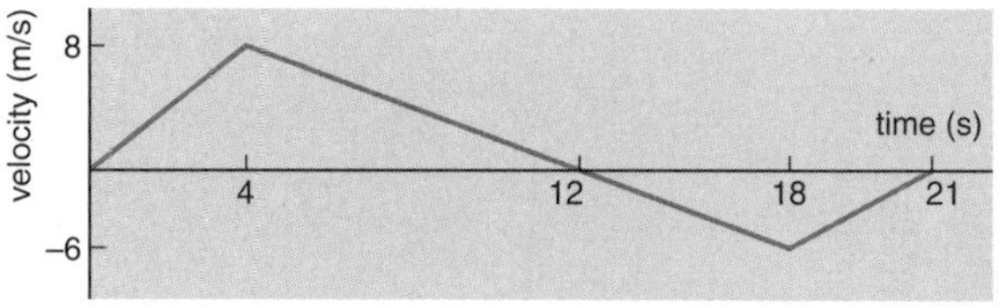

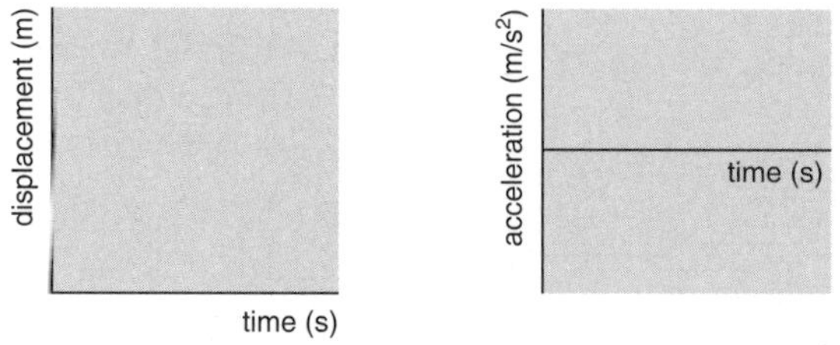

____

____

____

____

(8)

6 A satellite orbits a planet at 1 km/s. Draw speed–time and velocity–time graphs for its motion on the axes shown.

(4)

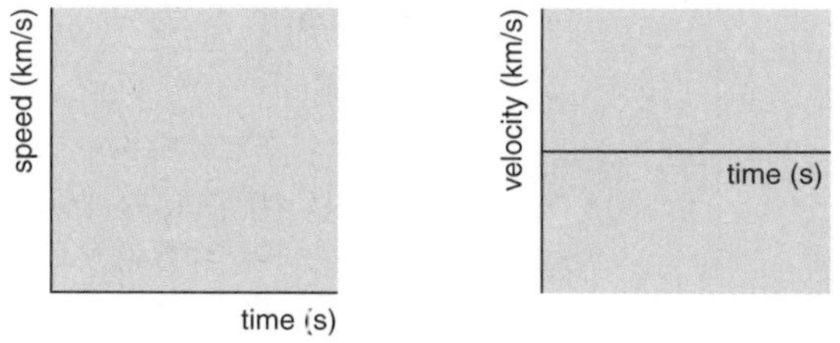

7 What is the average speed for the object in the distance–time graph shown in the figure on the right? Estimate the maximum velocity.

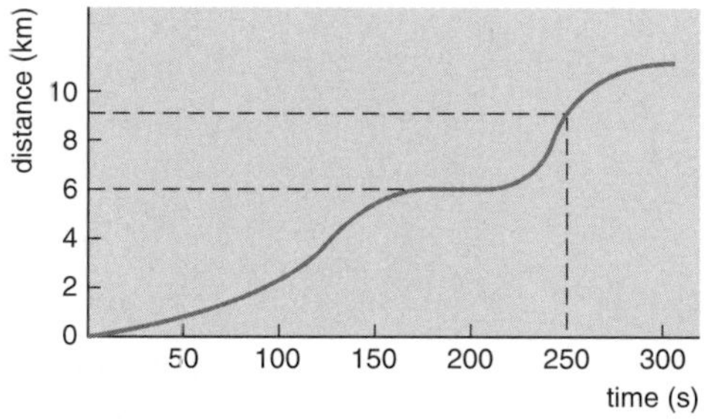

____

____

____

(8)

8 Draw a velocity–time graph for a football travelling at 20 m/s that bounces off a goal post back in the opposite direction with the same speed. Indicate on your graph the point where the bounce takes place. (3)

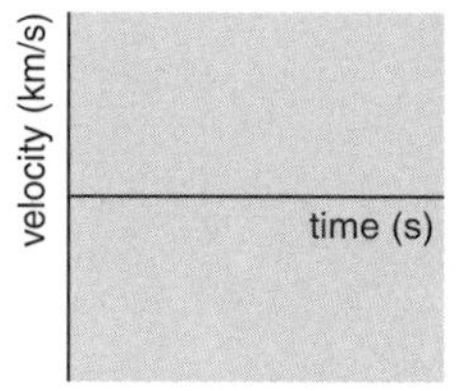

9 In the figure on the right there is a velocity–time graph showing the variation in velocity of a fish as it flips its tail whilst swimming upstream in a river. What is the maximum acceleration of the fish? Draw an acceleration–time graph for the time period shown in the figure below right and determine the displacement of the fish upstream after this length of time.

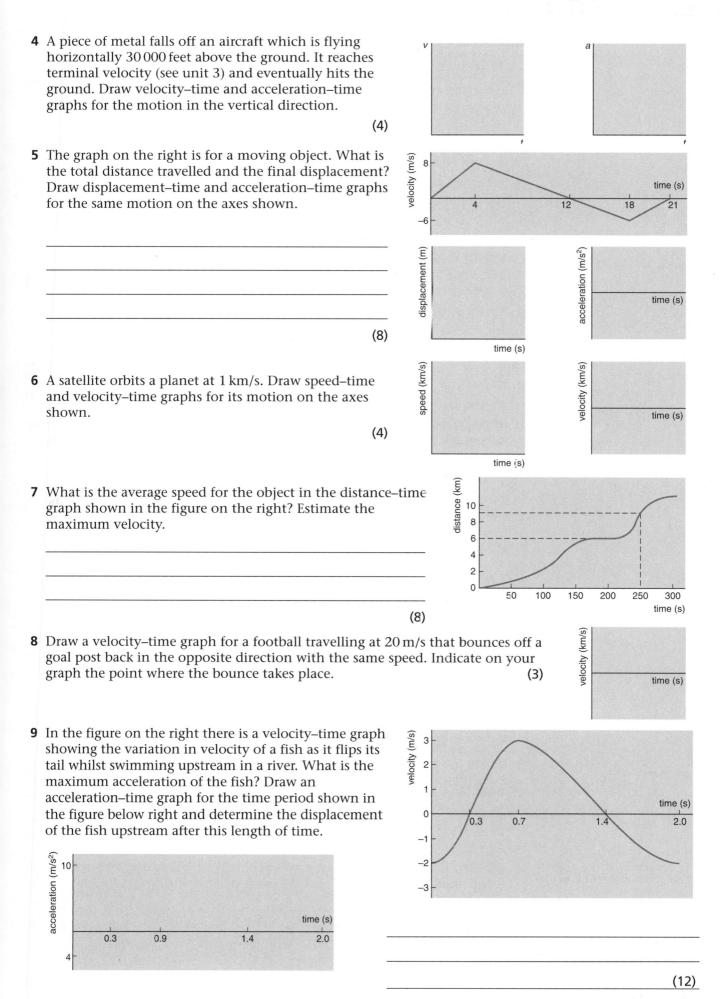

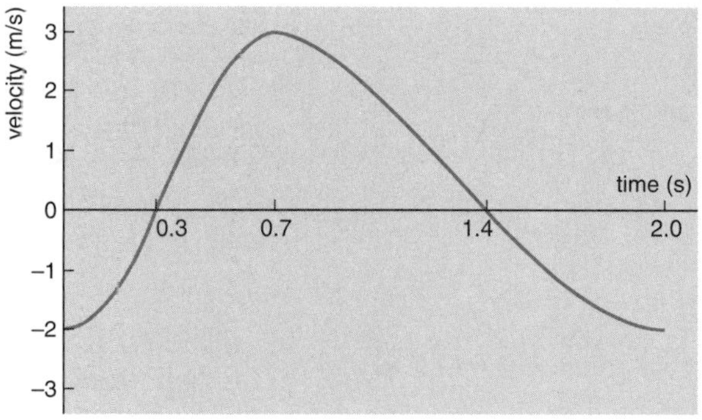

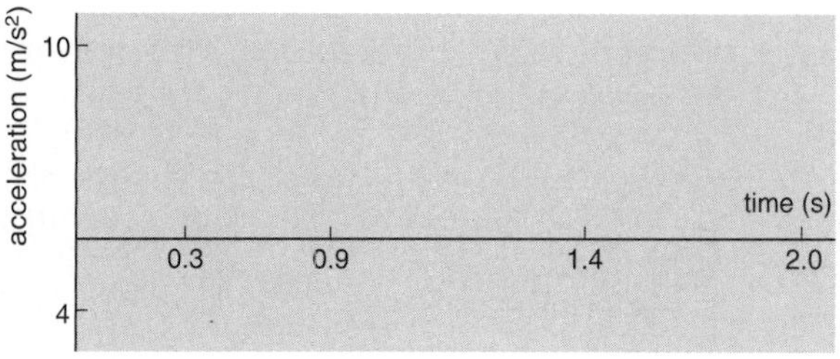

____

____

____

(12)

(Total 50 marks)

FORCES

- The two things that come into all areas of physics are forces and energy, so you can see how important forces are. We will deal with energy in unit 5. In this unit we are going to take a very simple approach to explaining what forces are and how we use them. We will go into a lot more detail later, in units 4 and 6.
- The first thing to do is explain what a **force** is. You should remember from GCSE that if a single force acts on an object, it can either push it away from a point or pull it towards that point.

  A **force** is something which will push or pull an object. It is a vector quantity, and is measured in newtons (N).

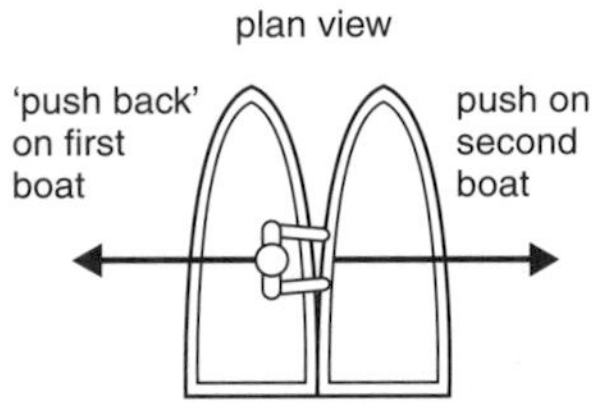

- Imagine you are in a boat. If there is another boat alongside, and you push it away, both boats will move off in opposite directions. You have pushed the other boat and it has 'pushed back' on you. So forces always occur in pairs.

  The push or pull of a force produced by one object acting on another will always produce an equal and opposite push or pull acting back on itself.

force = mass × acceleration
$F = ma$

- When you push the boat it is accelerating. If you push two boats of different masses with the same force, the one with the smaller mass will accelerate more; so mass determines the effect of a force.

  The force acting on an object is equal to the product of the mass of the object and the acceleration produced in the object ($F = ma$).

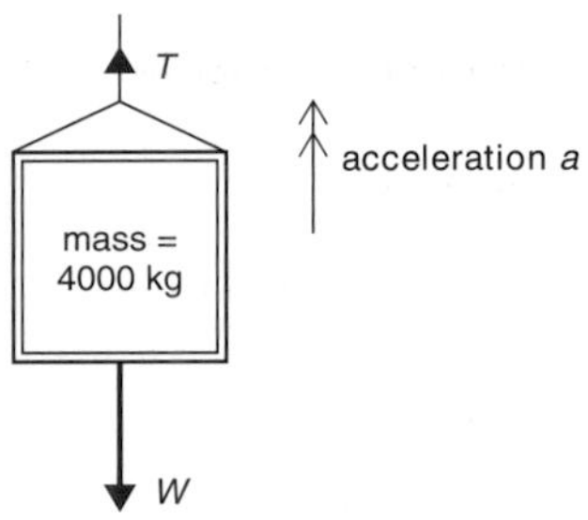

Example *A lift is accelerating upwards at 2.0 m/s². If the mass of the lift is 4000 kg, what is the tension in the cable pulling it up? (Take g as 10 m/s².)*

Start by drawing a diagram, taking upwards as positive. Then look at the forces in relation to each other.

Forces are vectors so $T - W = R$ where R is the resultant force.

If a is acceleration, m is mass, and g is gravity, $R = ma$ and $W = mg$ so

$T - mg = ma$ (rearrange the equation: $T = m(a + g)$)

Put in the numerical values: $T = 4000(2.0 + 10) =$ **48 kN**

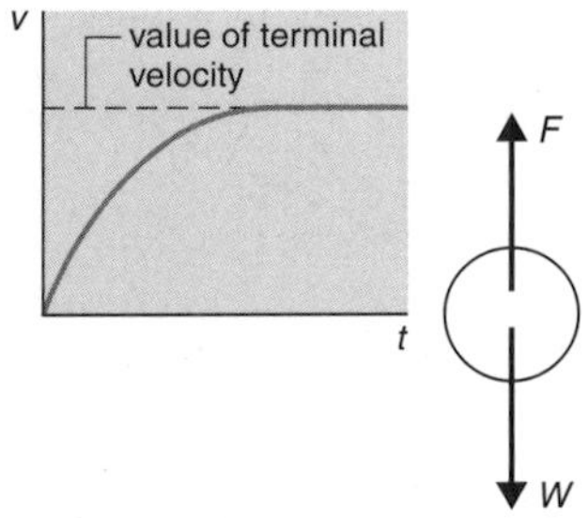

Forces at terminal velocity.

- When several forces act together they can cancel each other out and produce a resulting force of zero. This can result in two situations:
  1. The object is stationary.
  2. The object is moving at a constant velocity.

  The first case is obvious, but the second requires some explanation. When an object is released from rest in a fluid (liquid or gas), it initially accelerates because there is a resultant force acting on it. As its velocity increases the friction acting on it increases until it equals the weight and there is no resultant force; it then moves at a constant velocity. This is called **terminal velocity** and is illustrated in the diagram and graph shown on the left.

pressure = $\frac{\textbf{force}}{\textbf{area}}$

$P = \frac{F}{A}$

- When a force is applied to one object by another in contact with it, the effect depends upon the contact surface area. The larger the surface area the more spread out the force is. This is called the **pressure**, and is a measure of how concentrated the effect of a force is.

  Pressure is defined as force per unit area. It is measured in pascals (Pa) or N/m².

  An example of this is the way in which a drawing pin squeezed between two fingers pushes into the skin with its point but not with its flat head.

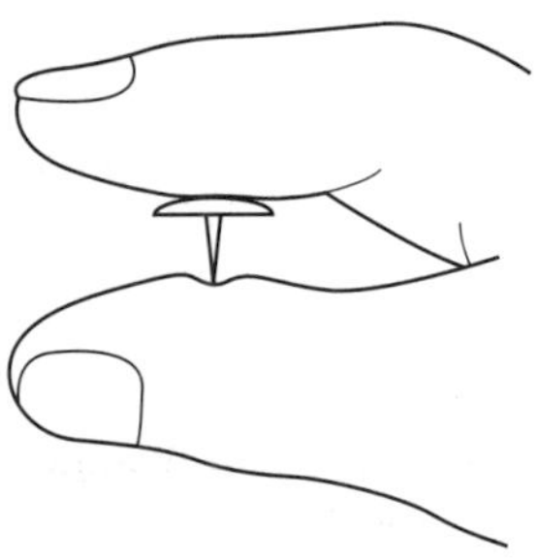
A drawing pin held between two fingers.

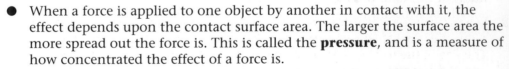

- Forces act in a particular direction, like velocity and acceleration, and are therefore vectors (see unit 1). When several vectors act together you can use one **resultant vector** to represent them all.

  A **resultant vector** is a single vector that produces the same effect as the combined effects of the individual vectors that it represents.

- It is possible to split a vector (***F***) up into two other vectors (***A*** and ***B***), at right angles to each other, which would produce the same effect as the original. This is called **resolving** a vector into two **components** (see right).

  From trigonometry:

$$\sin\theta = \frac{\text{opposite}}{\text{hypotenuse}} \qquad \cos\theta = \frac{\text{adjacent}}{\text{hypotenuse}} \qquad \tan\theta = \frac{\text{opposite}}{\text{adjacent}}$$

  $\sin\theta = A/F$, $\cos\theta = B/F$, $\tan\theta = A/B$, and so the components would be $\boldsymbol{A = F \sin\theta}$ and $\boldsymbol{B = F \cos\theta}$.

  The component next to the angle is always the cos component.

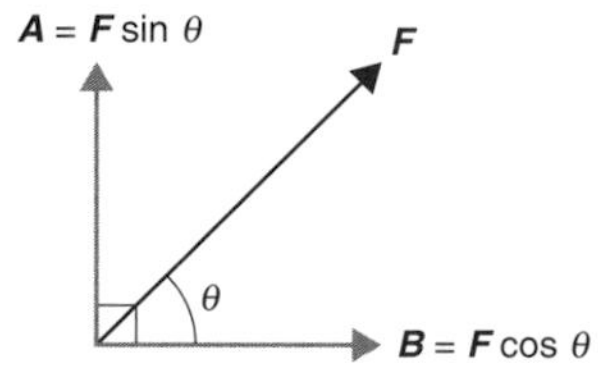

Resolving vectors.

vector components:
$\boldsymbol{A = F \sin\theta}$
$\boldsymbol{B = F \cos\theta}$

- It is also possible to combine two or more vectors together. If they are at right angles we use Pythagoras's theorem (see right).

  $\boldsymbol{A^2 + B^2 = C^2}$ and the angle between ***C*** and ***B*** would be given by $\boldsymbol{\tan\theta = A/B}$.

  If the vectors are not at right angles, the easiest way to work out the resultant is to construct a scale diagram where the lengths of the lines are proportional to the forces. A parallelogram is then constructed and the length of the diagonal shown is measured. This is proportional to the resultant force (see right).

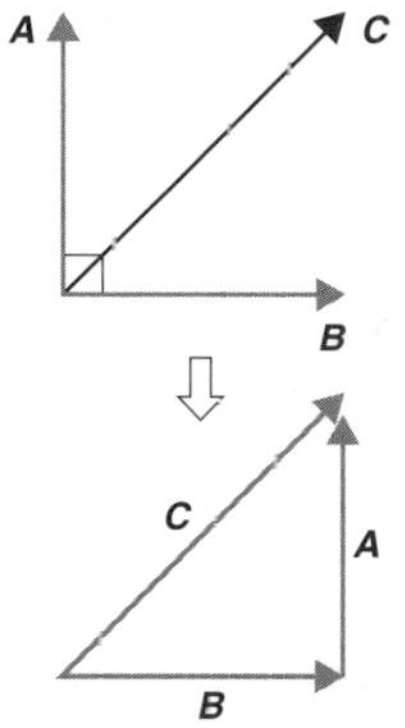

Combining vectors.

vector resultant:
$\boldsymbol{A^2 + B^2 = C^2}$
$\boldsymbol{\tan\theta = A/B}$

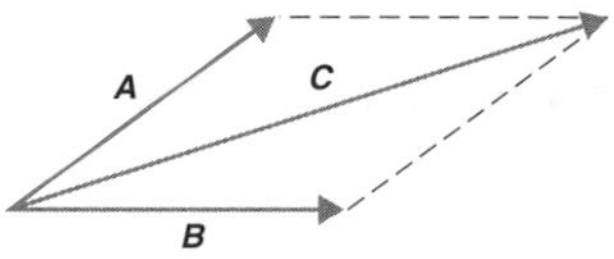

Parallelogram of forces.

- Forces can cause objects to rotate. If you push against a door, it is easier to make it move if you push it near the handle than near the hinge. This is the turning effect produced by a force and is called the **moment**.

  A **moment** (M) is the turning effect produced by a force, and equals the product of the force, F, and its perpendicular distance, s, from the axis of rotation. This is the shortest distance between the line of action of the force and the pivot point. The units of moment are N m.

  For an object not to be rotating (in rotational equilibrium), all of the moments taken about a point acting one way must balance with all the moments taken about that point acting the other way. This is called the **principle of moments**:

  For an object to be in rotational equilibrium the sum of the clockwise moments equals the sum of the anticlockwise moments, about any point.

moment = force × perpendicular distance to pivot

$\boldsymbol{M = F \times s}$

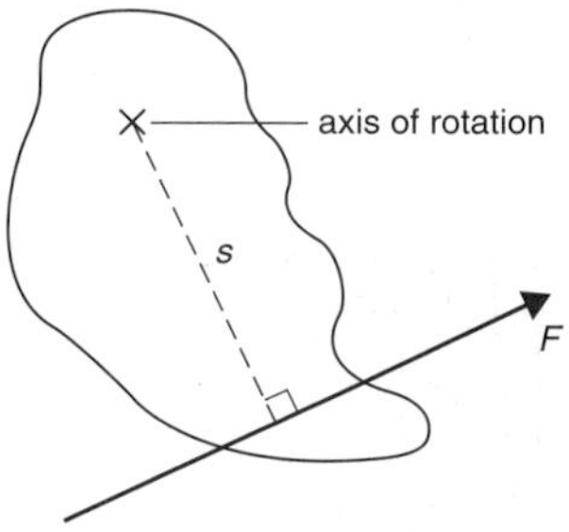

A moment acting on an object.

Σ (clockwise moments) = Σ (anti-clockwise moments)

Σ ↻ = Σ ↺

- When the moments acting on an object do not balance each other out, there is a resulting moment, and the object will rotate. You will often see the word 'torque' used in relation to resulting moments. **Torque** is defined as the resulting moment produced by a couple.

- When two identical anti-parallel forces act on an object, as in the figure on the right, they are called a **couple**. There is a turning effect but no resultant force. The size of the total moment or torque, T, produced is equal to the product of one of the forces, F, and the perpendicular distance between them, d.

- Some questions may ask you to calculate the torque produced by a single force. There is in fact an equal reaction force, not shown, acting at the pivot point, which together with the force shown constitutes a couple. In this situation the equation for a couple produces the same result as calculating the moment produced by the single force, so the reaction force is omitted from the diagram. If all this sounds confusing, just think of a torque as effectively the same as a moment, and work it out in the same way.

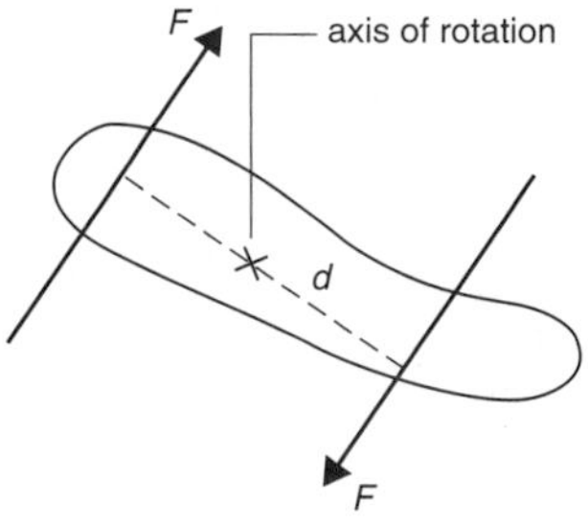

A couple.

moment of couple = one force × perpendicular distance between forces

$\boldsymbol{T = F \times d}$

TESTS

RECALL TEST

1 What is meant by 'terminal velocity'?

(2)

2 What are the two possible situations that may occur when all the forces acting on an object cancel each other out?

(2)

3 What is 'mass'?

(2)

4 Define 'moment'.

(2)

5 What is meant by the two 'components' of a vector?

(2)

6 What is meant by a 'resultant vector'?

(2)

7 What is the principle of moments?

(2)

8 Define 'pressure'.

(2)

9 What is a couple?

(2)

10 We talk of forces in terms of pushes or pulls. What effect will a single push or pull force have on an isolated object, and what does this depend upon?

(2)

(Total 20 marks)

CONCEPT TEST

Take $g = 9.8\ \text{m/s}^2$

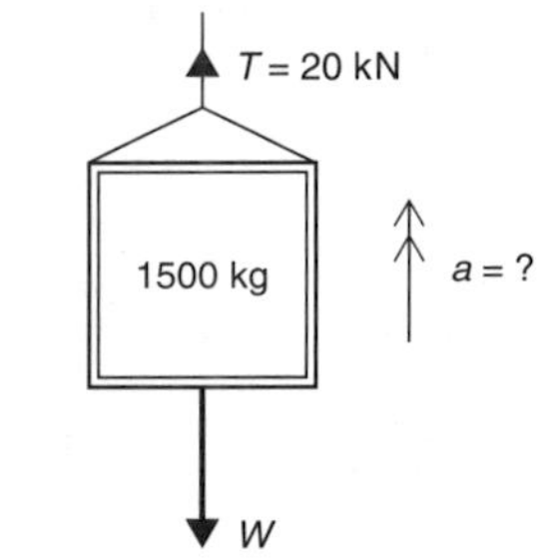

1 A car of mass 1500 kg is accelerated at $2.0\ \text{m/s}^2$. What force produces this acceleration?

(2)

2 A ball of mass 6 kg, travelling at 10 m/s, slows down to 6 m/s in 4 s. What is the frictional force (see unit 4) that produces this?

(2)

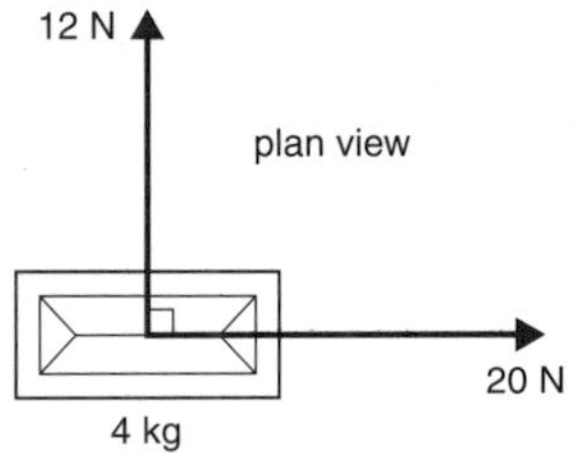

3 What is the acceleration of the lift above left? (Remember weight = *mg*.)

(2)

4 What is the acceleration of the brick with two forces acting on it shown left?

(2)

5 What is the unknown mass in the diagram below, if the rod is balanced?

(4)

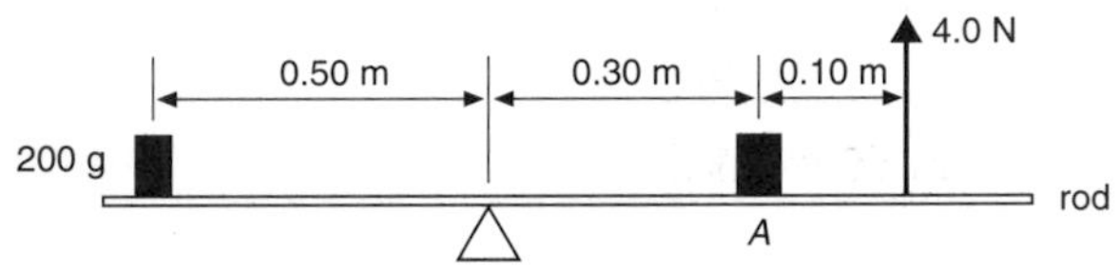

6 What is the linear acceleration of the sphere of mass 2 kg in the figure right?

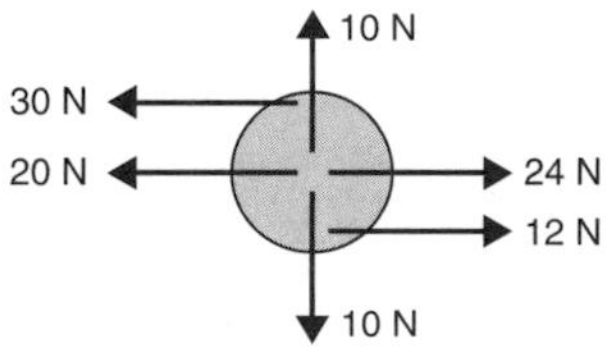

(4)

7 What is the tension (see unit 4) acting in the cable of the lift shown right?

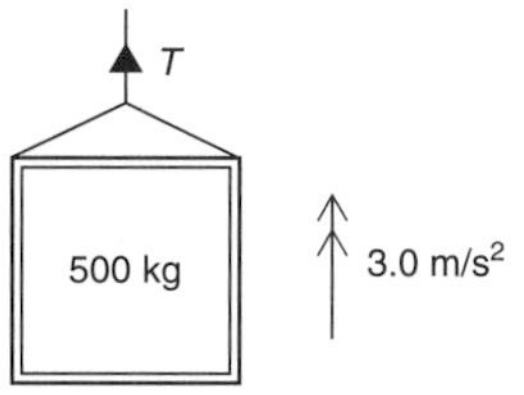

(4)

8 The figure below right shows a bicycle pedal. What is the moment produced by the cyclist in this position?

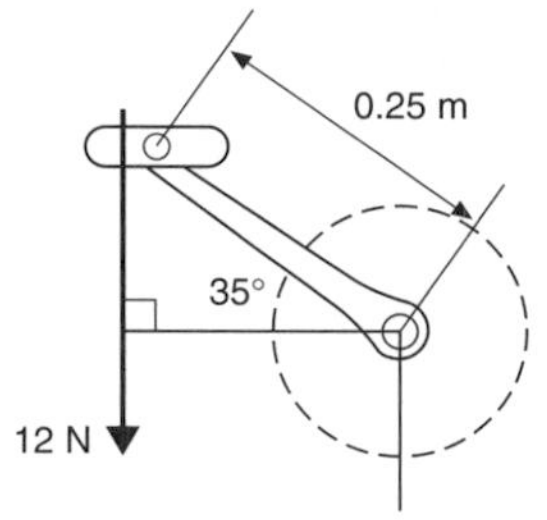

(4)

9 How far away from the end A is the centre of mass of the rod in the diagram below? Assume the rod is balanced.

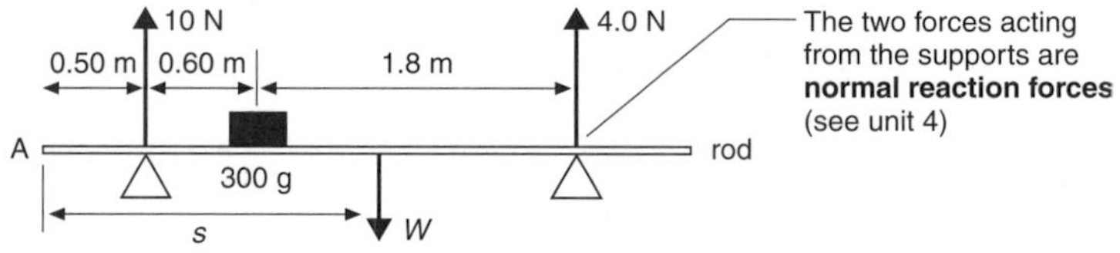

(4)

10 The diagram on the right represents the forces acting on a person's forearm. What is the maximum mass it can lift with the hand?

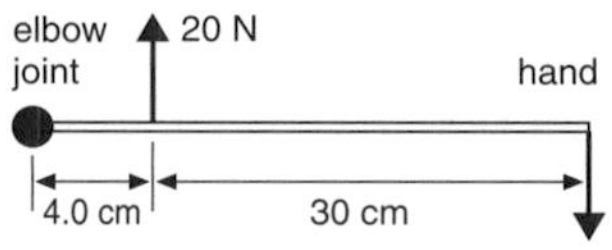

(4)

11 What is the moment produced by the spanner on the right if the force shown is applied?

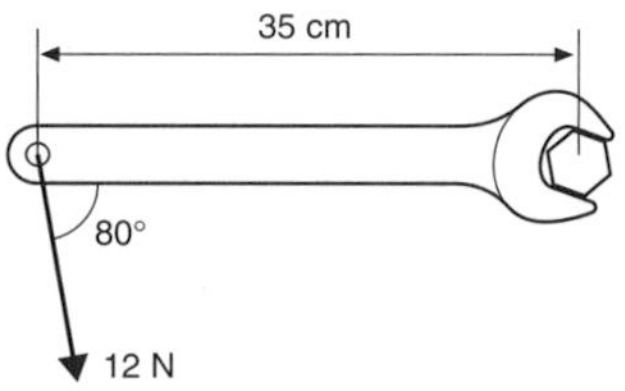

(2)

12 What is the resultant vector in this diagram? Give its size and direction.

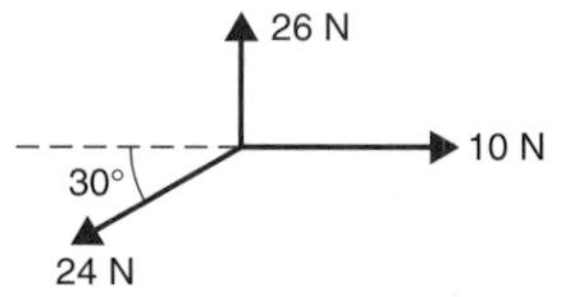

(4)

13 What force is required to lift the boulder with the lever system shown in the figure on the right?

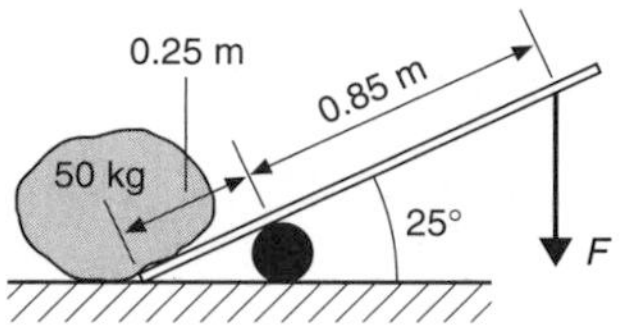

(4)

14 What is the resultant vector in the diagram on the right? Give its size and direction.

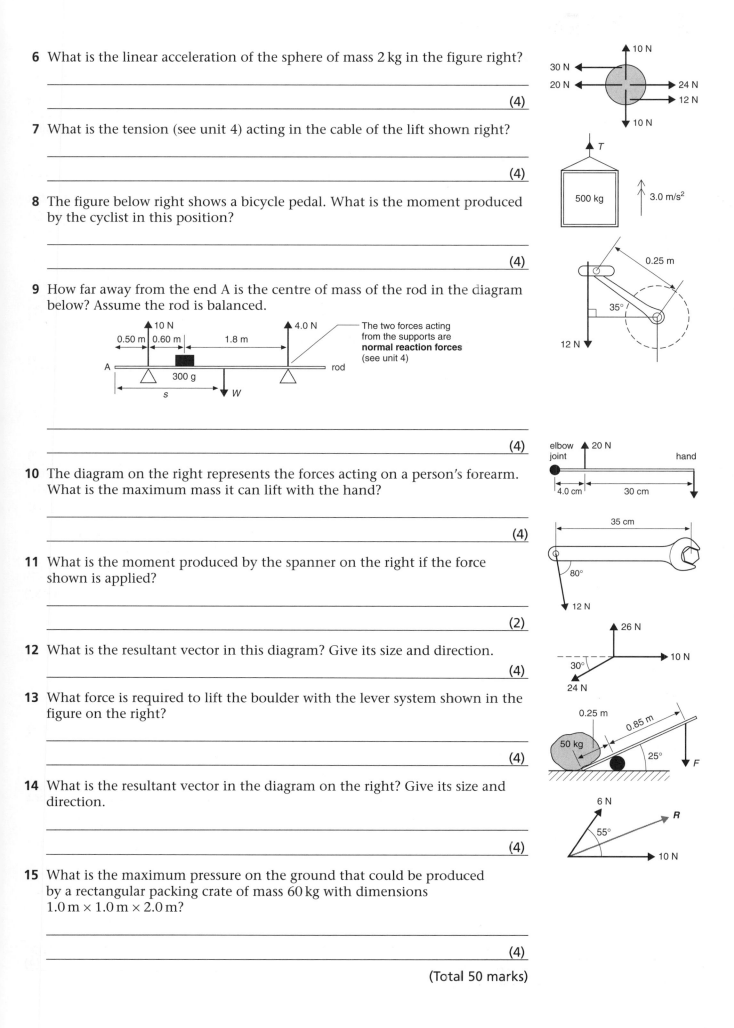

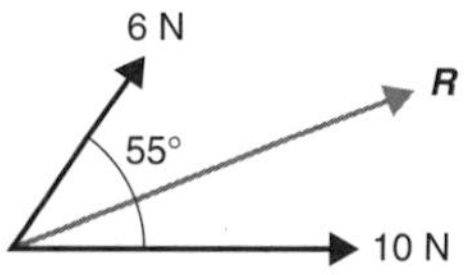

(4)

15 What is the maximum pressure on the ground that could be produced by a rectangular packing crate of mass 60 kg with dimensions 1.0 m × 1.0 m × 2.0 m?

(4)

(Total 50 marks)

Unit 4

TYPES OF FORCE

- Forces occur in many different forms and govern the behaviour of everything around us. To illustrate this, imagine yourself going through the following day.

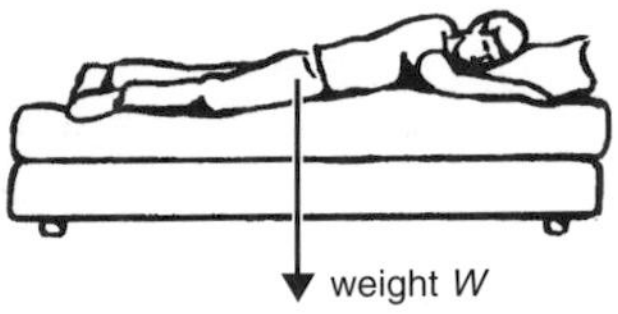

Gravity pulling downwards.

weight = mass × acceleration due to gravity

$$W = mg$$

- You wake up in the morning and push yourself upwards to get out of bed. This is because you have to overcome the force of gravity, more commonly referred to as your **weight**, which is pulling you downwards towards the Earth. The size of this force depends upon your mass.

  The **weight**, *W*, of an object is the force of the Earth's gravitational pull acting on its mass. It acts from the centre of gravity of the object towards the centre of the Earth.

  The **centre of mass** of an object is defined as the point at which if all the mass could be concentrated the object would still behave in the same way. The **centre of gravity** of an object is the point from which the weight may be considered to act. These are at the same position if gravity is constant; we can assume that gravity is constant near the Earth's surface. It has approximately the same value 100 m above the surface as it does on the surface.

The effect of removing friction.

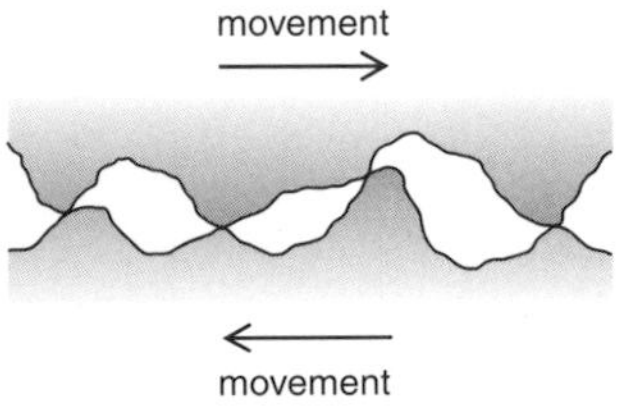

Surfaces in contact on a microscopic level.

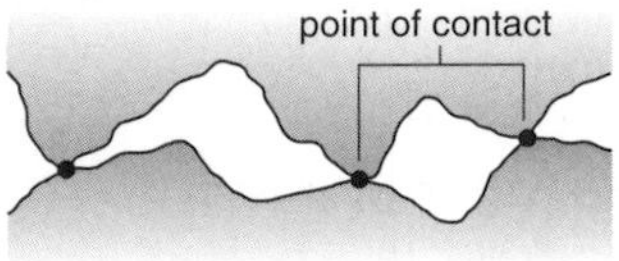

Surfaces bond together at a few points of contact.

- You then go into your bathroom, tread on a bar of soap left on the floor, and fall over. This brings us to one of the most important forces that we encounter in our daily lives, which we only really notice when it isn't there; namely that of **friction**.

  Friction, *F*, acts between surfaces or between fluids and surfaces, and opposes motion or probable motion. It acts parallel to and along the boundary between surfaces, from the point of contact.

  There is **static friction** and **dynamic friction**. The irregular nature of a surface on a microscopic level means that if two objects are in contact and moving relative to each other, bits of each object will get in the way of each other, causing **dynamic friction** (see left). If an object is placed on a surface, its contact forces are concentrated at a few points and the surfaces 'cold weld' together at these points (**static friction**). All of these things occur on a molecular and atomic level, and are due to interatomic forces that are electrostatic in origin.

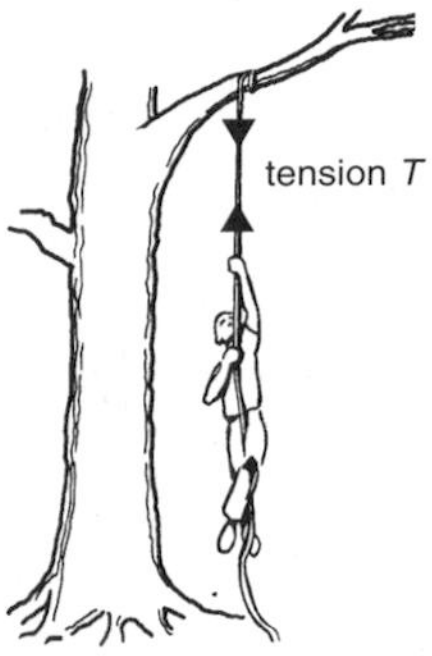

Tension in a rope being pulled.

- You decide to go for a walk. On the way you notice a rope hanging down from a tree, so quite naturally you decide to climb it. As you start to climb, the rope becomes stretched and taut. The rope is under **tension**.

  Tension is the force that occurs in any solid which is being stretched. It is electrostatic in origin and is due to the bonds between atoms in molecules. It opposes whatever force is stretching the object.

  Compression is similar to tension, and occurs when something is being squashed. Both these forces are important when considering the mechanics of the human body. Bones are designed to take compression and tendons are designed to take tension.

  A tension force is normally shown in a diagram by two arrows pointing in opposite directions, as shown left. To represent compression in something like a leg bone the arrows would be in the same positions, but each would be facing in the opposite direction. When drawing a free-body force diagram (see later) it is not always necessary to show both these arrows, as only one of them is acting on the object.

Normal contact force pushing up from the ground.

- You continue your walk along a road, past a field with a stone wall alongside. You jump over the wall. When you jump you push down on the ground and the ground pushes back with an equal and opposite force, enabling you to jump up. Although the force from the ground exists at all times and is necessary to balance the weight acting downwards, it is increased when you push downwards, and is known as the **normal contact force** or **normal reaction force**.

  Normal contact force (*N* or *R*) occurs between any two objects in contact. It acts at 90° to a surface from the point of contact.

On an atomic level most of matter is empty space. The normal contact force is the force that stops us sinking into the ground, or walking through walls (see unit 15). It is, in effect, the large-scale representation of all of the individual electrostatic interatomic forces between the two surfaces in contact.

The normal contact force N or R is linked to friction by the equation $\boldsymbol{F = \mu N}$ or $\boldsymbol{F = \mu R}$ where F is the **limiting friction** and μ is the **coefficient of static friction. Limiting friction** is the maximum value the friction can have before the object moves. The force of friction is independent of the contact surface area. In effect, friction and normal contact force are the two components of the actual force that acts between two surfaces (see right).

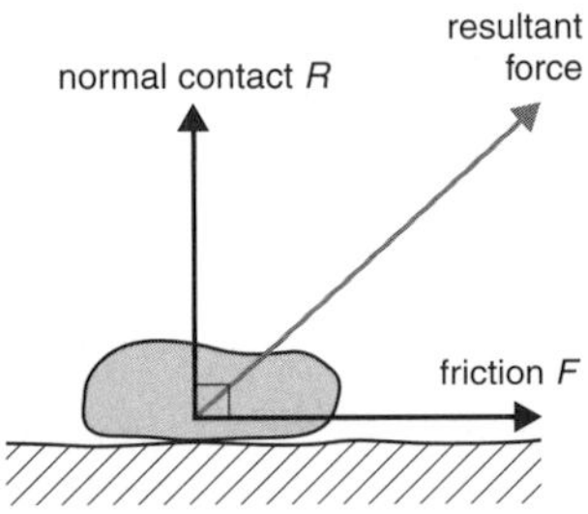

The relationship between F and R.

- You walk past a small river. It is a hot day, you see a friend having a swim, and decide to join him. You dive in and float on your back. The force that enables you to float is called **upthrust**.

  Upthrust occurs in fluids (liquids or gases), and is equal to the weight of the fluid displaced by the presence of an object. It is produced indirectly by the gravitational force acting on the displaced fluid, and directly by the interatomic forces due to the fluid pushing on the bottom of the object.

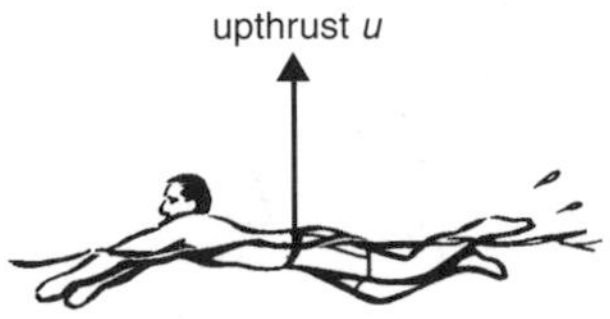

The upthrust force acting in water.

- There are several other types of force which you may meet: thrust, lift, drag, electrostatic force, magnetic force, and nuclear force. **Thrust** is the force generated by an engine or person pushing against something, which enables that object to move. **Lift** is the upward force produced by fluid flow over aerofoil shapes, like a wing. **Drag** is the force of friction found when an object moves through a fluid. **Electrostatic force** is the force experienced between charges; like charges repel and unlike charges attract. **Magnetic forces** are similar to electrostatic forces, except that we deal with north and south poles; like poles repel and unlike poles attract. **Nuclear forces** occur in the nucleus; see units 15 to 17.

- Forces surround us all of the time, on both a large and a small scale. It is necessary to classify them into general types in order to be able to model situations and predict the behaviour of objects. This is why we use weight, friction, tension, normal contact force, and upthrust. On a fundamental level of physics there are only actually four types of force: strong and weak nuclear forces, electromagnetic, and gravitational (see units 16–19).

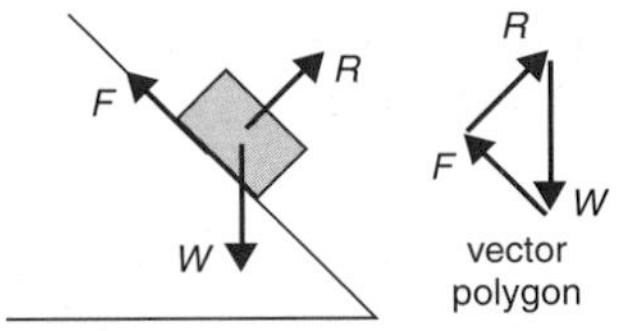

Object on a inclined plane.

- The forces previously described are used to model what is happening in real situations. This is done by drawing **free-body force diagrams**.

  A **free-body force diagram** treats an object as if it is isolated, and only shows the forces acting on the object. It does not show components or resultants unless they are specifically asked for. In diagrams, forces are usually represented by arrows with solid triangular heads. Resultant forces are usually indicated by using either a double or larger arrow head. In this text we are using a larger arrow head. In a similar way, in this text resultant accelerations are distinguished from velocities by using a double arrow head to represent them.

  The various forces acting in a free-body force diagram can be used to construct a vector polygon representing the situation. A closed vector polygon shows the object is in equilibrium, and hence can only be drawn for an object at rest or moving with constant velocity. If it is not closed, the resultant force is represented by a vector that would complete the polygon.

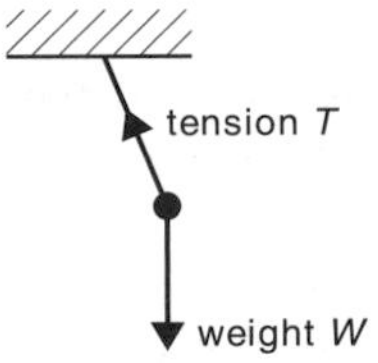

Simple pendulum.

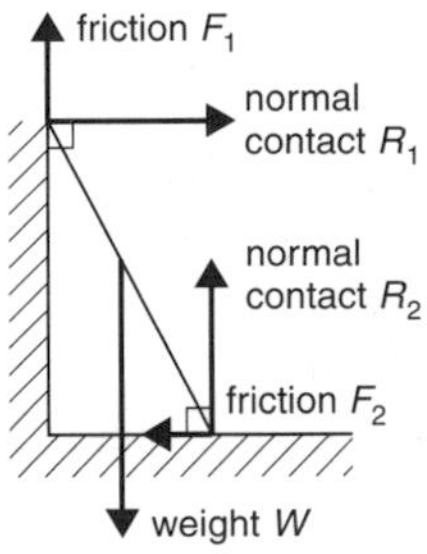

Ladder resting against a wall.

Example *Draw free-body force diagrams for an object on an inclined plane, a simple pendulum, a ladder resting against a wall and an aircraft in level flight. Draw a vector polygon for the first diagram.*

To draw the forces, just consider weight, friction, thrust, normal contact, and upthrust, and see which ones apply. For the vector polygon: take a force and draw it out roughly to scale in the right direction. Going around either clockwise or anticlockwise, add each force vector to the end of the previous vector.

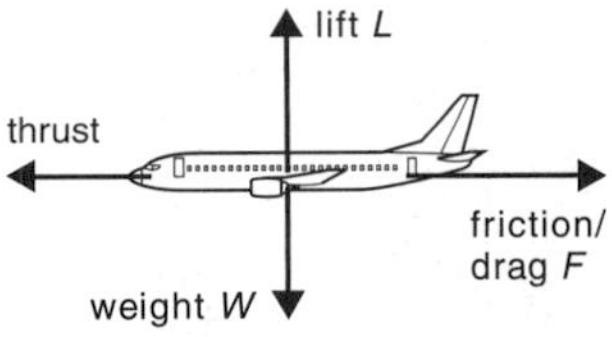

Aircraft in level flight.

Unit 4

TESTS

RECALL TEST

1 What is meant by 'weight'?

(2)

2 What is 'friction'?

.(2)

3 What is 'tension'?

(2)

4 What is meant by 'normal reaction force'?

(2)

5 What is 'upthrust'?

(2)

6 What is a free-body force diagram?

(2)

7 What is meant by the 'centre of mass' of an object?

(2)

8 Why do we use the forces weight, friction, normal contact, tension, and upthrust in diagrams?

(2)

9 Which of the forces mentioned in question **8** are electrostatic in origin, and which are gravitational in origin?

(2)

10 What are the four fundamental forces in physics?

(2)

(Total 20 marks)

CONCEPT TEST

Take $g = 9.8\,\text{m/s}^2$

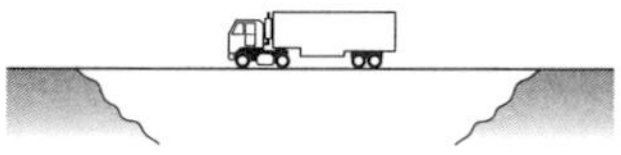

1 Draw on the diagram left the forces acting on a bridge when a lorry is one-third of the way across it. Write down one equation linking all the forces.

(4)

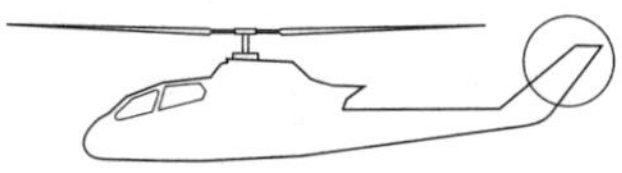

2 Draw the forces on the diagram of a helicopter moving with a constant velocity at a constant height above the ground. Write down two equations linking the forces.

(4)

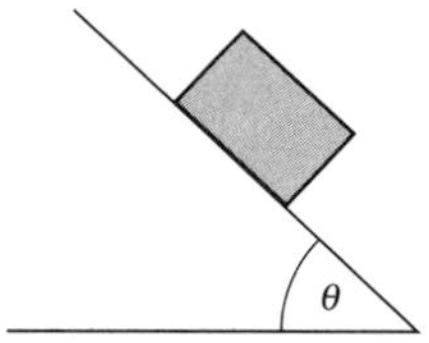

3 A rectangular box is resting on a rough inclined plane as shown left. Draw on the diagram the forces acting on the box, as well as a vector polygon of the forces. By resolving the weight force into two components, parallel and perpendicular to the plane, write down two equations linking the forces.

(6)

4 Sarah is abseiling down a rock face. She is momentarily at rest in the position shown right with the forces shown acting on her. The tension in the rope acts at 40° to the vertical. If her mass is 60 kg what are the values of tension, T, in the rope, and normal reaction force, R? Draw a vector polygon of the forces acting on her.

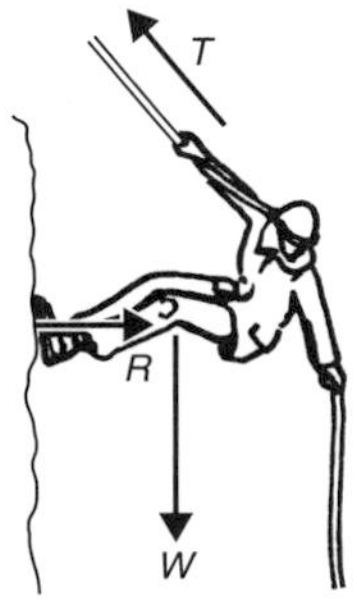

(6)

5 A man is moving upwards with constant velocity on a ski lift, as shown in the free-body force diagram. If the weight, W, of the man is 800 N, and the tension, T, acting downwards in the cable is 600 N, what is the force, F, pulling the man upwards in the direction of F?

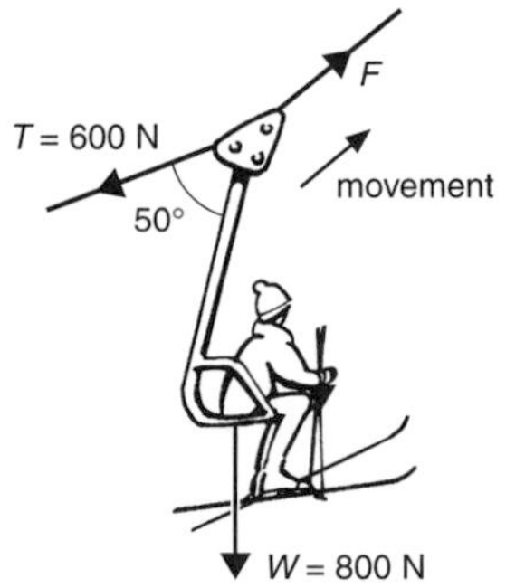

(6)

6 A tank is moving up a slope with uniform velocity as shown on the right, with the forces shown acting on it. Draw a vector polygon of the forces. If the mass is 5000 kg what is R? If Th is 20 000 N what is the value of F?

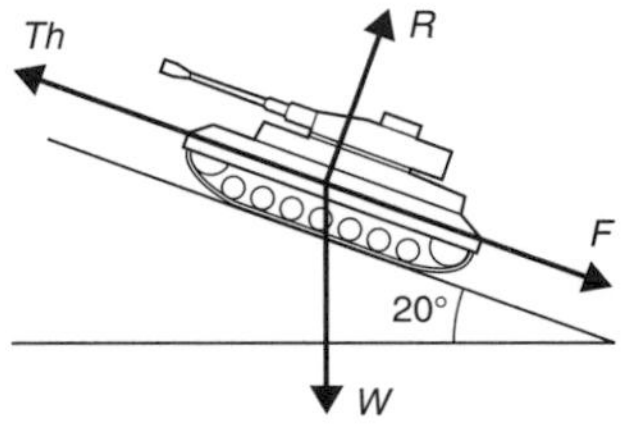

(6)

7 Josie is the captain of a boat which is being pulled along by two smaller boats called *George* and *Arthur* with a constant velocity against the tide as shown right. Draw a vector polygon to represent this situation. The tide then turns and the boats accelerate. Determine the total friction of the boats with the water at this moment if the resultant acceleration is 0.3 m/s^2.

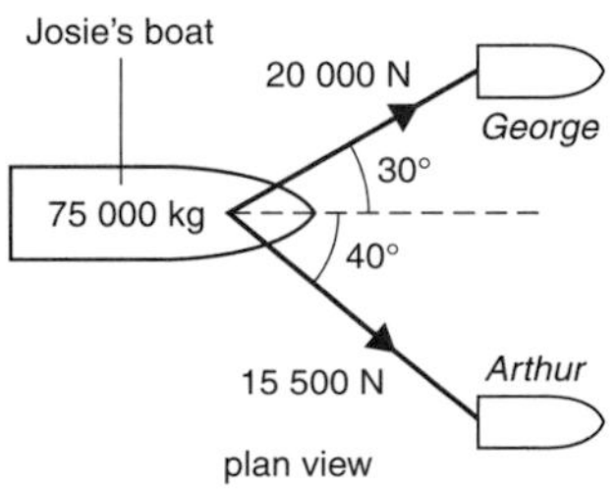

(8)

8 Tony is paragliding and being pulled along by a boat at constant velocity, as shown right. Draw a vector polygon of the forces acting on him. If Tony's mass is 70.0 kg, what are the values of the drag force, D, and the tension, T? Bearing in mind that forces always come in pairs, what is the driving force of the boat if the friction with the water is 1000 N?

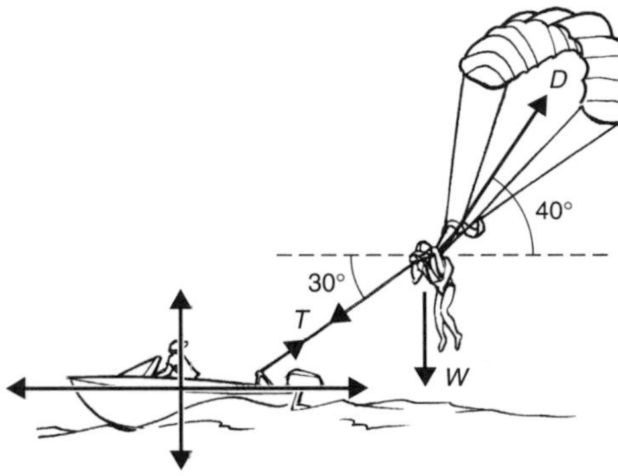

(10)

(Total 50 marks)

Unit 5 ENERGY

- Energy comes into virtually everything in physics, and is one of those underlying concepts that form the foundation of the subject. It is easier to say what it is not than what it is. It is not something tangible or physical in the sense that you can put it in a bottle or touch it, yet you can feel its effects. It can change form and pass from one object to another in several different ways. In a way, it is a measure of the effect that a force has on an object when it moves it. You can **quantify** (give a number to) this effect, and the change in energy that occurs is equal to the **work** that has been done on the object, or by it.

$\boldsymbol{W = F \times \Delta s}$

- **Work done** (W or ΔE) is defined as the product of force, F, and the distance moved, Δs, in the direction of the force. It is a scalar quantity, measured in joules (J), and is given by the equation $\boldsymbol{W = Fs}$ or $\boldsymbol{W = F\Delta s}$. If the force is not constant we use the average force.

  When an energy change takes place we say work is done. The work done is equal to the change in energy that has taken place; for instance, the change in kinetic energy or in potential energy (see opposite).

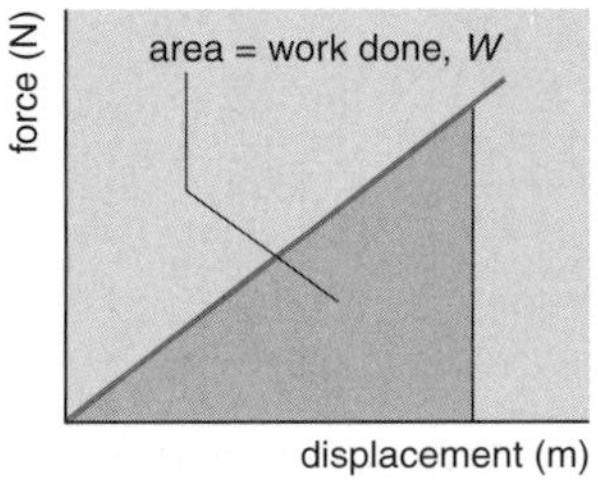

A graph of force versus displacement.

  The area under a force–displacement graph is the work done (see the section on graphs in unit 2 and the figure left). In nearly all cases work is only done on or by an object if it moves.

  The **energy** an object has is really the capacity of that object to do work, or the amount of work that has been done on it.

  Example *A packing case is dragged 6.0 m across a floor by a force of 200 N. What is the work done against friction, and into what form does the energy change?*

  Simply apply the equation.

$$W = F \times \Delta s \text{ so } W = 200 \times 6.0 = \mathbf{1200\,J}$$

  The energy is transformed into heat energy.

- One way of looking at energy is to think of it like money. If a group of people are given a fixed amount of money, that money can take several forms: different coins, notes, or cheques. It can be transferred from one form to another or from one person to another, yet the total amount is always fixed. This is in essence the **principle of conservation of energy**:

  The principle of conservation of energy states that energy cannot be destroyed or created, only changed from one form to another.

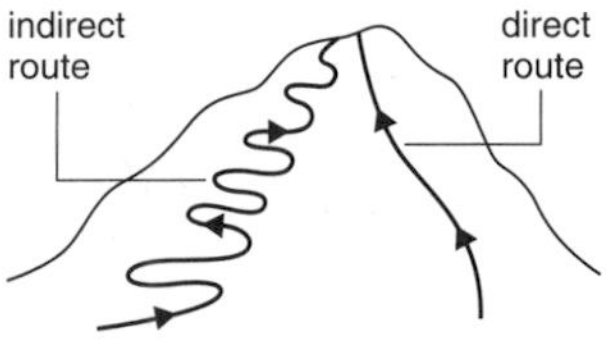

Two routes up a mountain.

- The time over which the energy change takes place is also important. For example, two people, Chris and Charlotte, take two different routes walking to the top of a mountain: Chris takes a short direct one straight up the mountain side, while Charlotte takes a longer, wiggly, indirect one. The gravitational potential energy change (see opposite) is the same, but if they both travel at the same speed, Chris takes less time to reach the top. He also finds it harder to climb than Charlotte. They are both using up the same amount of energy, but over different periods of time. Chris is using more **power** than Charlotte.

  Power is defined as the rate of doing work. It is a scalar quantity, measured in watts (W).

$\boldsymbol{P = \frac{W}{\Delta t}}$ or $\boldsymbol{P = Fv}$

$$P = \frac{W}{\Delta t}$$

  Also, $\Delta W = F \times \Delta s$ and $v = \frac{\Delta s}{\Delta t}$ so $P = \frac{F\Delta s}{\Delta t}$ or $P = Fv$.

  $P = Fv$ only gives you the instantaneous power being used at a particular velocity. This is sometimes called 'instantaneous motive power'. To calculate the average power of an object which is accelerating, use the equations $P = \Delta KE/t$ or $P = \Delta PE/t$ (KE and PE stand for kinetic energy and potential energy respectively – see opposite). The area under a graph of power against time is the work done.

- Lots of devices or machines convert energy from one form to another, but sometimes in the process some of the energy is converted into non-useful and unwanted forms. For example, an electric motor converts electrical energy into kinetic energy, but in the process some is converted into heat and sound energy. The effectiveness of the energy conversion into useful forms is called the **efficiency** of the device.

  Efficiency is the ratio of useful energy output to energy input. It is given as either a ratio or a percentage:

$$\textbf{efficiency} = \frac{\textbf{useful energy output}}{\textbf{energy input}} \times 100\% \text{ or } \frac{\textbf{useful power output}}{\textbf{power input}} \times 100\%$$

- On a fundamental level there are really only two forms of energy: **potential energy** (PE) and **kinetic energy** (KE).

- **Potential energy** is the energy that an object has due to its position or the arrangement of its atoms and molecules. This can take several forms:

  Gravitational PE is the energy that an object has because of its height above the ground (see right), given by the equations $\boldsymbol{PE = mgh}$ or $\boldsymbol{\Delta PE = mg\Delta h}$ where m is mass, g is acceleration due to gravity, h is height above the ground, and Δh is change in height. This only works as long as g can be considered constant, i.e. relatively close to the Earth's surface.

  Elastic PE is the energy stored in any stretched object. It is given by the equation $\boldsymbol{\Delta PE = \frac{1}{2}Fx}$ or $\boldsymbol{\Delta PE = \frac{1}{2}kx^2}$, where F is force, x is extension, and k is the spring constant (see unit 27).

  Chemical PE is the energy stored in interatomic and molecular bonds.

  Nuclear energy is the energy associated with the atomic nucleus seen in nuclear fission, nuclear fusion, and radioactivity. Energy $\boldsymbol{E = mc^2}$, where m is the mass converted into energy and c is the speed of light (see unit 15).

- **Kinetic energy** is the energy an object has because of its motion. It is given by the equation $\boldsymbol{KE = \frac{1}{2}mv^2}$, where m is mass, and v is velocity (see right). Several other types of energy are just different forms of kinetic energy.

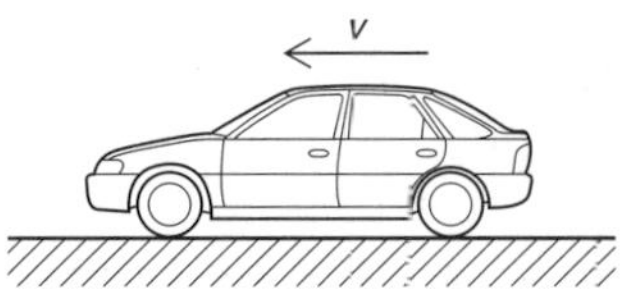

A moving car has kinetic energy.

kinetic energy: $\boldsymbol{KE = \frac{1}{2}mv^2}$

  Electrical energy is the energy associated with the movement of an electric charge as it flows around a circuit (see right). It is given by the equation $\boldsymbol{E = IVt}$ where I is current, V is voltage, and t is time. (See unit 9.)

  Heat energy is the thermal energy inside an object that gives rise to its temperature. This is the KE of the atoms vibrating or moving around. A change in thermal energy is given by the equation $\boldsymbol{\Delta Q = mc\Delta\theta}$, where m is mass, c is the specific heat capacity, and $\Delta\theta$ is the change in temperature (see unit 25).

  Sound energy is the energy associated with the sound waves that we hear. When it travels through a material the particles in that material move from side to side, passing energy to their neighbours, and so on through the material. (See unit 11.)

  Light energy, in a similar way to sound, is transmitted as a wave moves through a material. However, light is different in that it consists of photons of electromagnetic radiation whose energy is given by $\boldsymbol{E = hf}$ where h is the Planck constant and f is frequency. (See unit 13.)

  From the above you can see that it is usually possible to calculate changes in energy, whatever its form is, and express them as numbers.

$$\textbf{efficiency} = \frac{\textbf{output}}{\textbf{input}} \times 100\%$$

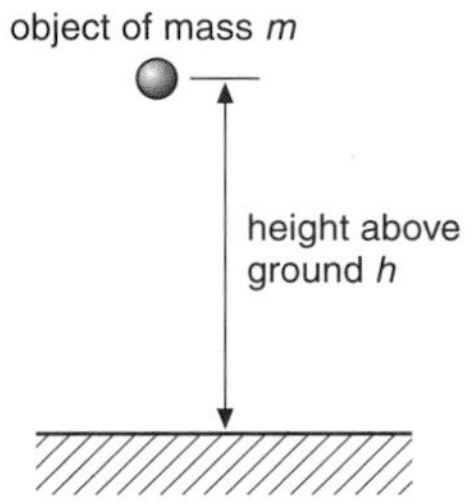

gravitational potential energy: $\boldsymbol{\Delta PE = mg\Delta h}$

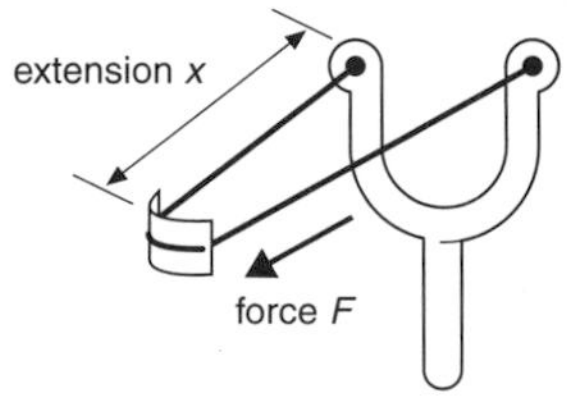

This catapult stores elastic potential energy.

elastic potential energy: $\boldsymbol{\Delta PE = \frac{1}{2}Fx}$

nuclear energy: $\boldsymbol{E = mc^2}$

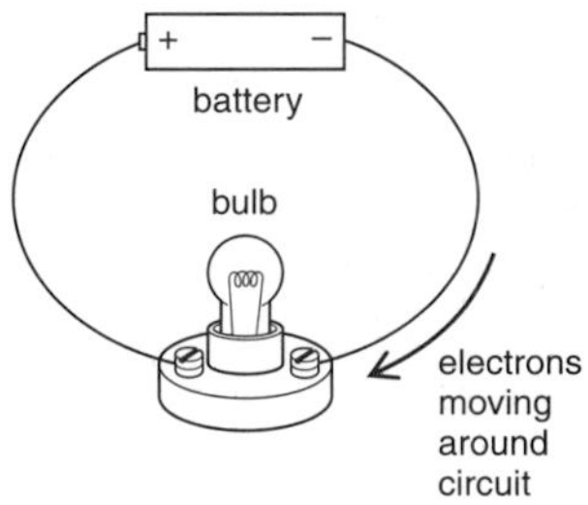

Electrical energy.

electrical energy: $\boldsymbol{E = IVt}$

heat energy: $\boldsymbol{\Delta Q = mc\Delta\theta}$

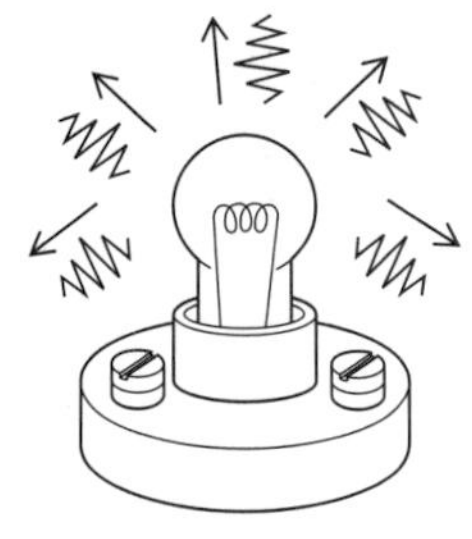

Photons of light being emitted.

light energy: $\boldsymbol{E = hf}$

TESTS

RECALL TEST

1 What is 'energy'?

(2)

2 What is the principle of conservation of energy?

(2)

3 What are the three types of potential energy?

(2)

4 Define 'work done'.

(2)

5 Define 'power'.

(2)

6 What is meant by the 'efficiency' of an engine?

(2)

7 What is meant by 'instantaneous power'?

(2)

8 Why does it feel harder to run up a flight of stairs than to walk up it, even though the change in PE is the same?

(2)

9 When a car applies its brakes and slows down what happens to its KE?

(2)

10 What is the area under a force–displacement graph equal to?

(2)

(Total 20 marks)

CONCEPT TEST

Take $g = 9.8\,\text{m/s}^2$

1 A ball of mass 4.00 kg falls from a height of 6.00 m to the ground.

a What is its original potential energy?

(2)

b What is its velocity just before impact with the ground?

(2)

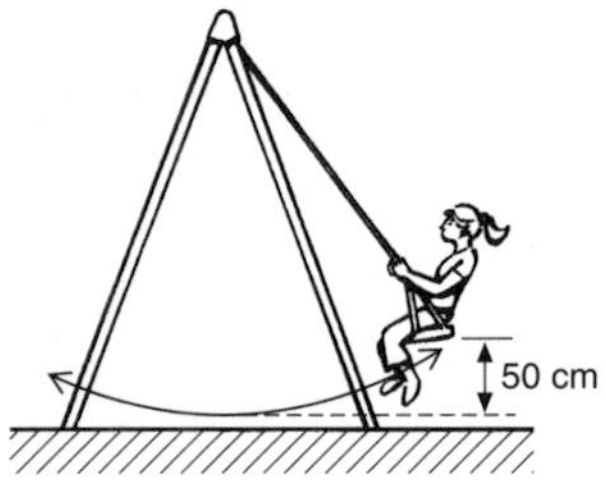

2 A child is on a swing as shown left. Outline the energy changes that occur, and calculate the maximum velocity.

(4)

3 Dom and Clare decide to run 1 km to get to a pub. Clare runs fast and covers the distance in a short time, whereas Dom runs more slowly and takes longer. Assuming that they use the same amount of energy in the process, discuss the advantages and disadvantages of each method in terms of power.

(4)

4 A ball of mass 200 g drops from a height of 10 m, and bounces to a height of 6.0 m. If the ball stays at a constant temperature, how much heat energy is gained by the ground?

____

______________________________ (4)

5 A sledge of mass 200 kg, with a man of mass 80 kg riding on it, is pulled by six dogs across the snow, with each dog exerting a force of 20 N (see right). If the sledge starts from rest, what will be its kinetic energy and velocity after it has moved 15 m?

____

______________________________ (4)

6 A engine used to raise a lift is 60% efficient, and the lift's mass is 2000 kg. How much energy would be needed to raise it 10 m?

____

______________________________ (4)

7 A toy car of mass 50 g follows the track shown right.

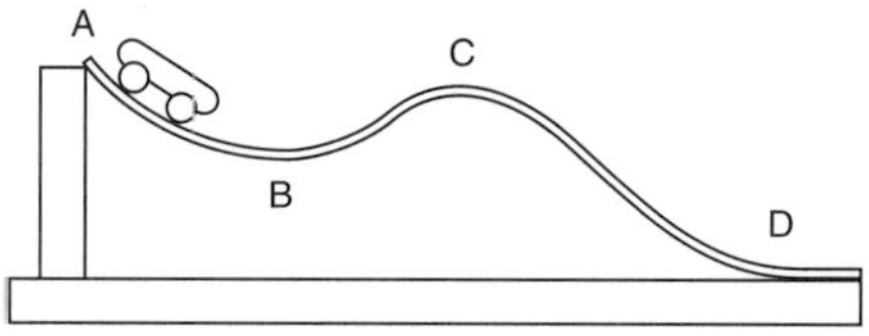

a Describe the energy changes that occur.

______________________________ (1)

b What will be the effect on the journey time from A to D if
i the mass of the toy is increased;

______________________________ (2)

ii the track is steeper between A and B?

______________________________ (2)

c If A is 1.2 m high and C is 0.9 m high what are the speeds at C and D?

______________________________ (4)

8 In the diagram on the right a container is being dragged up a slope in a loading bay. Calculate the energy dissipated against friction, and suggest a way of reducing it.

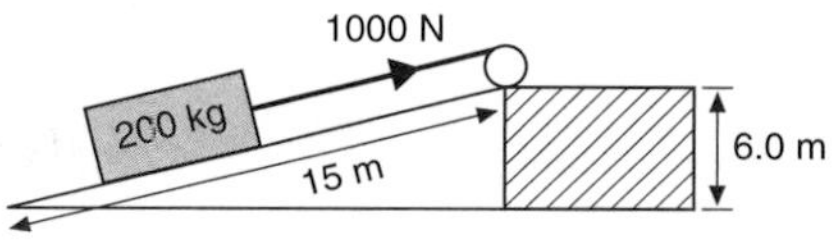

____

____

______________________________ (6)

9 A skier is going down a slope at a steady velocity of 20 m/s. If the frictional force opposing the motion is 15 N, what power is exerted in overcoming friction?

______________________________ (2)

10 The graph on the right shows a force acting on an object as it moves through a certain displacement. How much work is done during this motion?

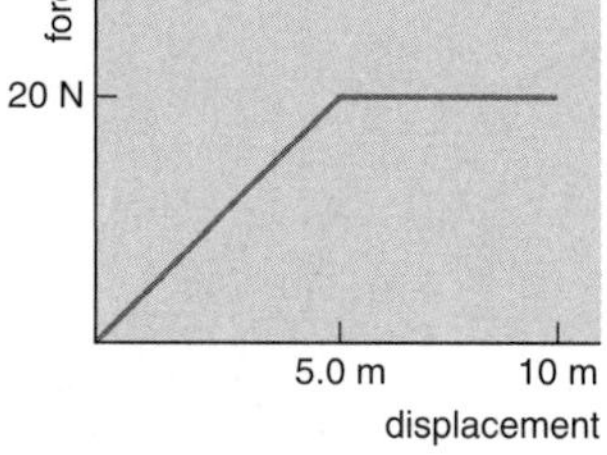

____

______________________________ (4)

11 An engine of a train locomotive produces a force of 2.00×10^8 N. Its mass is 5.00×10^6 kg, and its initial velocity is 4.00 m/s. What is the average frictional force acting on the engine when it reaches 25.0 m/s after travelling 300 m?

____

____

______________________________ (6)

(Total 50 marks)

FORCES IN ACTION (NEWTON'S LAWS)

Floating in space with a ball.

- Imagine you are floating in space, light-years away from anything, and in front of you is a ball which is stationary in relation to you. Assume that your presence does not affect the ball, so that there are no significant forces acting on it. If you exert a force on the ball by pushing with your hand, it will start to move in the direction of the force; remember force is a vector quantity.

  The ball's velocity is changing, so it must be accelerating throughout the time the force is applied. Once you stop pushing it with your hand it will stop accelerating and continue to move with a constant velocity equal to the value it had at the moment the force stopped. If you now apply another force to it by pushing it again, it will accelerate/decelerate, change direction, or both, depending upon the direction in which the force is applied. Summarizing:

  A force will cause an object to change its state of motion.

  This is harder to see on Earth, because most objects have many forces, such as friction, acting on them. The behaviour of forces is described using several simple rules, which are known as **Newton's laws of motion**.

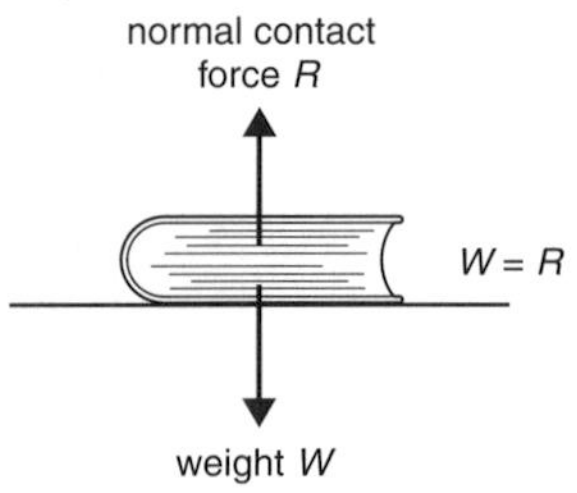

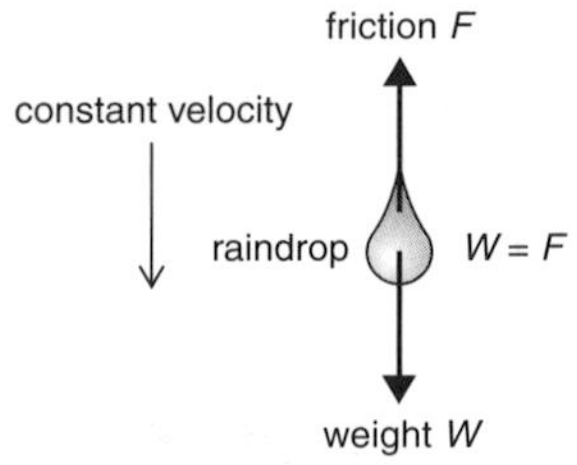

- Until a force acted on it, the ball in space stayed the way it was. More common everyday examples would be a stationary book lying on a table (left) or a raindrop falling with constant velocity through the air (below left).

  In each case the forces acting on the object cancel each other out, leaving a resultant of zero. This is **Newton's first law of motion** (**NI**), which is:

  An object will remain in a state of uniform motion or at rest, unless acted upon by a resultant non-zero force.

  'Resultant force' means the combined effect of all the forces acting.

- When the forces do not balance there is a resultant force. When the ball in space was initially pushed, it started to move, accelerating in the direction of the force. This is **Newton's second law of motion** (**NII**), which is:

  The resultant force acting on an object is directly proportional to the rate of change of momentum that it produces.

  Momentum is defined as the product of mass and velocity (see unit 7).

  So the law can be written as $\textbf{force} = \dfrac{\textbf{change in momentum}}{\textbf{change in time}}$

Newton's second law:
$F = \dfrac{mv - mu}{t}$
$F = ma$

This law can be put into several forms of equation:

$$F = \frac{mv - mu}{t} = \frac{m(v - u)}{t} = ma \quad \text{or } F = \frac{\Delta p}{\Delta t} = m\frac{\Delta v}{\Delta t} = ma$$

where mass is m, initial velocity is u, final velocity is v, time is t, a is acceleration, p is momentum, and force is F.

Its two most common forms are $F = \dfrac{mv - mu}{t}$ and $F = ma$.

These equations are used in situations where the mass being accelerated is constant. If it is not, we use

$$F = \frac{v(m_1 - m_2)}{t} \quad \text{or } F = \frac{m_1 u - m_2 v}{t} \quad \text{or } F = \frac{v\Delta m}{\Delta t}$$

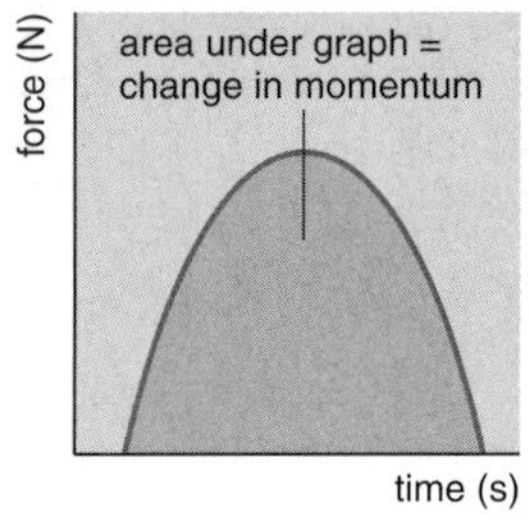

- The graph shows a typical variation of the force acting on an object during a collision with something else. When the object initially makes contact it is still moving and will both compress whatever it is hitting and be compressed by it. The greater the compression, the greater the repulsive forces produced between atoms. These collectively make up a force acting on the object that opposes the compression. The force reverses the object's direction of motion, thereby reducing the compression and the resulting force. The area under the graph equals the change in momentum or the impulse (see unit 7).

Example *A car of mass 2000 kg is travelling at a velocity of 34 m/s and hits a brick wall, stopping in 0.50 of a second. What is the average force acting on the car?*

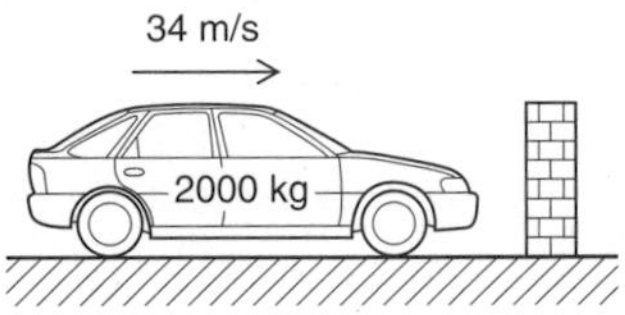

This is an NII situation because it involves momentum change. Draw a diagram (see right).

We use the equation $F = \frac{\Delta p}{\Delta t} = \frac{mv - mu}{t}$

$$F = \frac{2000 \times (34 - 0)}{0.50} = 136\,000 \quad \boldsymbol{F = 140\,000\,\mathrm{N}}$$

Example *Rain is falling on a flat roof. If each drop has a velocity of 20 m/s and a mass of 0.1 g, and the force acting on the roof is 4 N, how many rain drops hit the roof each second?*

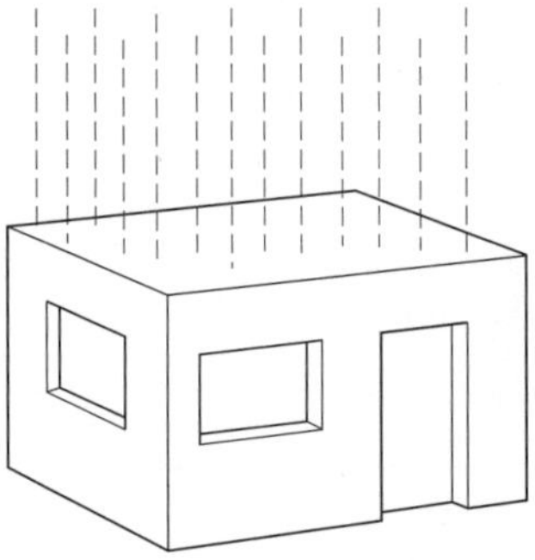

Assume that the rain drops do not bounce. This is similar to the previous example.

Start as before: $F = \frac{\Delta p}{\Delta t} = \frac{N(mv - mu)}{t}$ where N is the number of drops.

$$4 = \frac{N(20 \times 0.1 \times 10^{-3})}{1} \qquad N = \frac{4}{2 \times 10^{-3}} = \mathbf{2000}$$

- From the equation $F = ma$ we can define **1 newton**.
  1 newton is the force that will accelerate a mass of 1kg by 1m/s^2.
- When you push your finger against a table, you can feel the table pushing back against your finger. When you drag a chair across the floor you can feel it pulling against you. The conclusion from this is that forces do not act in isolation. When a force is applied to an object, that object exerts a force back in opposition to the original force. This is **Newton's third law (NIII)**:

  Every force that acts on an object has an equal and opposite force reacting to it.

  So all forces occur in action–reaction pairs which *must* be
  1 equal in magnitude (size);
  2 opposite in direction;
  3 the same type of force;
  4 acting on different objects.

Example *What is the reaction force to the gravitational pull, W, of the Earth acting on a falling skydiver?*

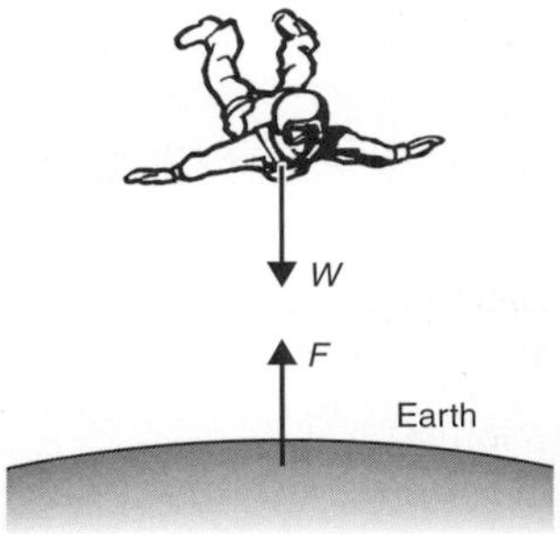

Once again, start by drawing a diagram (right).

The reaction force, F, is the gravitational pull of the skydiver acting on the Earth.

Example *The diagram on the right shows the forces acting on a book lying on a table. Use Newton's third law to determine the reaction forces that pair off with the two forces shown.*

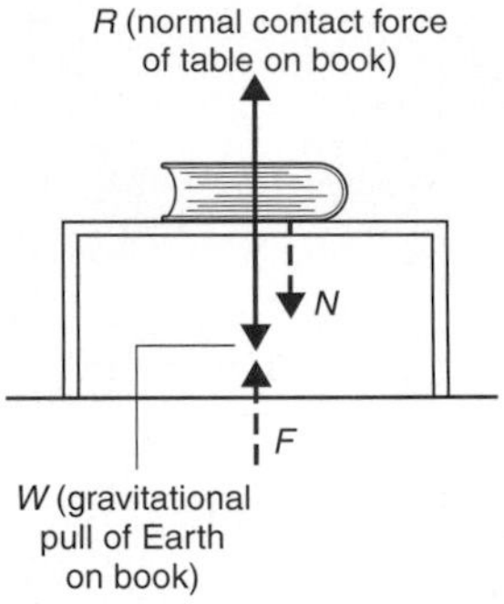

To get the reaction force to a given action force simply swap the positions of the two objects in the sentence.

R is the normal contact force of the table on the book.
N is the normal contact force of the book on the table.

W is the gravitational pull of the Earth on the book.
F is the gravitational pull of the book on the Earth.

TESTS

RECALL TEST

1 In what situations will no resultant force act on an object?

(2)

2 If there is a resultant force acting on a stationary object what will it do?

(2)

3 Using Newton's third law determine the reaction force that goes with the frictional force of the ground acting on a case being dragged across the floor.

(2)

4 What is equal to the area under a force–time graph?

(2)

5 What is meant by a 'resultant force'?

(2)

6 What is Newton's second law?

(2)

7 What is Newton's first law?

(2)

8 What is Newton's third law?

(2)

9 An object is moving at constant velocity. A force acts on it and, depending upon the direction it acts, could have several different effects. What are they?

(2)

10 Define the unit of one newton.

(2)

(Total 20 marks)

CONCEPT TEST

Take $g = 9.8$ m/s^2

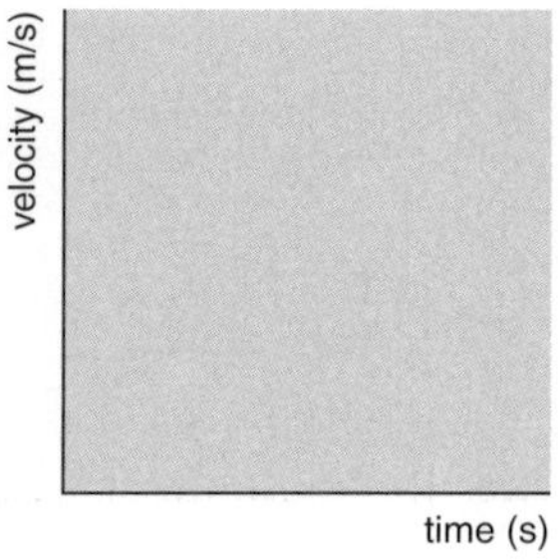

1 A ball bearing is held just beneath the surface of the water in a container. It is released and falls through the water. Initially it accelerates but then it reaches a steady velocity. With reference to what you know about forces or Newton's laws, explain why this happens. Draw a velocity–time graph for the motion.

(4)

2 A box of mass 8.0 kg is being dragged along the ground by a force of 30 N.

a If the friction force is 26 N what is the resulting acceleration?

(3)

b If the frictional force is one-quarter of the normal reaction force, what is the acceleration?

(3)

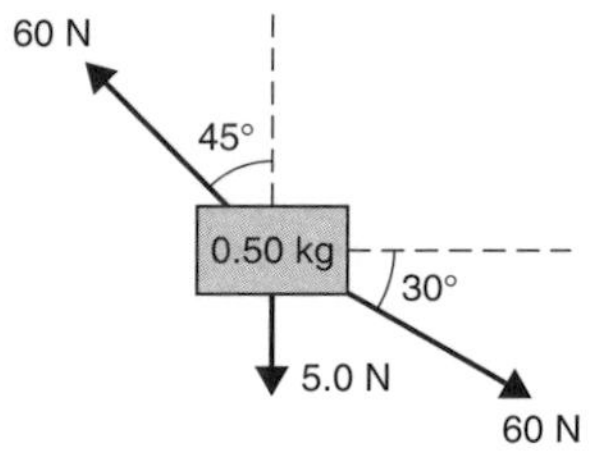

3 What is the resultant force acting on the object in the figure left, and what is its consequent acceleration? (Here g is taken as 10 m/s^2.)

(4)

4 What is the mass of the lift accelerating upwards in the figure on the right?

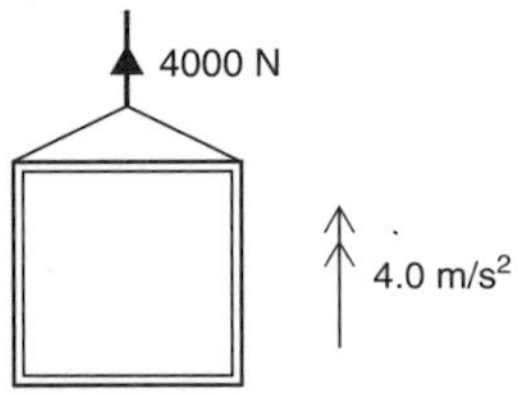

(4)

5 Explain how Newton's laws apply to the fish on the right as it swims with constant velocity at a constant depth. Draw the forces acting on the fish.

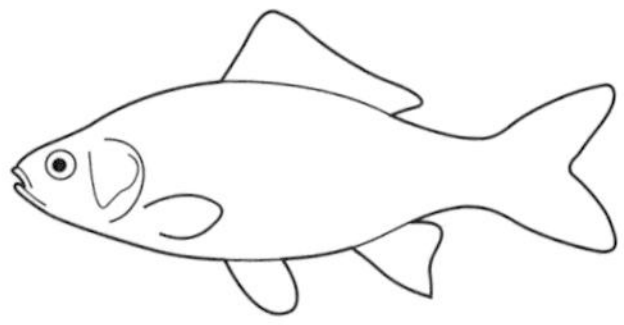

(4)

6 A squash ball of mass 0.20 kg travelling at 15 m/s hits the front wall of the court at 90° to the surface and bounces back in the opposite direction with a velocity of 8.0 m/s. What is its change in momentum and the average force exerted on it if the duration of the collision is 0.10 s? Draw force–time, acceleration–time, and velocity–time graphs for the duration of the impact on the axes below. Take the original direction of motion as negative.

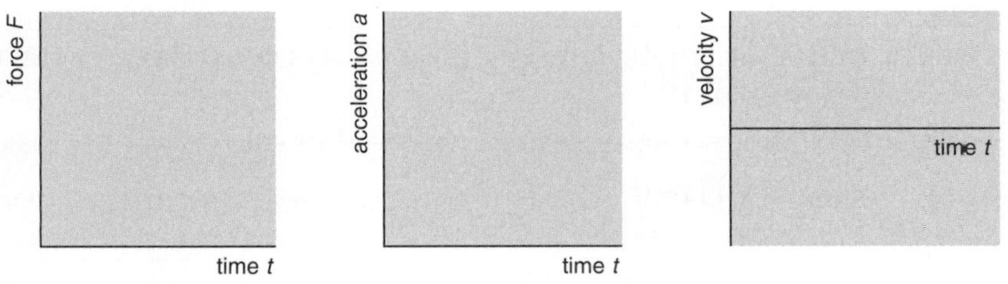

(8)

7 A spaceship of mass 2000 000 kg is moving with a constant velocity of 3.0×10^4 m/s as shown in the figure on the right. It is suddenly hit by a mysterious force of 8.0×10^{10} N for 10 s in the opposite direction to its motion. What is its final velocity and direction of motion?

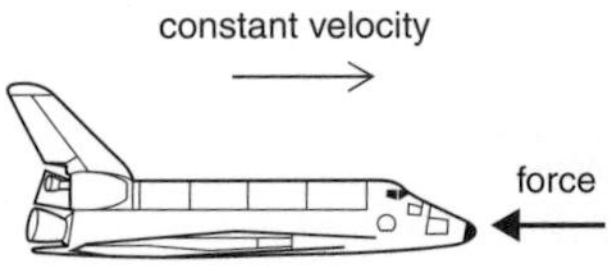

(6)

8 A car is driving through a rainstorm. Each raindrop of mass 0.000 05 kg hits the windscreen with a velocity of 16 m/s. If 500 raindrops hit the glass and run down the windscreen each second without bouncing, what force is exerted on the windscreen?

(4)

9 A brick is lying on the ground as shown in the figure on the right, with the forces shown acting on it. Write down what these forces are and with them their corresponding reaction pair force.

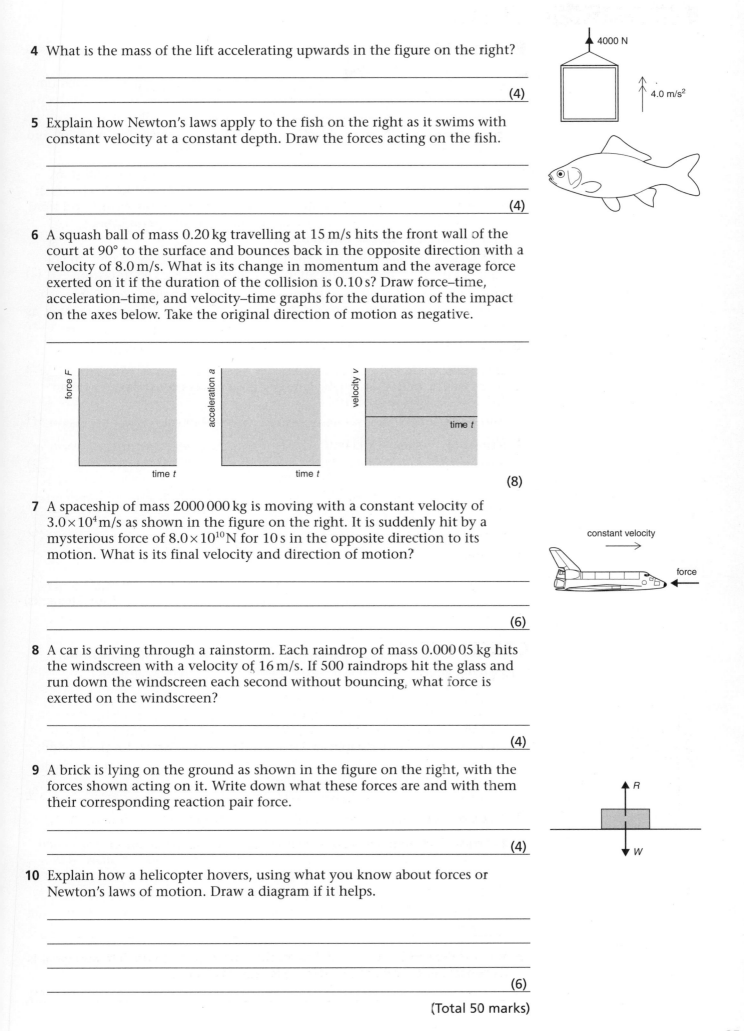

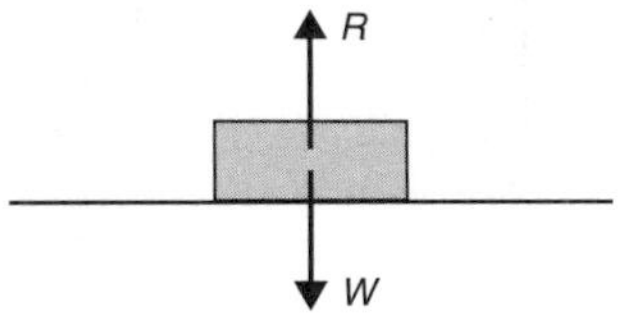

(4)

10 Explain how a helicopter hovers, using what you know about forces or Newton's laws of motion. Draw a diagram if it helps.

(6)

(Total 50 marks)

Unit 7 CONSERVATION OF MOMENTUM

- Momentum has already been mentioned in several units. It is necessary to expand upon the basic idea of momentum as there are many useful applications of it. People often talk about things having a certain 'momentum', and what they mean is that a moving object has a particular quality that keeps it moving in a particular direction or makes it hard to stop it moving.

momentum = mass × velocity
$p = mv$

The **momentum**, p, of an object is defined as the product of its mass, m, and velocity, v. It is a vector quantity, measured in kg m/s.

- The idea of momentum is particularly useful when considering collisions between objects, and is linked with the effects of forces acting on objects. It is related to Newton's laws of motion. If you take Newton's second and third laws and combine them you get what is known as **the principle of conservation of linear momentum**:

In a system of colliding bodies the total momentum before a collision is equal to the total momentum after the collision, in a given direction.

From this you can work out the velocities of objects before and after collisions. 'Conservation' means that the total amount is always the same, as with energy.

- When things collide they can do it in two different ways: **elastic** and **inelastic**.

In an **elastic collision** kinetic energy (KE) is always conserved; but in an **inelastic collision** KE is not conserved.

Both total energy and momentum are conserved in each type of collision.

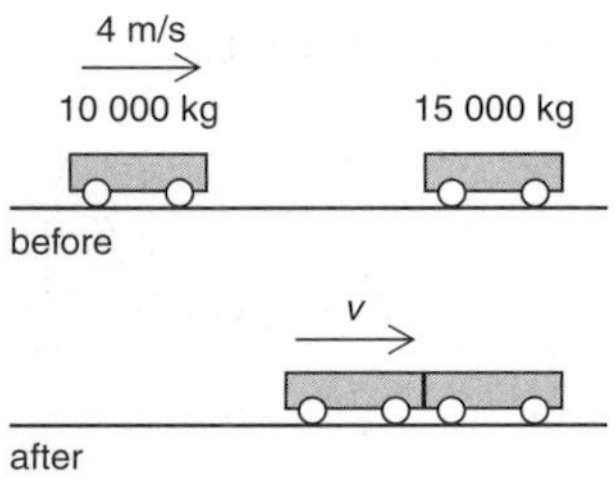

Example *A train carriage of mass 10 000 kg travelling at a constant velocity of 4.0 m/s in the sidings of a railway station hits a stationary carriage of mass 15 000 kg and couples to it. What is the final velocity of the combined carriages?*

Draw a diagram and then apply the principle of conservation of momentum.

momentum before = momentum after
$(10\,000 \times 4.0) + (0) = (25\,000)v$
$\mathbf{v = 1.6\ m/s}$

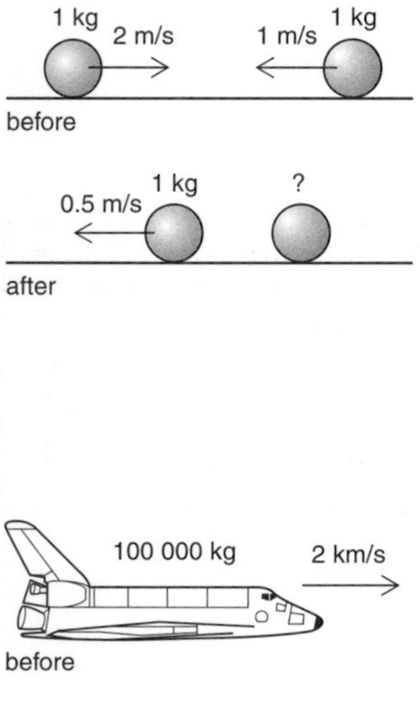

Example *A snooker ball of mass 1.0 kg, travelling at 2.0 m/s, collides head on with another identical ball travelling in the opposite direction with a velocity of 1.0 m/s. If the first ball bounces back with a velocity of 0.50 m/s, what does the second ball do? Is this collision elastic?*

Draw a diagram and apply the principle of conservation of momentum. Remember, momentum is a vector quantity. We will take moving to the right as positive.

momentum before = momentum after
$(1.0 \times 2.0) - (1.0 \times 1.0) = (-1 \times 0.50) + (1.0)v$
$v = 1.5$ m/s. **The second ball bounces back with a velocity of 1.5 m/s.**

If elastic, KE before = KE after, so $(0.5 \times 1.0 \times 2.0^2) + (0.5 \times 1.0 \times 1.0^2)$ should equal $(0.5 \times 1.0 \times 0.50^2) + (0.5 \times 1.0 \times 1.5^2)$
KE before = 2.5 J   KE after = 1.3 J

so it is **not elastic**.

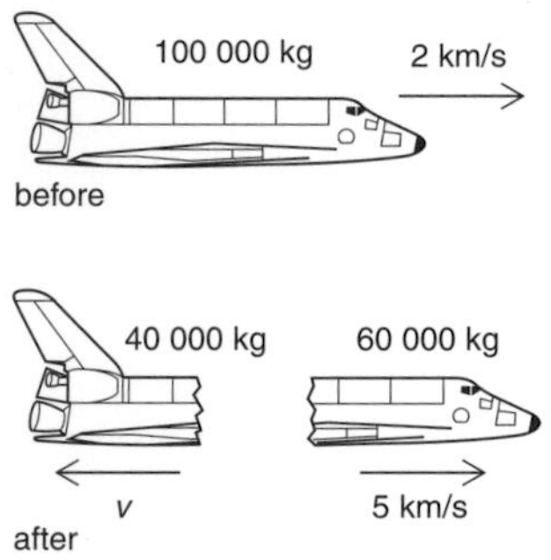

Example *A space ship of mass 100 000 kg and velocity 2.0 km/s explodes and breaks up into two pieces. One piece of mass 60 000 kg continues to move in the original direction with a velocity of 5.0 km/s, the remainder moves backwards in the opposite direction. What is its velocity?*

Draw a diagram and apply conservation of momentum. We will take moving to the right as positive.

momentum before = momentum after
$(100\,000 \times 2000) = (60\,000 \times 5000) + (40\,000)v$
$2.0 \times 10^8 = (3.0 \times 10^8) + (4.0 \times 10^4)v$
$\mathbf{v = -2.5\ km/s}$

- When an object's momentum changes, a force is required to make it happen. The effect of that force depends upon how long it is applied for. This leads us into the idea of **impulse.**

  Impulse is defined as the product of force and the time over which the force is applied. This is also equal to the change in momentum.

  Taking Newton's second law and rearranging $F = (mv - mu)/t$ gives us

  impulse = $\boldsymbol{Ft = (mv - mu)}$ or $\boldsymbol{Ft = \Delta p}$ Units are N s or kg m/s.

  m is mass, v and u are velocities, F is force, t is time, and Δp is change in momentum.

impulse = change in momentum

$\boldsymbol{Ft = mv - mu}$

- From this equation you can see that if the change in momentum, Δp, is fixed, then if you increase time, t, the force, F, is reduced. Examples of this include:

  1 the way in which you bend your knees when landing on the ground after jumping off a high wall, in order to prolong the duration of the impact with the ground and so not break your legs;

  2 the way in which you move your hands in the direction a cricket ball is moving when you catch it, to increase the time of the impact and so stop it hurting;

  3 crumple zones in cars, which fold up like paper, increasing the time of any impact, and so reducing injury to the people inside.

A child jumping off a wall.

Catching a cricket ball.

A car hitting a wall.

Example *A large oil tanker of mass* 2.5×10^6 *kg is leaving a dock. It is attached to a tug boat to pull it clear of the harbour. It is initially at rest, and receives an impulse of* 3.0×10^6 *N s. What is its final velocity? If the impulse acts for 2.0 s what force is applied?*

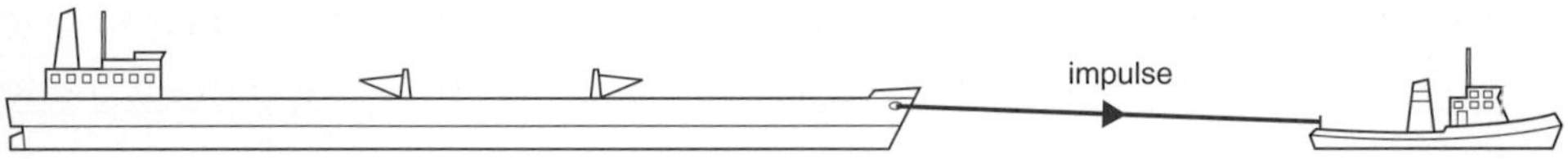

Start by drawing a diagram (see above).
Take the equation

$$\text{impulse} = mv - mu$$
$$3.0 \times 10^6 = (2.5 \times 10^6)v - 0$$
$$\boldsymbol{v = 1.2\ \text{m/s}}$$

$$\text{force} = \text{change in momentum} \div \text{time}$$
$$F = (3.0 \times 10^6)/2.0$$
$$\boldsymbol{F = 1.5 \times 10^6\ \text{N}}$$

- The relative velocity with which two objects separate from each other after a collision is proportional to their relative velocity of approach. The constant of proportionality is called the **coefficient of restitution**, e, which depends upon the elastic properties of the objects and their type of surface.

  relative velocity of separation = $\boldsymbol{e \times}$ **relative velocity of approach**

  The values of e are linked to the type of collision (see right). This is not on all A-level syllabuses so check yours before worrying about it.

Collision	e
completely inelastic	0
inelastic	<1
elastic	1

TESTS

RECALL TEST

1 What is 'momentum'?

(2)

2 What is the principle of conservation of momentum?

(2)

3 What is the difference between an elastic and an inelastic collision?

(2)

4 Define 'impulse'.

(2)

5 What are the units of impulse?

(2)

6 Why does a parachutist bend her knees and roll to land on the ground?

(2)

7 Is an explosion an elastic or inelastic situation?

(2)

8 Why is it necessary to define the direction as positive or negative when dealing with momentum, but not when dealing with kinetic energy?

(2)

9 Is it true that true elastic collisions only occur on an atomic level?

(2)

10 Explain how crumple zones in cars reduce injury to passengers.

(2)

(Total 20 marks)

CONCEPT TEST

Take $g = 9.8\,m/s^2$

1 A gun of mass 3.0 kg fires a bullet of mass 30 g at 400 m/s. What is the recoil velocity of the gun?

(4)

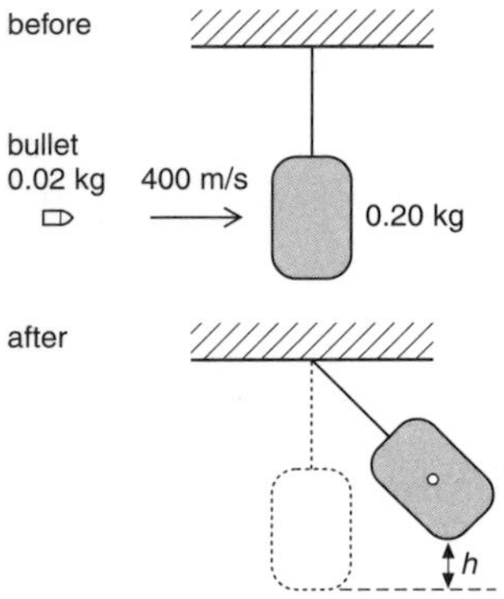

2 An arrow of mass 400 g travelling at a velocity of 25 m/s hits and sticks into an apple of mass 1.0 kg. What is their combined velocity immediately after impact?

(4)

3 A bullet of mass 20 g, travelling at 400 m/s, hits a sand bag, mass 0.20 kg, suspended by a string (see left). If the bullet sticks in the bag, through what height does the bag swing up?

(6)

4 A boy is sitting in a boat with a combined mass of 200 kg moving at 2.0 m/s directly towards a friend standing on the bank. His friend throws a ball to him, which gives him an impulse of 750 N s in the opposite direction. What is his final velocity?

(4)

5 A railway car of mass 2500 kg, moving at a velocity of 4 m/s, collides with and sticks to a stationary identical car. What is the resultant velocity?

(4)

6 A meteor of mass 20 000 kg, travelling at a velocity of 500 m/s, hits another object which gives it an impulse of 2.00×10^6 kg m/s in the opposite direction. What is its final velocity?

(4)

7 A squash ball of mass 20 g travelling at 12 m/s hits a wall and returns in the opposite direction with a velocity of 9.0 m/s (see right). If the duration of the impact is 0.15 s, what is the impulse given to the ball and the force acting on it? Draw a graph of momentum against time for the collision of the ball with the wall, on the axes shown right.

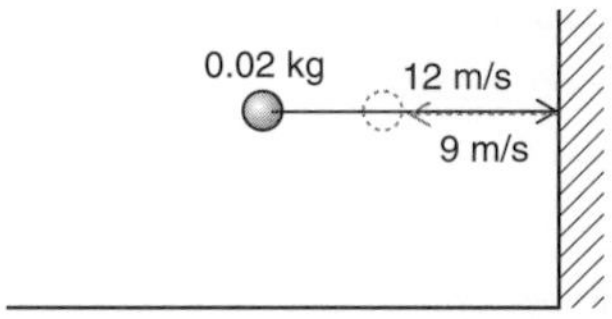

(6)

8 In a gas, atoms hit the side of a container elastically. Using the fact that the atoms will move faster when the temperature is increased, explain what will happen to the gas pressure when the gas is heated. What will happen to the gas pressure if we decrease the volume of the container while keeping the temperature constant?

(4)

9 A spaceship of mass 4000 kg is moving with a constant velocity of 3.12×10^5 m/s, as shown right. It is suddenly hit by a mysterious impulse of 1.65×10^9 N s in the opposite direction to its motion. What is its final velocity and direction of motion?

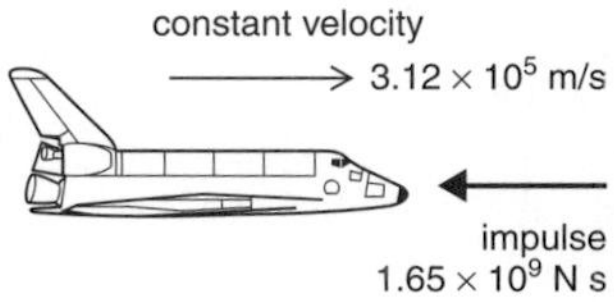

(4)

10 You are fishing by the side of a river. A small boat of mass 40 kg has broken free of its moorings and is drifting towards your fishing tackle at 0.5 m/s. You decide to try and stop it moving by throwing lumps of earth at it. If an average lump of soil has a mass of 1 kg and hits the boat with a velocity of 2.5 m/s, how many lumps of soil are required to stop the boat moving?

(4)

11 Explain why parachutists bend their knees and roll when they hit the ground. If a parachutist's mass is 80 kg, and the velocity on impact is 12 m/s, what is the minimum time required for the impact if the parachutist does not break a leg? Take the force required to break a leg as 600 N. (Don't forget you have two legs, and you have a weight.)

(6)

(Total 50 marks)

Unit 8a CIRCULAR MOTION

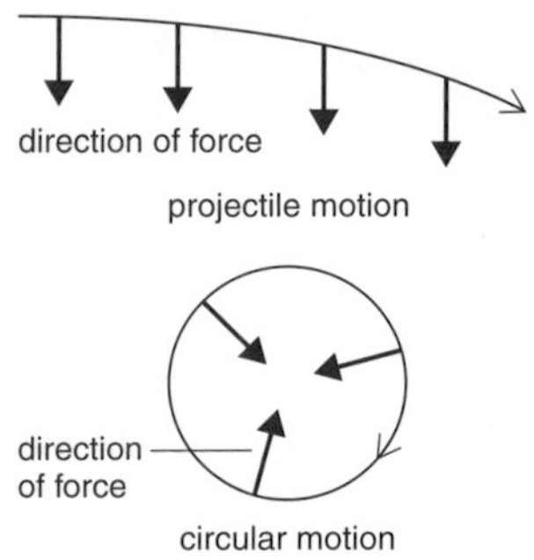

Projectile motion and circular motion. If an object has a constant speed and a force acting at 90° to the direction of motion it will move in a circular path.

- Units 1 and 2 dealt with motion in a straight line and parabolic (projectile) motion. Projectile motion requires a constant velocity in one direction and a constant acceleration in some other direction. This unit deals with what happens if the acceleration changes direction, and specifically if it is always at 90° to the direction of motion.

- From Newton's first law, an object continues to move in a straight line unless there is a force acting on it. An object moving around a circle is quite clearly changing its direction so there must be a force acting on it. From NII, $F = ma$; if there is a force there will also be an acceleration. The speed, however, is constant, so the force must act in such a direction as not to increase the speed. Therefore the force *must* act towards the centre of the circle, for if it was acting in any other direction a component of it would increase or decrease the speed. This is called the **centripetal force.**

  The **centripetal force** is the force acting towards the centre of a circle that enables an object to move in a circle. It is given by the equation

  $$F = \frac{mv^2}{r}$$

  where m is mass, v is velocity, and r is the radius of the circle.

centripetal force:

$$F = \frac{mv^2}{r}$$

- Centripetal force is not a force in its own right, but is provided by the resultant of other forces. Consequently, it should not be drawn in free-body force diagrams unless asked for. 'Centrifugal force' does not exist; it is just a term used to describe the sensation of being pushed outwards as you go around a circle, when in fact all you are doing is obeying NI.

- As an object turns through a circle it sweeps out an angle. The angle it has turned through is called its **angular displacement,** θ. The rate at which it does this is called **angular velocity**, ω.

  Angular velocity is the rate of change of angular displacement. Units are radians/second. In one complete revolution of a circle the angle turned through is 2π radians. The time taken to complete one revolution is the **period**, T.

angular velocity = $\dfrac{\textbf{angle turned through}}{\textbf{time taken}}$

$$\omega = \frac{\Delta\theta}{\Delta t} \text{ or } \omega = \frac{2\pi}{T}$$

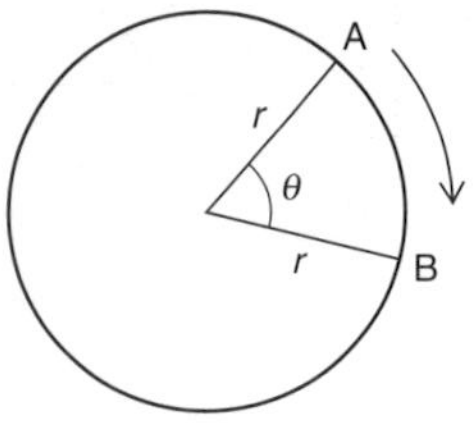

- If an object is moving around a circle as shown below left, the average speed in going from A to B is given by

  $$\text{speed} = \frac{\text{length of arc AB}}{\text{time}} \qquad v = \frac{r\theta}{t}$$

  We know that $\omega = \dfrac{\theta}{t}$ so $\boldsymbol{v = r\omega}$

  If we substitute this into $F = \dfrac{mv^2}{r}$ we get $\boldsymbol{F = mr\omega^2}$

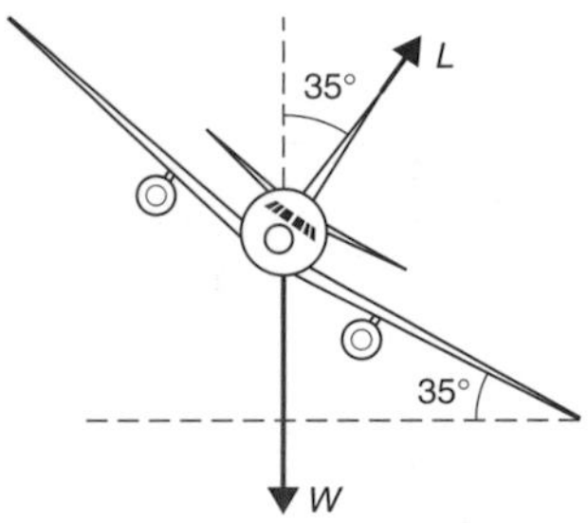

Example *An aircraft is banking as it turns, as shown left. What is the radius of curvature of the turn if the aircraft's velocity is 200 m/s and it is banked at 35°?*

Resolve the lift force L into two components: vertical and horizontal. The vertical component equals the weight and the horizontal component provides the centripetal force. This gives two equations:

$L \cos\theta = mg$ and $L \sin\theta = mv^2/r$; putting these together gives

$\tan\theta = v^2/rg$ so $\tan 35° = 200^2/9.8r$ and $\boldsymbol{r} = \textbf{5.8 km}$

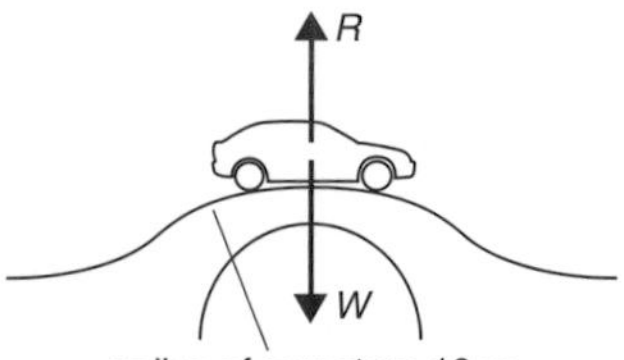

radius of curvature 10 m

Example *What is the maximum speed a car can go over a hump-backed bridge without losing contact with the ground, if the radius of curvature is 10 m?*

The resultant of the forces in the vertical direction provides the centripetal force; $W - R = mv^2/r$. If it just loses contact, $R = 0$ so

$mg = mv^2/r$ so $v^2 = rg = 10 \times 9.8$ $\quad \boldsymbol{v} = \textbf{10 m/s}$ (This is the same whatever the mass.)

STATICS

- Statics deals with objects that are stationary with respect to their surroundings. For an object to be stationary there must be no resulting forces acting on it, and no resulting moment. These are the **conditions for static equilibrium**.

  For an object to be in static equilibrium there must be no resulting force and no resulting moment acting on it.

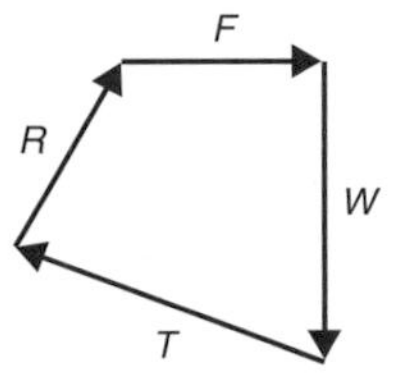

A closed vector polygon of forces.

- This can be expressed in other ways. Instead of saying there is no resultant force we could say that all the forces form a closed vector polygon (see right). Instead of saying there is no resultant moment we could say that the principle of moments must apply (see unit 3), or, in the case of an object acted upon by three forces, that they must all act through the same point (see right).

  Moments may be taken about any point in an object. If it is in rotational equilibrium the principle of moments applies. To make calculations involving an unknown force as simple as possible, take moments about a point through which it passes. The moment due to this force will be zero, as its perpendicular distance to the pivot is zero. When solving statics problems, equate all forces or their components in two directions at right angles to each other (usually vertically and horizontally) and take moments about one or more points.

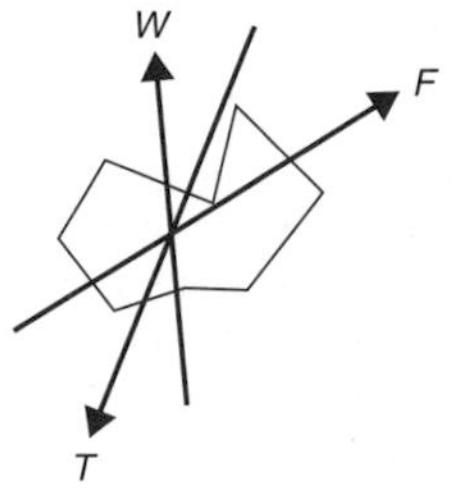

Three forces acting through the same point.

Example *A lorry of mass 10 000 kg is crossing a bridge. The bridge is uniform and has a mass of 100 000 kg. What are the forces on the two ends of the bridge when the lorry is 3/4 of the way across?*

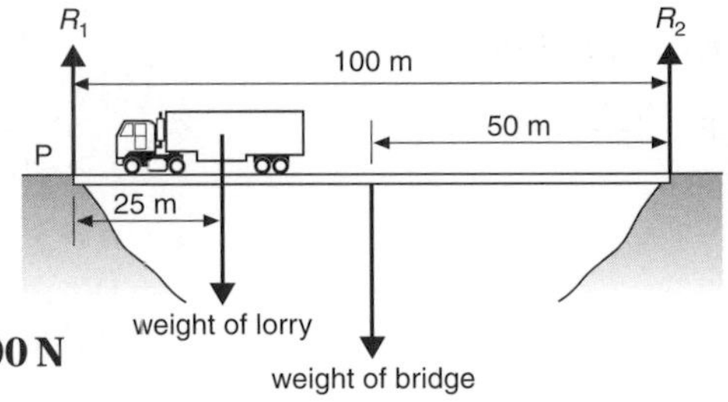

Take moments about P: $\Sigma\circlearrowright = \Sigma\circlearrowleft$

Σ (clockwise moments) = Σ (anti-clockwise moments)
$(100\,000\times9.8\times50) + (10\,000\times9.8\times25) = R_2\times100$ so $\boldsymbol{R_2} = \mathbf{510\,000\,N}$

Equating forces in the vertical direction:

$R_1 + R_2 = (10\,000\times9.8) + (100\,000\times9.8)$ so $\boldsymbol{R_1} = \mathbf{560\,000\,N}$

Example *A pub sign is supported by a chain as shown right. If the mass of the sign is 20 kg what is the tension in the chain and the force at the hinge A?*

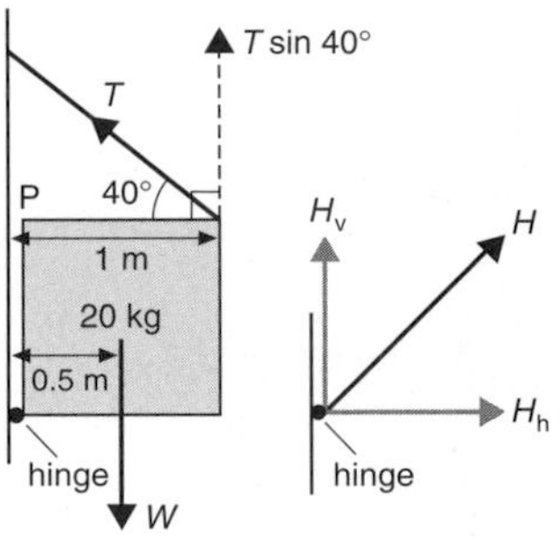

Take moments about P: $\Sigma\circlearrowright = \Sigma\circlearrowleft$

Σ (clockwise moments) = Σ (anti-clockwise moments)
$(20\times9.8\times0.5) = ((T\sin 40)\times1)$ $\quad \boldsymbol{T} = \mathbf{150\,N}$

Equating forces horizontally: $H_h = T\cos 40$ so $H_h = 115$ N
Equating forces vertically: $H_v + T\sin 40 = 20\times9.8$, $H_v = 100$ N

The force at the hinge is given by $H^2 = H_v^2 + H_h^2$ so $\boldsymbol{H} = \mathbf{150\,N}$

Example *The power is delivered to an electrical train by the overhead pylon shown in the upper figure right. Points X, Y, and Z are all freely pivoted. The force acting down due to the conductor is 600 N. What are the forces acting in the struts A and B? Are they compressive or tensile? What force acts on the base of the pylon?*

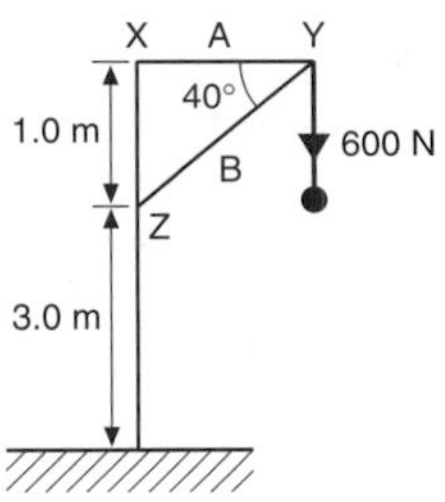

Take point Y. There is a downwards force of 600 N, so there must be an upwards force required to balance it. This can only come from a force in B acting diagonally upwards, as A is horizontal. Therefore B must be under compression with a component of the 600 N force pushing down on it. This produces a reaction force from Z acting upwards. The horizontal component of this force will place A under tension. So, equating forces vertically gives

$B\sin 40 = 600$ so $\boldsymbol{B} = \mathbf{930\,N}$ (compression)
Equating horizontally: $930\cos 40 = A$ so $\boldsymbol{A} = \mathbf{710\,N}$ (tension)

Look at the forces acting on the vertical strut. To find R, you need its vertical and horizontal components. Equating vertically gives

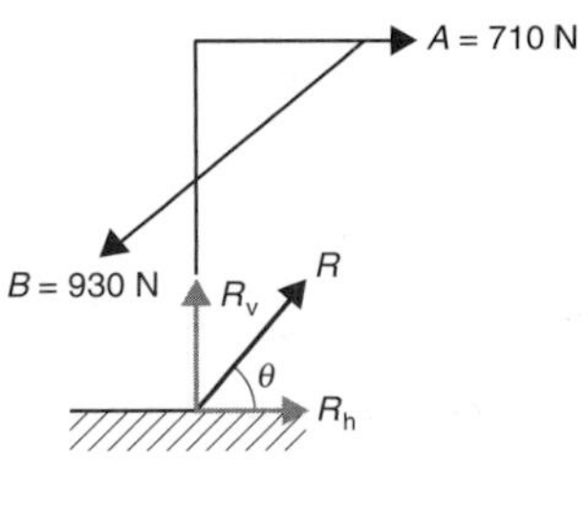

$R_v = B\sin 40$ so $R_v = 600$ N. Taking moments about Z: $A\times1.0 = R_h\times3.0$
so $R_h = 240$ N, $R^2 = R_h^2 + R_v^2$ so $R^2 = 240^2 + 600^2$ and $\boldsymbol{R} = \mathbf{650\,N}$

$$\tan\theta = \frac{R_v}{R_h} \text{ so } \theta = \tan^{-1}\frac{600}{240} = \mathbf{68°}$$

TESTS

RECALL TEST

1 What is meant by 'centripetal force'?

(2)

2 What is 'centrifugal force'?

(2)

3 In what direction does a centripetal force act?

(2)

4 What is 'angular velocity'?

(2)

5 Why are centripetal forces not put in free-body force diagrams?

(2)

(Total 10 marks)

CONCEPT TEST

Take $g = 9.8\ \mathrm{m/s^2}$

1 A car of mass 800 kg is turning a corner with a radius of curvature of 25 m. Where does the centripetal force come from, and what is the maximum speed it can take through this corner if the centripetal force is 3000 N?

(4)

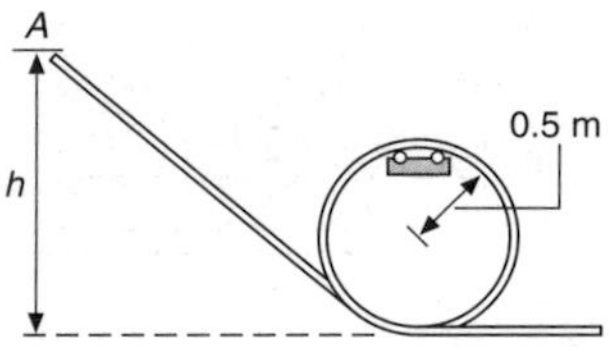

2 A toy car does a 'loop-the-loop' (left). What is its minimum velocity if it just remains in contact with the track at the top of the circle, if the circle has a radius of 0.50 m? If the car starts from rest at point A, what would be the height of point A above the ground?

(4)

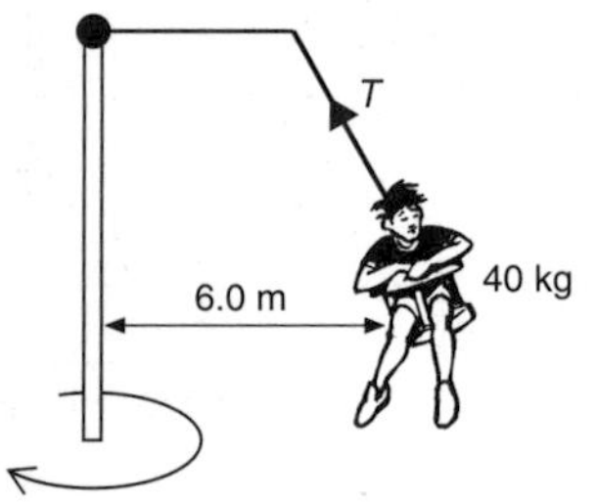

3 A child is sitting on a fairground ride, as shown left. The ride turns through one complete revolution every four seconds. If the combined mass of the child and the seat is 40 kg, and the radius of the circular path is 6.0 m, calculate the tension, T, in the support.

(4)

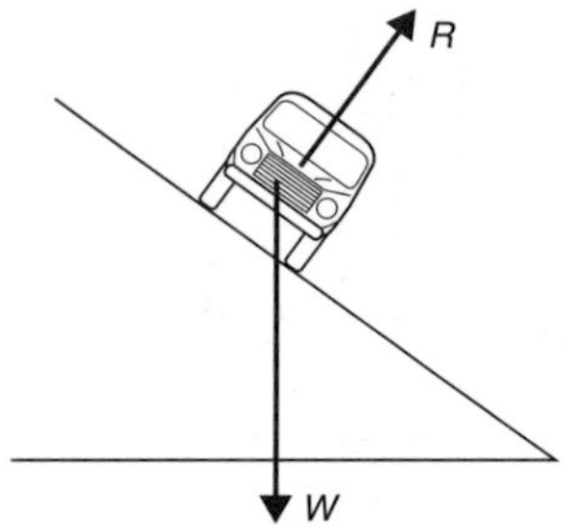

4 A van is going around a corner with a banked track, as shown left. If friction is negligible where does the centripetal force come from, and what is the angle of banking required if the speed is 14 m/s and the radius of curvature is 30 m?

(4)

5 The blades of a helicopter droop as shown. The maximum angular velocity they can rotate at with this droop before breaking is 5.0 rev/s. If the length of the blades is 5.0 m, and their mass is 30 kg, what force is needed to break the blades?

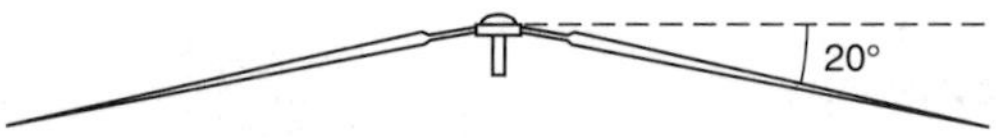

(4)

(Total 20 marks)

TESTS

RECALL TEST

1 What are the conditions for static equilibrium?

(2)

2 What is a vector polygon of forces?

(2)

3 What is the principle of moments?

(2)

4 What point would you normally take moments about?

(2)

5 What do you do to solve a static equilibrium situation?

(2)

(Total 10 marks)

CONCEPT TEST

Take $g = 9.8\,\text{m/s}^2$

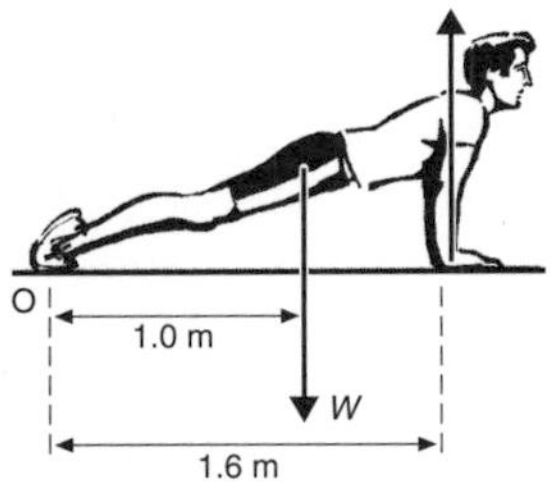

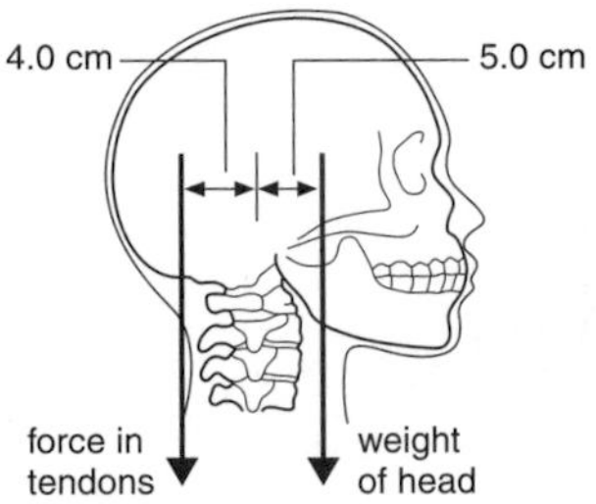

1 A man of mass 82 kg is doing press-ups as shown above right. What is the force in each arm?

(2)

2 A person's skull is pivoted about the top of the spine as shown above right. If the mass of the head is 10 kg, what is the force in the tendons at the back of the neck?

(2)

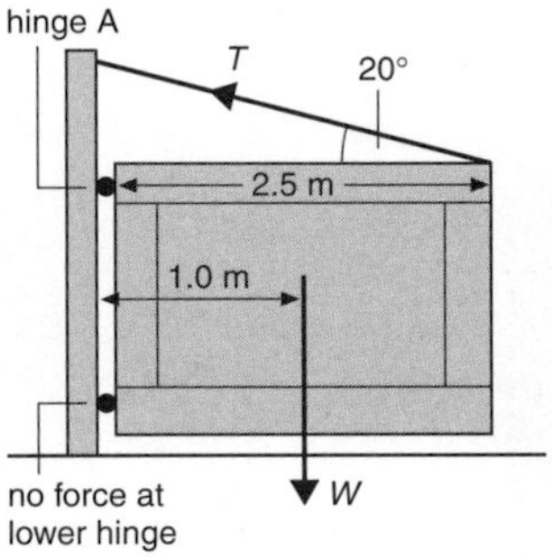

3 The gate to a field is shown above right. It is supported by a rope as shown. If the gate's mass is 30 kg, what is the tension T, and what is the force at the hinge A? Assume there is no force acting at the bottom hinge.

(4)

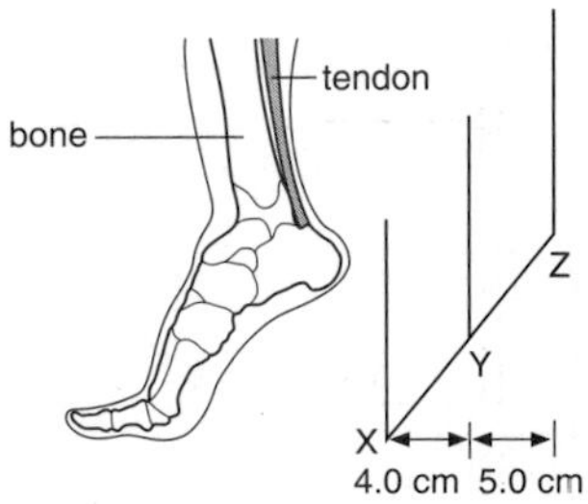

4 A dancer is standing on his toe as shown right. If the mass of the dancer is 75 kg, what are the values and directions of the forces acting through points X, Y, and Z? Are the forces at Z and Y compressive or tensile?

(4)

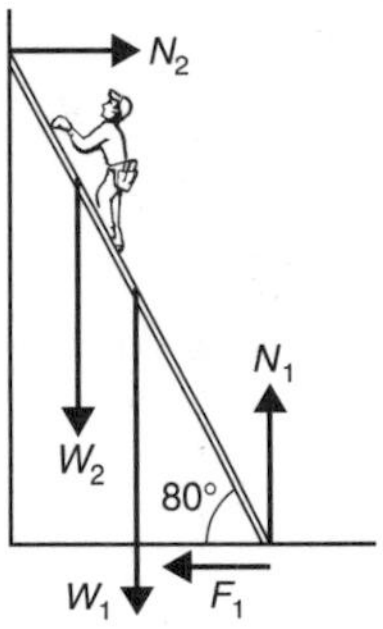

5 A ladder of mass 25 kg and length 4.0 m is resting against a wall (see right). A man of mass 70 kg is 3/4 of the way up the ladder. Taking the wall as smooth and the ground as rough, determine the size and direction of the force acting on the bottom of the ladder.

(4)

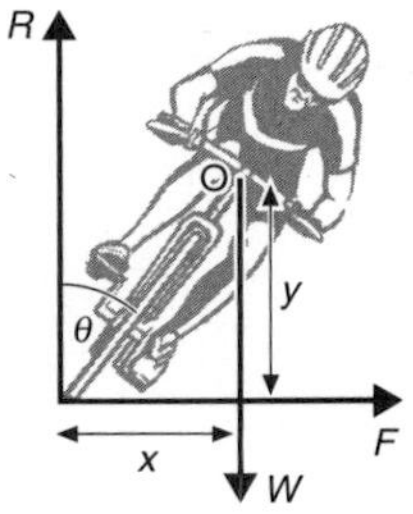

6 A cyclist is turning a corner on flat ground, as shown right. If the mass is 80 kg, the speed is 20 m/s, and the radius of curvature is 15 m, what is the angle of inclination from the vertical?

(4)

(Total 20 marks)

Unit 9

ELECTRICITY

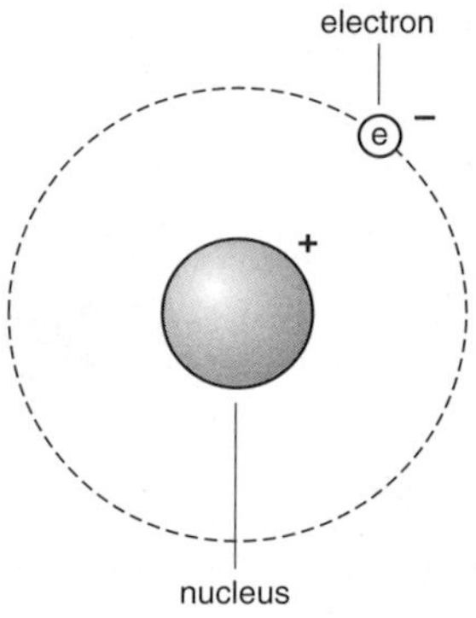

Simple atomic structure.

- There are two fundamental ideas in electricity: **current** and **potential difference**. Electric current deals with the movement of charge, potential difference deals with the energy used in moving the charge.
- Consider the simple model of atomic structure shown left. You should know that in an atom negatively charged electrons are present around a positively charged nucleus, producing a net atomic charge of zero. On a larger scale, an object becomes negatively charged when it has additional electrons present, and positively charged when some of its atoms have lost electrons. **Charge**, *Q*, is a property that electrons have, and like energy it is always conserved. The amount of charge present is always some multiple of the charge on an electron. It can be thought of as a measure of the presence or absence of electrons. Charge is measured in **coulombs** (C).

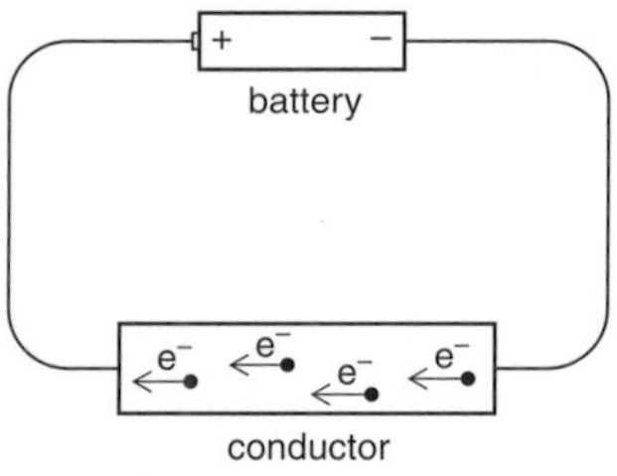

- In a good **conductor** there are many **free electrons**. (In a solid an atom's outer electrons are used in chemical bonding. A **free electron** is one that is not required by its atom for bonding with other atoms.) In an **insulator** there are no free electrons. From GCSE you will know that unlike charges attract and like charges repel. These free electrons will move through a conductor when a battery is connected across it. The battery supplies energy to the free electrons and attracts them to its positive terminal.

  The movement of these electrons is the electric current, *I*. The size of the current depends upon the number of electrons and how quickly they move.

  Current is the rate of flow of charge. It is measured in **amperes** (A). The units of current and charge are related as follows: 1 C is the charge that will flow when there is a current of 1 A for 1 s. 1 C = 1 A s.

current = $\dfrac{\textbf{charge}}{\textbf{time}}$

$$I = \frac{\Delta Q}{\Delta t}$$

- The battery supplies the energy and the electrons use up this energy when moving through the conductor. The amount of energy used in moving between two points is represented by the potential difference or voltage.

  Potential difference, p.d. or *V*, is defined as the energy used in moving unit charge (1 coulomb) between two points. It is measured in volts (V).

potential difference = $\dfrac{\textbf{energy}}{\textbf{charge}}$

$$V = \frac{\Delta E}{\Delta Q}$$

- The energy supplied by the battery is defined in similar terms, but has the unusual name of **electromotive force**, and is more commonly known as **e.m.f.**

  Electromotive force (**e.m.f.**) is defined as the energy given to each unit of charge as it passes through a cell, and is equal to the p.d. across a cell when no current is drawn from it.

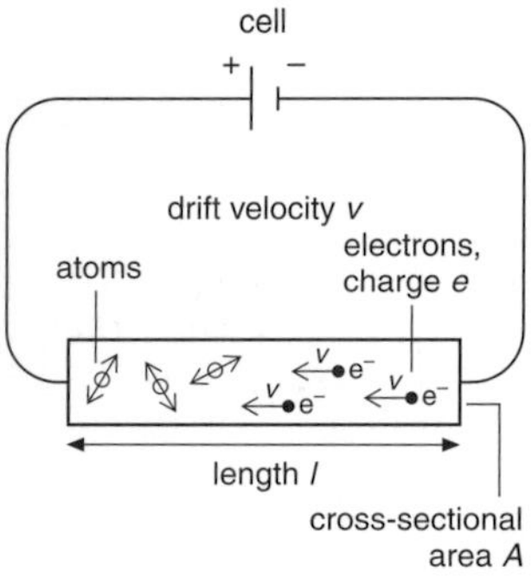

- When a cell is connected across the ends of a conductor, the conductor's free electrons start to accelerate and their kinetic energy (KE) increases. They collide with atoms, giving energy to them, and then move off again to repeat the process. These collisions represent the resistance of the conductor. The atoms vibrate more because of the increase in their KE, which represents the heating effect of the current. The chance of a collision is then increased. This represents the way that resistance increases with temperature for conductors.

  Look at the diagram left. If the volume of the conductor is *lA*, the number of electrons is *nlA*, and the total charge *Q* is *nlAe*, then the current is

$$I = \frac{Q}{t} = \frac{nlAe}{t} \text{ and } v = \frac{l}{t} \text{ so } \boldsymbol{I = nAve}$$

  where *n* is the number of charge carriers (electrons) per unit volume, *e* is the charge on an electron, and *v* is the **drift velocity** of the electrons.

drift velocity equation:
$\boldsymbol{I = nAve}$

- A typical drift velocity in a conductor is only around 10^{-3} m/s, because of the stop–start nature of the electrons' motion. If *n* and/or *A* are increased, *I* is greater. *e* is fixed and *v* depends upon the type of material. In a **semiconductor**, a material which only partially conducts, *n* is much lower, but as temperature increases more free electrons are released, increasing *n*. All insulators are semiconductors at high temperatures, and all semiconductors are insulators at low temperatures. Whether a material is categorized as one or the other depends upon its behaviour at room temperature.

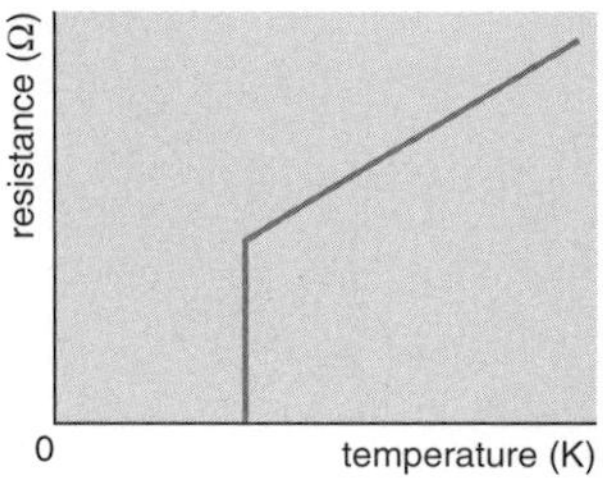

A graph of resistance against temperature for a superconductor. (Temperature is measured in kelvin: 0 K = –273 °C (see unit 25).)

- **Superconductors** are materials which have zero resistance at very low temperatures. This means that when the drift electrons pass through a material no energy is passed to the atoms. New research is producing new types of material which superconduct at higher and higher temperatures, and at some point room-temperature superconductors may be produced.
- The collisions of electrons with atoms are the **resistance** of the conductor.

  Resistance, R, is defined as the ratio of potential difference to current. It is the opposition to the flow of charge and is measured in ohms (Ω).

  Resistance is greater the longer a piece of conductor is, because there are more electron collisions. It is less if the cross-sectional area is larger, because it is easier for the electrons to get through with fewer collisions. The type of material involved also affects the resistance, because the atoms of different elements may have different sizes and spacings, and so affect the frequency of collisions. This is summarized in the equation

  $\boldsymbol{R = \frac{\rho l}{A}}$ where l is length, A is cross-sectional area, and ρ is the resistivity of the material.

  So $\rho = RA/l$, which gives us the definition of **resistivity**.

  Resistivity is defined as the resistance of a unit cube, and is measured in Ω m. **Conductivity**, σ, is related to resistivity by the equation $\boldsymbol{= 1/\rho}$.

  Example *A heating coil of resistance 40 Ω is made from wire of diameter 2.0 mm and resistivity 4.0×10^{-4} Ω m. How long is it?*

  Rearrange the equation $R = \rho l/A$:

  $l = RA/\rho$

  area = πr^2 so $A = \pi \times (1.0 \times 10^{-3})^2 = 3.14 \times 10^{-6}\,\text{m}^2$

  $l = (40 \times 3.14 \times 10^{-6})/(4.0 \times 10^{-4}) = \mathbf{0.31\,m}$

$$R = \frac{\rho l}{A}$$

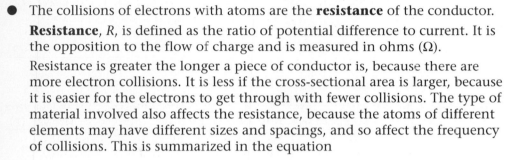

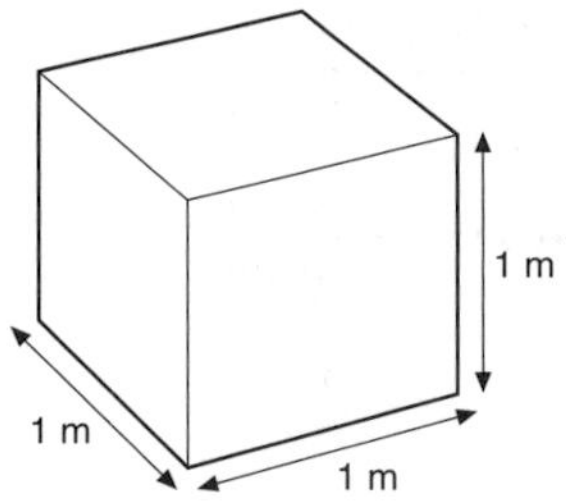

A unit cube.

voltage = current × resistance

$\boldsymbol{V = IR}$

- So far we have effectively been looking at three concepts: current, potential difference, and resistance. They are all linked together in Ohm's law.

  Ohm's law states that the current through a conductor is directly proportional to the p.d. across its ends, if temperature is constant.

  p.d. ∝ current or **p.d./current = a constant**

  This constant is the resistance of the conductor, so $\boldsymbol{V/I = R}$ or $\boldsymbol{V = IR}$.

  By combining these ideas together we get some useful equations:

  Take $I = Q/t$ and rearrange to get $Q = It$. (1)
  Take $V = E/Q$ and rearrange to get $E = VQ$. (2)

  Substitute (1) into (2) to give $\boldsymbol{E = VIt}$
  so **energy** = **voltage** × **current** × **time**.

  Furthermore, power = $\frac{\text{energy}}{\text{time}}$ so $P = \frac{E}{t} = \frac{VIt}{t} = VI$.

  $\boldsymbol{P = IV}$

  If we now substitute in Ohm's law ($V = IR$), we get $\boldsymbol{P = I^2R}$ and $\boldsymbol{P = V^2/R}$, which are useful for determining the power dissipated by electrical components.

energy = voltage × current × time

$\boldsymbol{E = VIt}$

power = current × voltage

$\boldsymbol{P = IV}$

$\boldsymbol{P = I^2R = V^2/R}$

- The domestic unit of electrical energy is the kilowatt hour (kW h). It is the energy transformed when a power of 1 kW is used for one hour.
  1 kW h = 1000 × 60 × 60 = 3.6×10^6 J.

TESTS

RECALL TEST

1 What is 'charge'?

(2)

2 Define 'electric current'.

(2)

3 Define 'potential difference'.

(2)

4 What is 'resistance'?

(2)

5 What is Ohm's law?

(2)

6 What is a semiconductor?

(2)

7 What is a superconductor?

(2)

8 Define 'resistivity'.

(2)

9 What four things affect the resistance of a conductor?

(2)

10 What is 'e.m.f.'?

(2)

(Total 20 marks)

CONCEPT TEST

Take $e = 1.6 \times 10^{-19}$ C

1 A 3.0 A current flows for 2.0 s. How much charge flows in this time? How many electrons is this?

(4)

2 The potential difference across a resistor is 4.0 V. How much energy is dissipated when 3.0 C of charge passes through it?

(2)

3 40 C of charge flows in 5 s, dissipating 200 J of energy. What are the current and voltage?

(4)

4 A 20 Ω resistor is connected to a 6 V power supply. What are the current through the resistor and the power dissipated by it?

(4)

5 An electric motor draws a current of 6.0 A from a 24 V power supply when operating normally. What is its resistance and power consumption? When it is pulling a heavy load its power consumption is trebled. If the voltage is fixed, what is the new current?

(6)

6 A string of Christmas tree lights is connected to a supply voltage of 240 V, and has 20 bulbs each with a resistance of 10 Ω, connected up so the resistances add together. What is the current through the circuit and the power consumption of the whole string? If one bulb short circuits and the circuit remains intact, what is the new power consumption?

(4)

7 Explain in terms of electrons and atoms what is meant by resistance, the heating effect of a current, and how this affects resistance. Explain why a bulb lights immediately when a circuit is switched on even though its electrons are only moving at 2 mm/s.

(6)

8 A power transmission cable has a resistance of 12 ohms per metre. If it loses 500 watts per km of length to heat, what is the output power if it delivers a p.d. of 200 000 volts?

(6)

9 The filament of a bulb is to emit 60 W of light energy. If the bulb is 40% efficient at converting electric energy into light energy, what is its resistance when connected to a 240 V supply? What is its length if its resistivity is 2.0×10^{-3} at its normal operating temperature and it is 0.50 mm thick?

(8)

10 The flow of current in a wire is given by the equation $I = nAve$ where n is the number of charge carriers per unit volume, e is the charge on each charge carrier, A is the cross-sectional area, and v is the drift velocity. Estimate the drift velocity of the electrons in a torch bulb given that n is 5.0×10^{25}.

(6)

(Total 50 marks)

Unit 10

ELECTRIC CIRCUITS

- In any practical work with electricity, batteries (which are made up of cells) are connected to various components with conducting copper wire. A continuous conducting path between all the components produces an **electric circuit**.

  An **electric circuit** is an unbroken conducting path in a continuous loop.

  We represent this by drawing a circuit diagram. This represents each of the components with a particular symbol, which you need to know (see left).

  Conventional current is taken as flowing from a point of high potential to one of low potential, from the positive terminal to the negative on a battery, and this is usually what is shown in circuit diagrams. Electrons actually flow the opposite way. (See unit 9.)

- The number of free electrons in a circuit is fixed, so the total amount of charge present is constant. We call this the **principle of conservation of charge**.

  The **principle of conservation of charge** is that the total amount of charge present is always the same.

- Electric circuits are governed by this principle. The total amount of charge flowing into a junction must be the same as the total amount flowing out. Current is the rate of flow of charge, so the total current going into a junction will be the same as the total coming out. This is known as **Kirchhoff's first law**.

  Kirchhoff's first law (**KI**) states that the total current at a junction is always zero. This comes from the principle of conservation of charge. Take current flow into the junction as positive and out of it as negative.

- From the definitions of potential difference (p.d.) and e.m.f. in the last unit it follows that the e.m.f. is the energy being supplied by the cell to the charge moving in a circuit, and the p.d. is the energy being used up as the charge moves through various components. So from the principle of conservation of energy, the e.m.f. in a series circuit must equal the sum of the p.d.s. This is **Kirchhoff's second law**.

  Kirchhoff's second law (**KII**) states that the sum of the e.m.f.s will equal the sum of the p.d.s in any series closed loop of a circuit. This is based upon the principle of conservation of energy .

- There are basically two types of electric circuit: series and parallel.

  Two rules apply to series circuits (see left):

  1 The current is the same at all points; so using I for current gives $I_1 = I_2 = I_3$ (from KI).

  2 The p.d.s add up to equal the cell e.m.f. so $V_T = V_1 + V_2$ (from KII).

  From $V_T = V_1 + V_2$ and Ohm's law $V = IR$ we get
  $IR_T = IR_1 + IR_2$; I is constant so $\boldsymbol{R_T = R_1 + R_2}$.

  Two rules apply to parallel circuits (see left):

  1 The p.d.s across conductors connected in parallel are equal, so $V_T = V_1 = V_2$, which is derived from KII.

  2 The current splits at a junction, with most of it taking the path of least resistance, so $I_1 = I_2 + I_3$, which is derived from KI.

  From $I_1 = I_2 + I_3$ and $I = V/R$ we get

  $$\frac{V}{R_T} = \frac{V}{R_1} + \frac{V}{R_2}$$

  V is constant so $\dfrac{1}{R_T} = \dfrac{1}{R_1} + \dfrac{1}{R_2}$

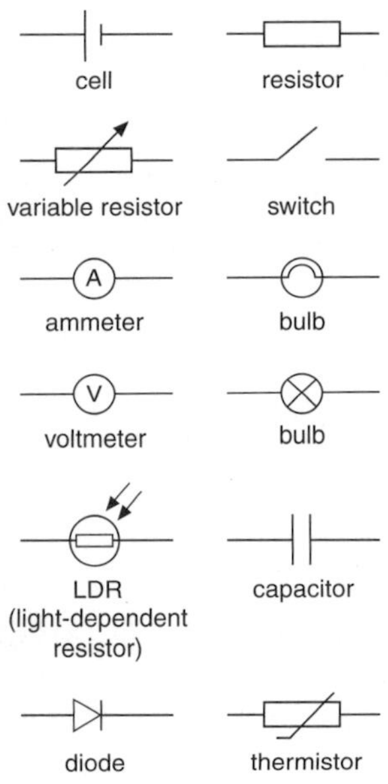

Common electrical component symbols.

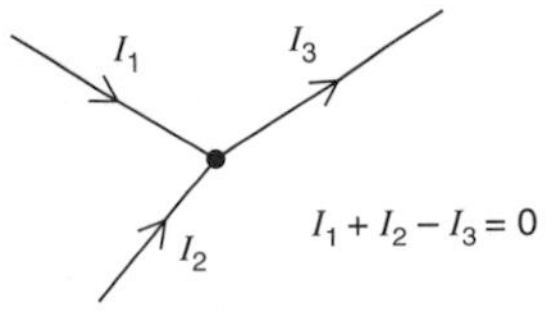

Kirchhoff's first law.

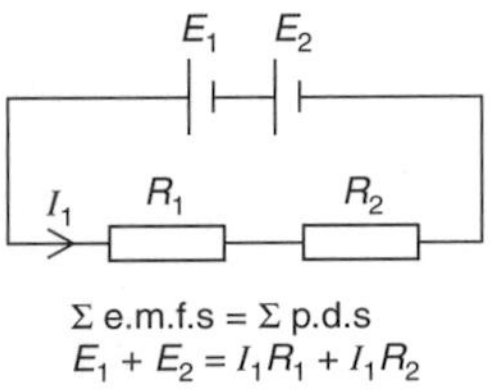

Kirchhoff's second law.

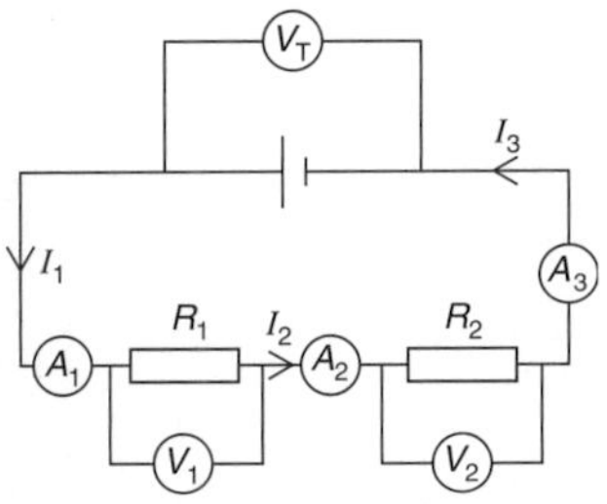

A series circuit.

resistors in series:
$\boldsymbol{R_T = R_1 + R_2}$

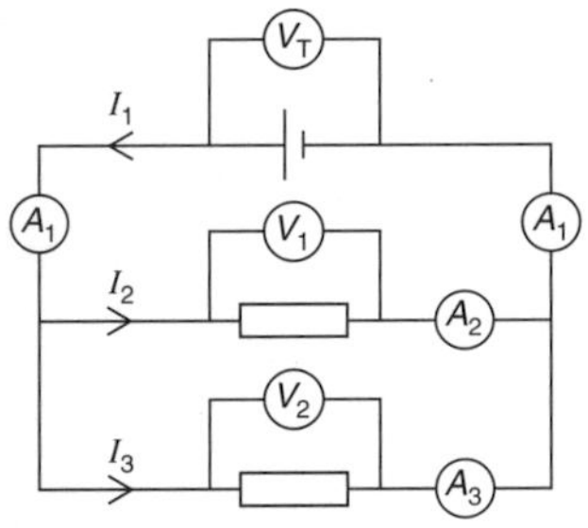

A parallel circuit.

resistors in parallel:
$$\boldsymbol{\frac{1}{R_T} = \frac{1}{R_1} + \frac{1}{R_2}}$$

- Electrical meters are used in a circuit to measure current and potential difference and their presence should affect the circuit as little as possible. Ammeters are placed in series, and must have low resistance, so as not to reduce the current. Voltmeters are placed in parallel, and must have high resistance, so as not to draw much current from the circuit (see right).

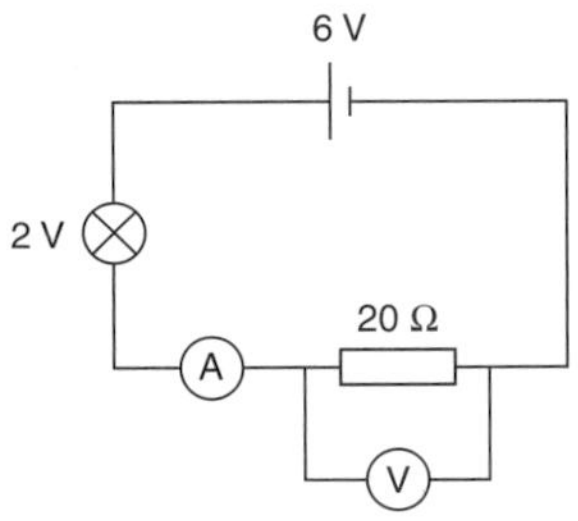

Example *What are the meter readings in this circuit?*

If there are 2 V across the bulb then from KII there must be 6 – 2 = 4 V across the resistor. Then, using Ohm's law:

$I = V/R$ so $I = 4/20 =$ **0.2 A** and $\mathbf{V = 4\ V}$

- As mentioned before, a cell or battery supplies the energy for a circuit. Cells are made up of various chemical compounds which can store energy. They also have a resistance, which is called the **internal resistance**, *r*, of the cell.

The internal resistance of the cell is due to its internal chemical structure, and accounts for the difference between e.m.f. and terminal p.d. (the voltage across the terminals of the cell) (see right).

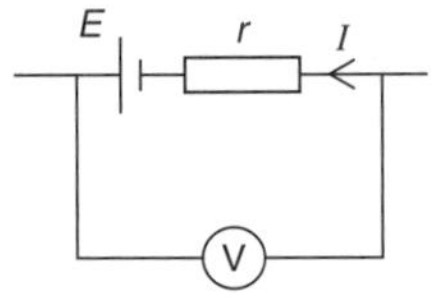

A cell's internal resistance can be depicted as a resistor in series with the cell.

The internal resistance, *r*, uses up some of the energy supplied by the e.m.f., *E*, when current, *I*, flows through it, so the terminal p.d., *V*, is less than the e.m.f. These are called the '**lost volts**' of a cell. Some high-voltage power supplies have an increased internal resistance to reduce the current delivered, for safety reasons.

internal resistance:

$$V = E - Ir$$

Example *A 10 000 V power supply delivers 200 V to a 400 Ω resistor. What is the current flow in the circuit and the internal resistance of the supply?*

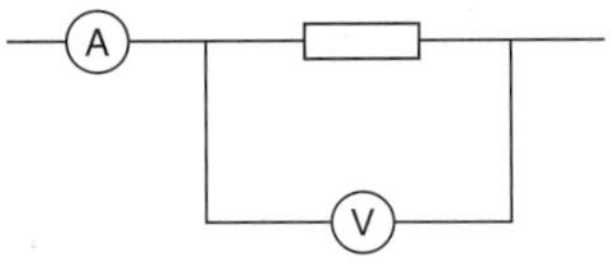

Use Ohm's law on the resistor to find the current, then $V = E - Ir$ to calculate the internal resistance.

For the resistor: $V = IR$ so $I = V/R = 200/400 = 0.500$ A

For the power supply: $V = E - Ir$ so $r = (E - V)/I = 9800/0.500 = \mathbf{1.96 \times 10^4\ \Omega}$

- When a graph of the variation of current against voltage is plotted for a particular component you get its **characteristic curve**. In a graph of *V* against *I* the gradient equals the resistance, and in a graph of *I* against *V* the gradient is the inverse of the resistance. On a curve, use the gradient of the tangent at a point. A component which produces a straight-line graph is described as an **ohmic conductor** because it obeys Ohm's law.

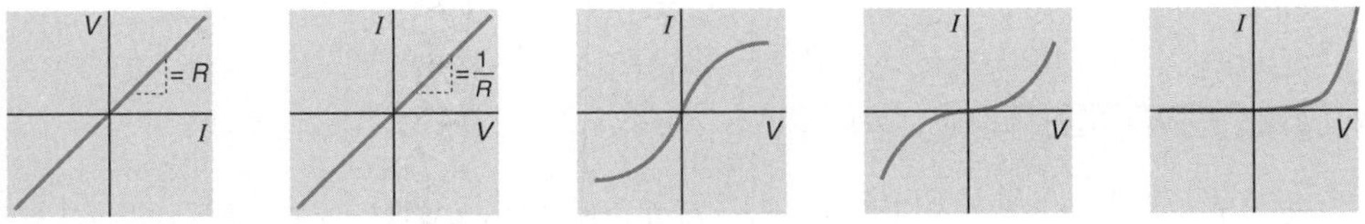

(a) carbon resistor (obeys Ohm's law)
(b) bulb (resistance increases as temperaure rises)
(c) thermistor (resistance decreases as temperature rises)
(d) diode (only lets current flow one way)

Characteristic curves.

- A **potential divider** is used to vary the p.d. across a particular component (see right). For example, you may have a 6 V battery and a 3 V motor that you wish to run off it. A potential divider can also be used in circuits which switch on other circuits, for example a circuit containing an LDR which switches on a porch lamp when it gets dark.

Total resistance of circuit is $R_T = R_1 + R_2$

Current in circuit, $I = V/R = V/(R_1 + R_2)$

p.d. across resistor $R_1 = V_1 = IR_1 = R_1V/(R_1 + R_2)$ so $\mathbf{V_1 = V\left(\frac{R_1}{R_1 + R_2}\right)}$

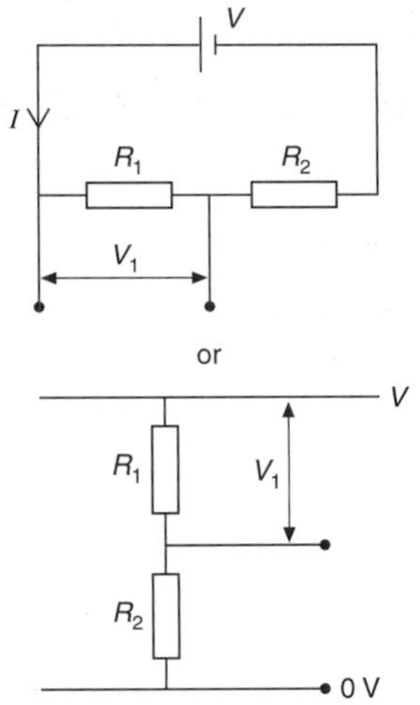

Potential divider circuits.

potential divider:

$$V_1 = V\left(\frac{R_1}{R_1 + R_2}\right)$$

TESTS

RECALL TEST

1 What is an electric circuit?

(2)

2 What is the principle of conservation of charge?

(2)

3 What are the two rules that always apply to a series circuit?

(2)

4 What are the two rules that always apply to a parallel circuit?

(2)

5 What is 'internal resistance'?

(2)

6 What is meant by the 'lost volts' of a cell?

(2)

7 What is a potential divider?

(2)

8 What is the gradient of a voltage–current graph equal to?

(2)

9 What is Kirchhoff's first law, and what principle is it based upon?

(2)

10 What is Kirchhoff's second law, and what principle is it based upon?

(2)

(Total 20 marks)

CONCEPT TEST

1 Describe how the principle of conservation of energy applies to the voltages around an electric circuit. A current of 6 A passes through a resistor of 5 Ω in 4 s. What is the voltage, and how much energy is dissipated?

(4)

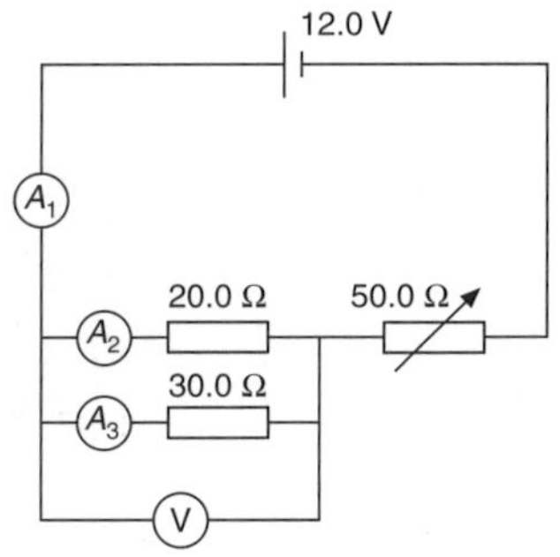

2 What are the meter readings in the circuit on the left?

(4)

3 In a circuit the current in a wire is 10 A. If the resistance of the wire is 500 Ω, what is the power consumption of the wire? If an identical wire is placed in parallel with the first, what is the new power consumption of the pair?

(6)

4 A power transmission cable has seven strands of aluminium of resistance 5.0 Ω per metre, and twenty strands of steel of 10 Ω per metre. What is the resistance of a 1 km length?

(6)

5 In a circuit a voltmeter of resistance 10 kΩ, measuring the p.d. across a resistor of 200 Ω, reads 10 V. What is the current drawn from a cell supplying the circuit, and the power dissipated by the meter? If the meter develops a fault and its resistance becomes 500 Ω what are the new values of current and power?

(6)

6 A 12.0 V bulb is connected across a 12.0 V supply voltage. A voltmeter placed across the bulb reads 8.60 V. Explain why it does not read 12.0 V. If it is a 10.0 W bulb, what is the total resistance of the circuit?

(4)

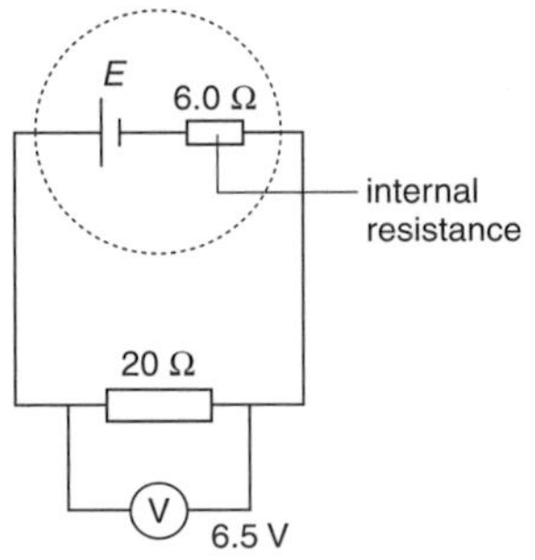

7 People have received serious burns by placing a bunch of keys in a pocket with a battery. Explain why this might happen. What is the e.m.f. of the cell in the circuit shown right?

(4)

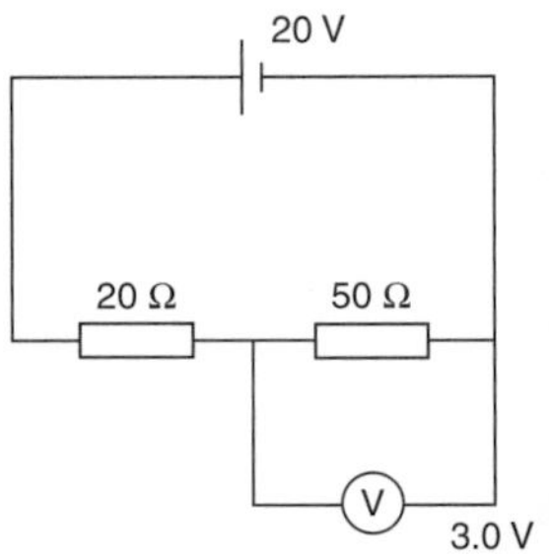

8 In the circuit shown right, the meter reads 3.0 V. What is the resistance of the meter?

(4)

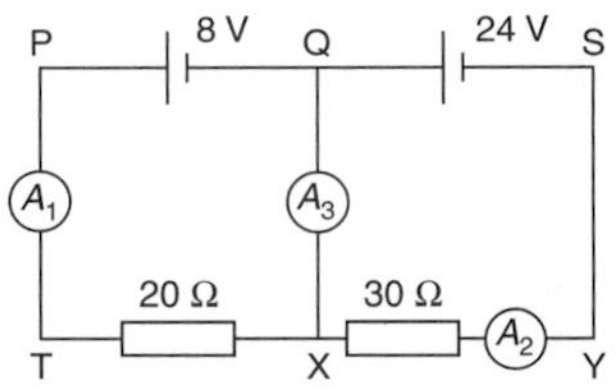

9 Using Kirchhoff's laws calculate the currents A_1, A_2, and A_3 in the circuit on the right. Also, determine the power dissipated in each resistor.

(6)

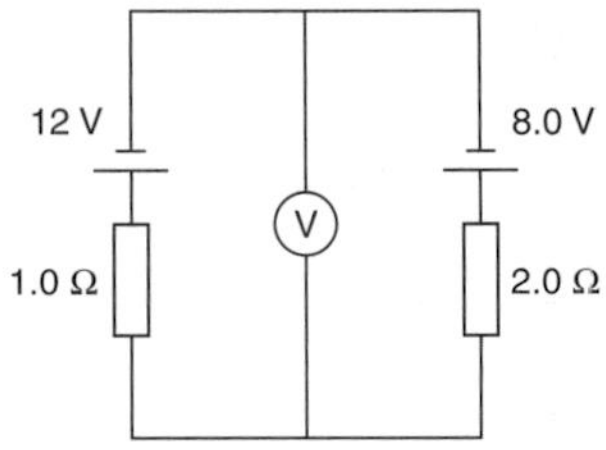

10 Assuming that no current is drawn by the voltmeter in the figure on the right, what would its reading be?

(6)

(Total 50 marks)

Unit 11 WAVES

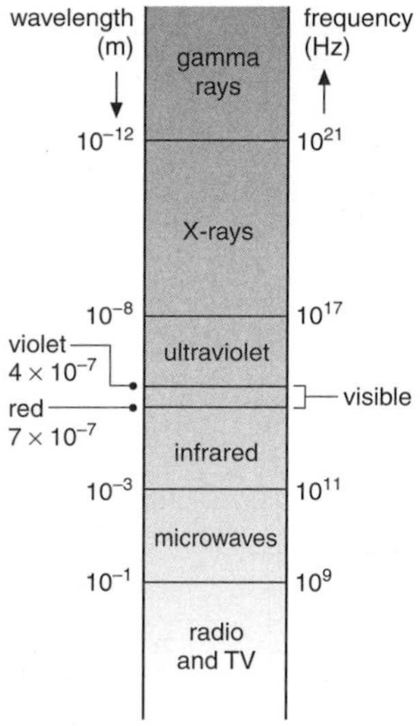

The electromagnetic spectrum.

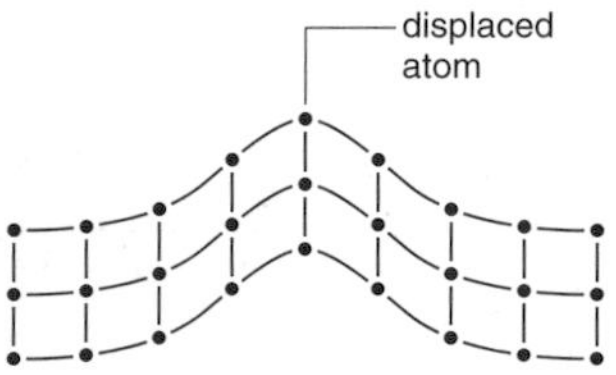

In a solid a displaced atom will drag its neighbours along with it.

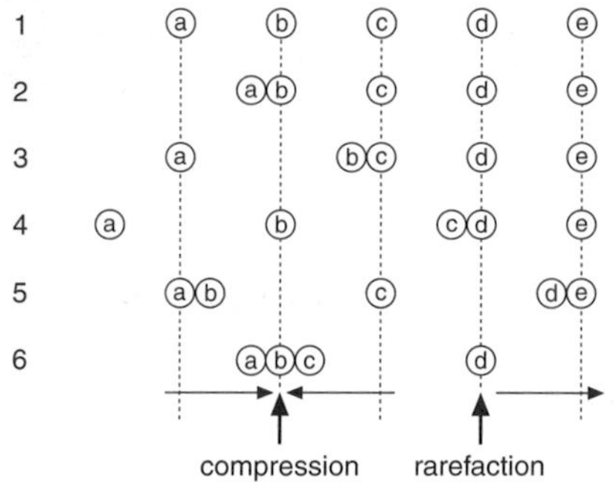

These six 'snapshots' show the different stages in setting up a longitudinal wave in a gas.

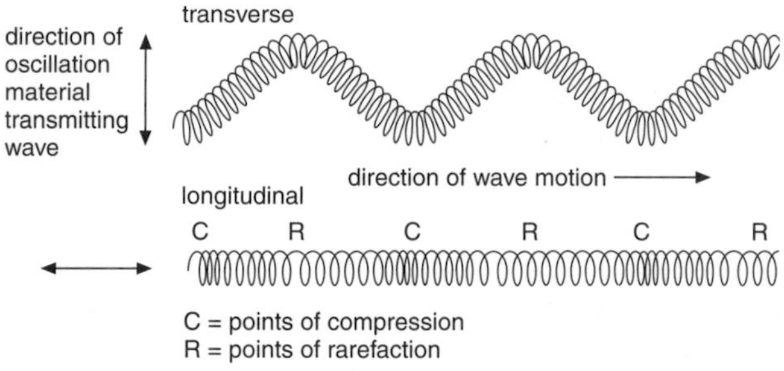

- Waves are classified as either **mechanical** or **electromagnetic**. The electromagnetic spectrum on the left shows all the different types of electromagnetic waves. They all travel at the same speed in a vacuum. **Mechanical waves** are caused by the oscillation of matter, and **electromagnetic** waves by the oscillation of electromagnetic fields.

- When a water wave out at sea passes a point, the surface of the water moves up and down. This type of wave is called a **transverse** wave. (Particles of water actually move in a vertical circle as a wave passes.) It does *not* move horizontally, even though the wave appears to; what actually moves is energy. A wave which transmits energy is called a **progressive** wave, and the direction it moves in is called the direction of **propagation**.

  In a **transverse wave** the direction of oscillation of the medium transmitting the wave is perpendicular to the direction of wave propagation. All forms of electromagnetic radiation are transverse waves. There are also mechanical transverse waves, such as seismic S waves (see later).

- Any gas consists of lots of atoms or molecules moving around randomly (see unit 25). These atoms exert no force on each other unless they are close together. As there are no bonds between atoms, if one is pushed in a certain direction, it does not drag its neighbours along with it as in a solid (see left); instead it will just keep on moving on its own until it meets another atom. The electrostatic forces between the two will then force the first atom back in the opposite direction, and displace the second atom in the original direction of motion (see below). The process continues with more atoms, causing them to oscillate and produce wave motion, but unlike transverse waves, the oscillation is in the same direction as the wave motion. This type of wave is called a **longitudinal** wave, and this is how sound travels through air.

  In a longitudinal wave the direction of oscillation of the medium transmitting the wave is the same as the direction of wave propagation.

  A 'slinky' spring can be used to demonstrate the two different types of wave motion (see left). You are often asked to classify waves as either transverse or longitudinal. Most waves are transverse, sound and seismic P waves are longitudinal.

- Mechanical longitudinal waves can travel through fluids and solids whereas mechanical transverse waves can only travel through solids and along the surface of fluids. A good example of this is in seismic waves: longitudinal **pressure** (P) waves travel through the liquid outer core of the Earth and transverse **shear** (S) waves do not (see diagram). The paths are curves because of refraction due to changes in density in the rock.

- When a wave passes through a material each particle in the material oscillates as the wave passes, and undergoes repeated cycles of motion that take the same time. The time taken for each of these oscillations is called the **time period**. The inverse of this is **frequency**.

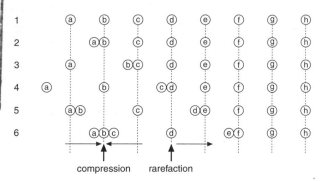

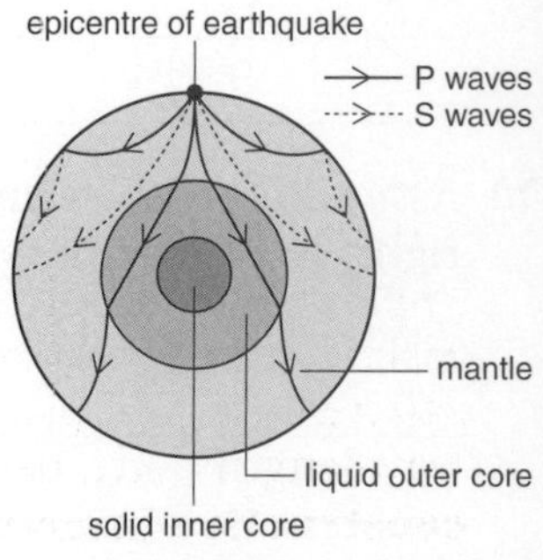

The paths of P and S waves in the Earth after an earthquake.

The **period**, T, is the time taken for one complete oscillation or for one complete waveform to pass a point. It is measured in seconds.

- **Frequency**, f, is the number of waves passing point each second or the number of oscillations occurring each second. Its units are hertz (Hz).

- When a wave passes a particle, that particle is displaced from its equilibrium position. The **amplitude**, A, of a wave is defined as its maximum displacement from its equilibrium position (see right).

  The amplitude depends upon the energy being carried by the wave. The intensity, I, is directly proportional to the square of the amplitude.

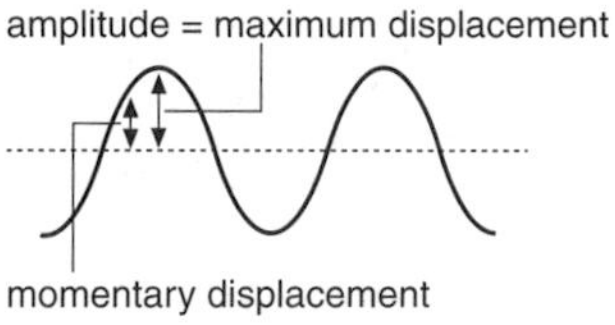

$$f = \frac{1}{T}$$

$$I \propto A^2$$

- When a wave passes a particle, the particle will move up and down and back again to its original position as the wave passes (see right). The particle goes through one wave cycle. What stage in a wave cycle a point is is represented by **phase** or **phase angle.**

  The **phase angle** represents where a particle is in terms of its wave cycle. One complete wave cycle is equivalent to 2π rad.

  A graph of displacement against time for such a particle can be mapped onto a point moving around a circle with constant speed, as shown right. The angle the point has turned through is the phase angle. The difference between two phase angles is the **phase difference**. The phase difference between two particles on the same wave can be determined using ratios.

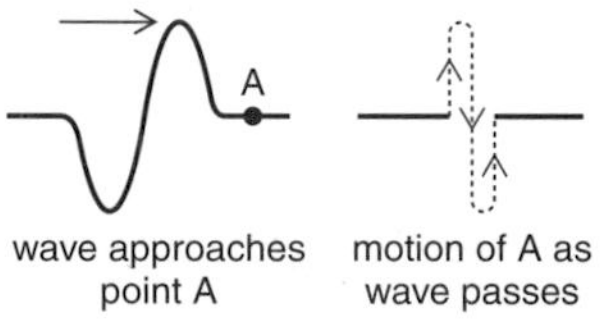

The movement of a particle as a wave passes.

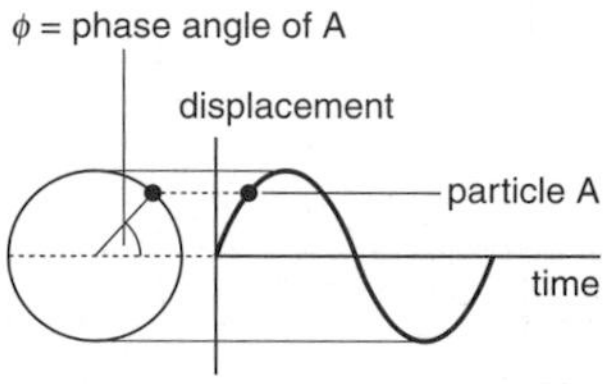

- The distance between two particles in a wave at the same stage of the wave cycle, for example, wave crest to wave crest, is known as the **wavelength.**

  The **wavelength**, λ, is defined as the distance between two adjacent positions in phase on a progressive wave, and is measured in m.

  Wavelength and frequency are linked, together with the velocity of the wave motion, through $v = s/t$. If $s = \lambda$, and $f = 1/t$, then obviously $\boldsymbol{v = f\lambda}$. The velocity of the wave is the velocity with which the energy is transferred.

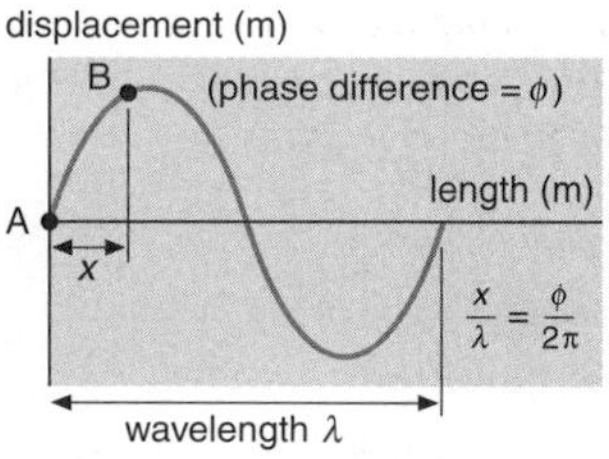

Phase angle and phase difference.

wave equation: $v = f\lambda$

- It is quite common to see two water waves meet, combine together for a moment, pass through each other, and then reform as they were before. You can see this in the sea or in your own bath. This particular effect is called **superposition**, and it deals with the interference between waves.

  The **principle of superposition** states that if two or more waves of the same type are in the same place at the same time, their wave displacements add. When the waves completely add together we get **constructive interference**; when the waves completely cancel each other out we get **destructive interference** (see diagrams).

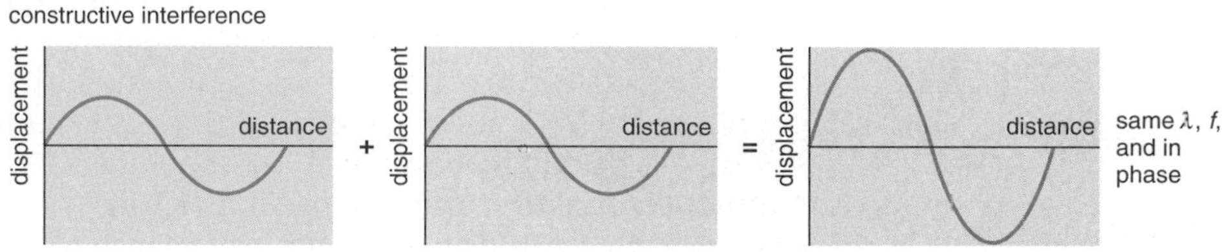

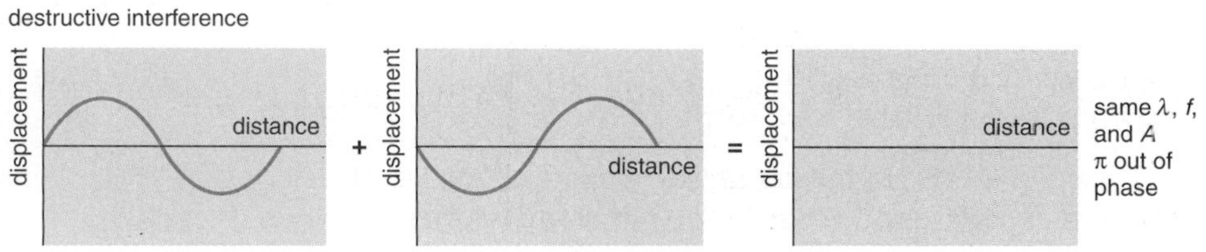

- Two wave sources, A and B, are emitting identical in-phase waves, which arrive together at C, and produce interference. If the difference in length of the two paths (AC – AB), the **path difference**, is equal to a whole number of wavelengths, the waves arrive in phase and produce constructive interference. If the path difference is $\frac{1}{2}$ a wavelength, $1\frac{1}{2}$ wavelengths, $2\frac{1}{2}$ wavelengths, etc., destructive interference occurs. **If path difference = $n\lambda$ constructive interference occurs** where n is a whole number. **If path difference = $(n + 1/2)\lambda$ destructive interference occurs.**

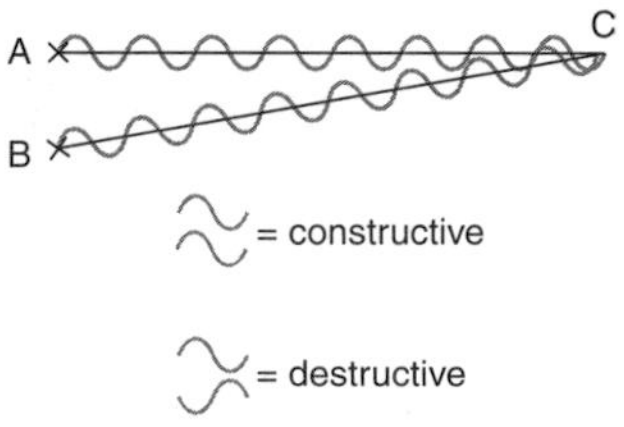

Path difference and interference.

Unit 11 TESTS

RECALL TEST

1 What are the two different types of wave?

______________________________ (2)

2 What is the difference between transverse and longitudinal wave motion?

____

______________________________ (2)

3 What moves in a progressive wave?

______________________________ (2)

4 Define 'frequency'.

______________________________ (2)

5 Define 'wavelength'.

______________________________ (2)

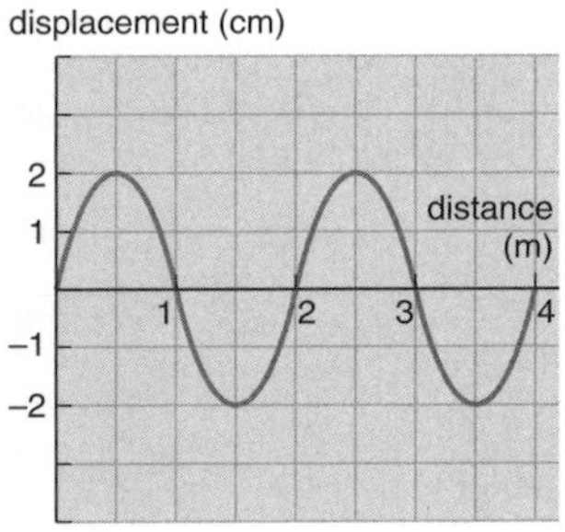

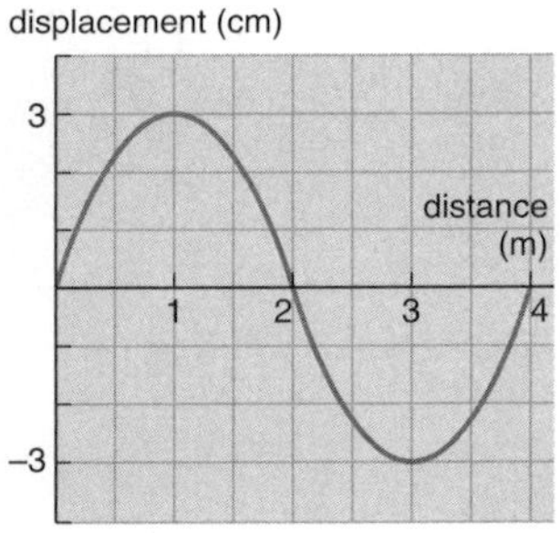

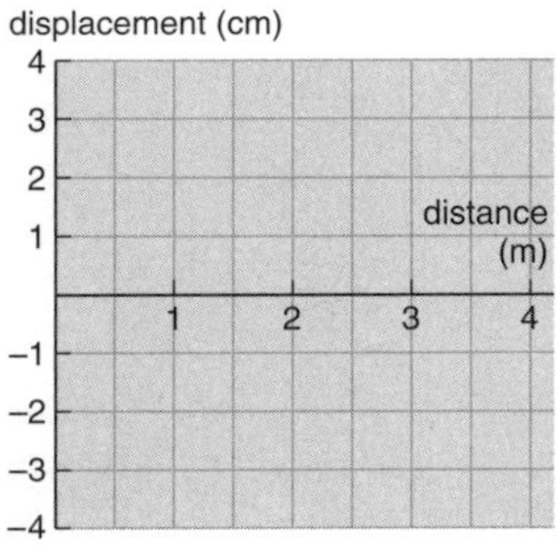

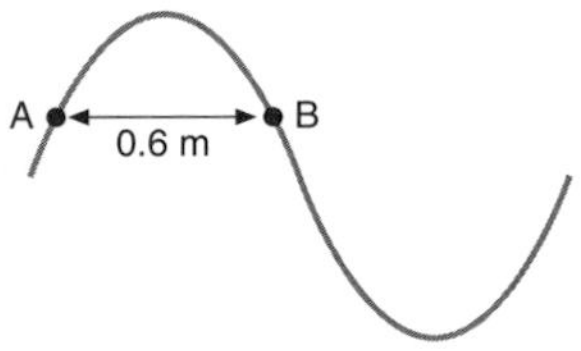

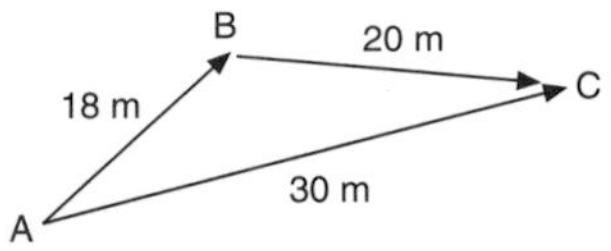

6 Define 'amplitude'.

______________________________ (2)

7 What is 'phase'?

______________________________ (2)

8 What is the principle of superposition?

______________________________ (2)

9 What conditions are necessary for the constructive interference of two waves?

______________________________ (2)

10 What conditions are necessary for the destructive interference of two waves?

______________________________ (2)

(Total 20 marks)

CONCEPT TEST

Take speed of light in a vacuum $c = 3 \times 10^8$ m/s

1 As a wave passes a point, a particle moves through 6 oscillations in 2 seconds. What is the time period and frequency of the particle?

______________________________ (2)

2 The two waveforms above left arrive in the same place at the same time. Sketch the resulting waveform on the axes beneath them. (4)

3 The wave left is travelling at 300 m/s, and has a frequency of 140 Hz. What is its wavelength, and what is the phase difference between A and B?

____

______________________________ (4)

4 The wave shown left is travelling at 30 m/s, and has a frequency of 50 Hz. A is the wave source, and the waves travel by two routes to C as shown. What is the wavelength? What are the path difference and phase difference of the waves at C?

____

______________________________ (6)

5 a Explain why, when an earthquake happens in Japan, only longitudinal (P) waves will be detected in South America.

____

______________________________ (2)

b A small fishing boat out at sea bobs up and down 10 times a minute. A fisherman calculates the speed of one of the waves by measuring the time it takes to travel between two buoys 100 m apart, and finds it to be 20 s. What is the wavelength of the wave?

____

______________________________ (4)

6 Light hits a lens (see right). Some of it is reflected from the surface of the lens, but most is transmitted through the lens. The reflected light can cause problems in some optical devices, so to get rid of it the lens is coated with a thin film. Light is then reflected from the top and bottom surfaces of the film. The thickness of the film is just sufficient to produce destructive interference between the two reflected waves and thereby dispose of them. If the coating is 1.4×10^{-7} m thick, what frequency of reflected light is cancelled out by the film?

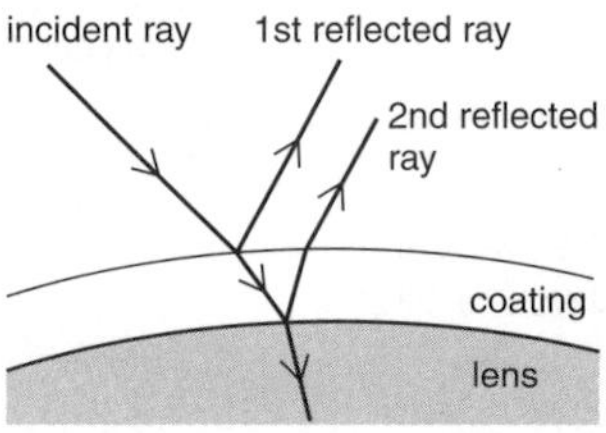

____

______________________________ (6)

7 An aircraft is flying overhead while you are watching TV. The waves from a transmitter reflect off the aircraft and produce a second, 'ghost' image on the screen, 3.0 cm to the right of the normal image, and a flickering screen. Explain why this happens, using the figure on the right. The screen is 40 cm wide, and the electron beam in the TV crosses the screen 625 times a second. By how much further has the wave reflected off the aircraft travelled than the wave direct from the transmitter?

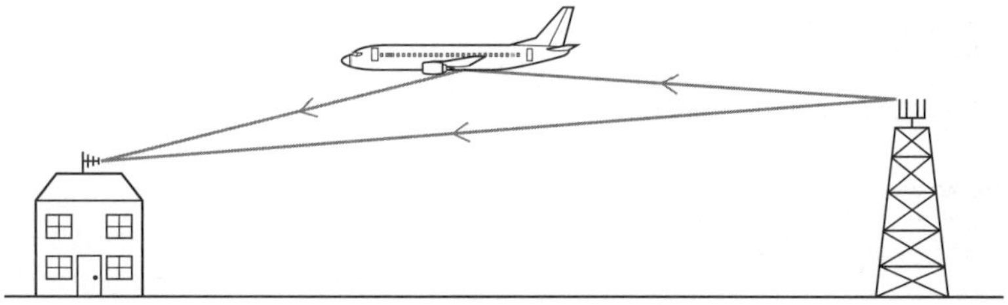

____

______________________________ (8)

8 A man is walking towards a radio as shown right. Radio waves are emitted by the mast, absorbed by the man, and then re-emitted, so that he acts as a second source. The sound from the radio varies in loudness. Explain why this happens. If the distance he walks between points where the loudest sound and the quietest are produced is 2 m, what is the frequency of the emitted radio wave?

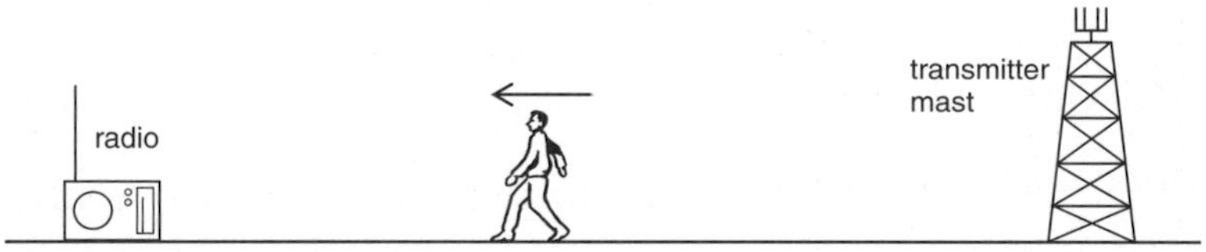

____

____

______________________________ (6)

9 A wave source is emitting waves equally in all directions. If the intensity at the source is 2×10^{-2} W, what are the intensity and amplitude at a distance of 5 m from the source? Hint: treat the source as if it is at the centre of a sphere with the energy spread over its surface. Amplitude at source = 3×10^{-10} m. (Surface area of sphere = $4\pi r^2$.)

____

______________________________ (8)

(Total 50 marks)

WAVE PHENOMENA

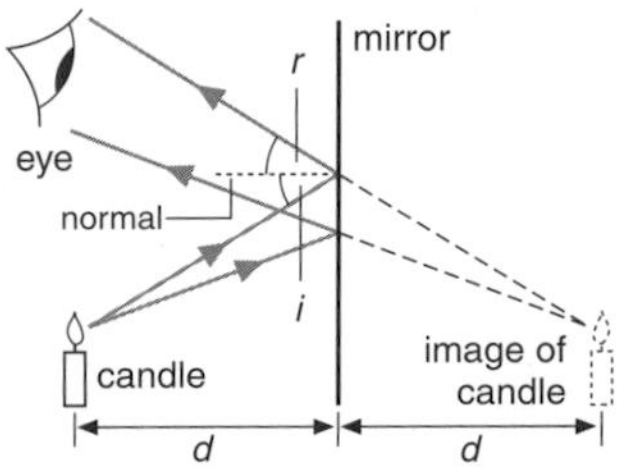

reflection: $i = r$

- One of the most common phenomena that we see is **reflection** (see left).

  When a wave is reflected from a surface the **angle of incidence**, *i*, will equal the **angle of reflection**, *r*.

  The angles *i* and *r* are always measured from the normal (the **normal** is a line drawn from the point where the wave hits the reflecting surface, perpendicular to the surface). The image of the object lies the same distance behind the reflecting surface as the object lies in front of it.

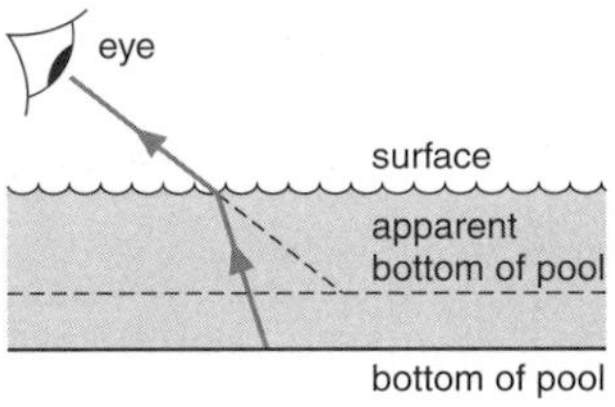

Refraction in a swimming pool.

refraction: $\frac{\sin i}{\sin r} = {}_1n_2$

$n_1 \sin \theta_1 = n_2 \sin \theta_2$

- Another wave effect is illustrated by the way that the bottom of a swimming pool appears shallower than it actually is (see left). This effect is produced because light changes speed when it travels from one material to another, causing it to change its direction of motion. This effect is called **refraction**.

  Refraction is the change in direction of motion of a wave as it travels from one material to another, due to a change of speed. The angles of incidence (*i* or θ_1) and refraction (*r* or θ_2) are related by the equations

  $$\frac{\sin i}{\sin r} = \frac{\sin \theta_1}{\sin \theta_2} = \frac{c_1}{c_2} = \frac{\lambda_1}{\lambda_2} = {}_1n_2$$

  $$n_1 \sin \theta_1 = n_2 \sin \theta_2$$

  where c_1 and c_2 are the speeds in the respective materials, λ_1 and λ_2 are the wavelengths, and ${}_1n_2$ is the **refractive index** for a wave travelling from the first material to the second (see left). Frequency is constant. ${}_1n_2 = n_2/n_1 = c_1/c_2$ where n_1 is the **absolute refractive index** (refractive index with respect to a vacuum, or air) of the material the wave is coming from, and n_2 is the absolute refractive index of the material it is moving into. The diagram on the left applies these equations to a wave that slows down as it travels from one material to another ($c_1 > c_2$).

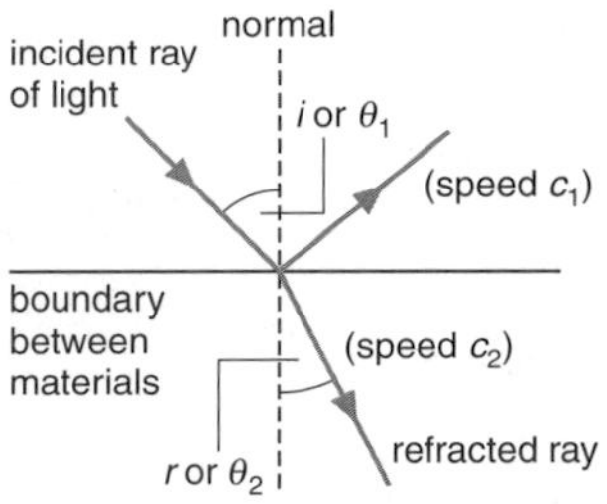

Refraction of light.

- When a wave travels from a fast to a slow medium it bends towards the normal. When it travels from a slow to a fast, it bends away from the normal. The diagram below left shows a wave that speeds up as it passes from one material to another ($c_1 < c_2$). At a particular angle of incidence the refracted wave passes along the boundary between the two materials. This angle is called the **critical angle**, i_c.

  The critical angle of incidence produces an angle of refraction of 90°.

  $$\frac{\sin i}{\sin r} = \frac{c_1}{c_2} = {}_1n_2 = \frac{1}{{}_2n_1} \text{ and } \sin 90 = 1 \text{ so } \mathbf{\sin i_c = \frac{1}{{}_2n_1}}$$

  If the 'faster' medium is air then ${}_2n_1$ becomes n_1 (the absolute refractive index of the 'slower' medium), from ${}_2n_1 = n_1/n_2$ and $n_2 = 1$. So $\mathbf{\sin i_c = 1/n_1}$.

  If the angle of incidence is greater than the critical angle **total internal reflection** occurs. Some light is always reflected from the boundary when refraction occurs.

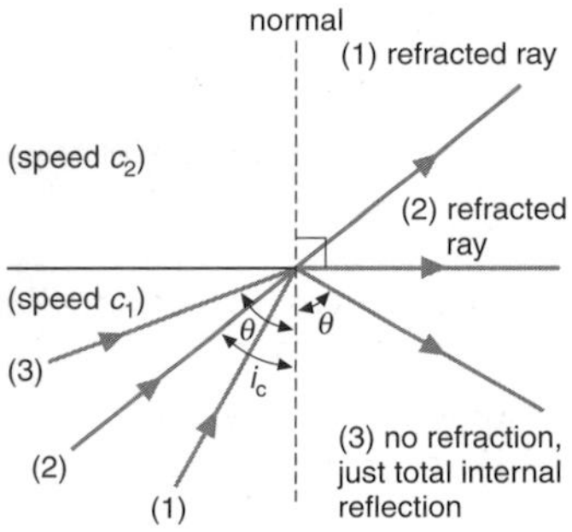

critical angle: $\sin i_c = 1/n_1$

- **Young's double-slit experiment** shows the interference of light, and that it is a wave. If you have two monochromatic, coherent sources of light close together, distance *a* apart, you will get an interference pattern of light and dark fringes an equal space *y* apart on a screen distance *D* away (see left).

  (**Monochromatic** means of a single wavelength and **coherent** means having a constant phase difference.)

  Two slits act as a coherent source if light from a single light source passes through them. Light photons need to have come from the same original source to be coherent.

  If each slit is a wave source, at the central point (O) waves from each slit have travelled the same distance, the path difference is zero so they are in phase,

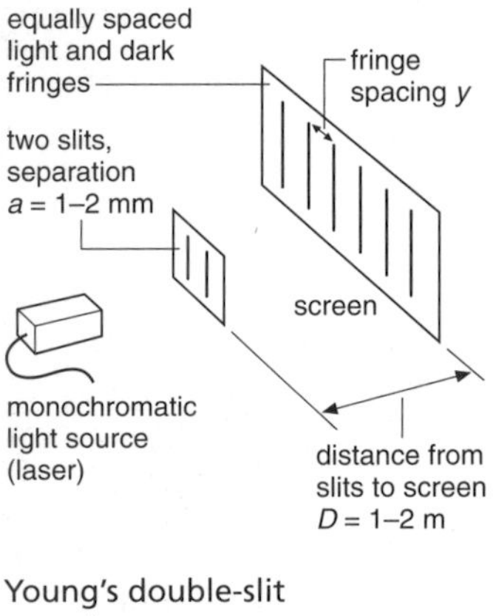

Young's double-slit experiment.

Young's fringes: $y = \frac{\lambda D}{a}$

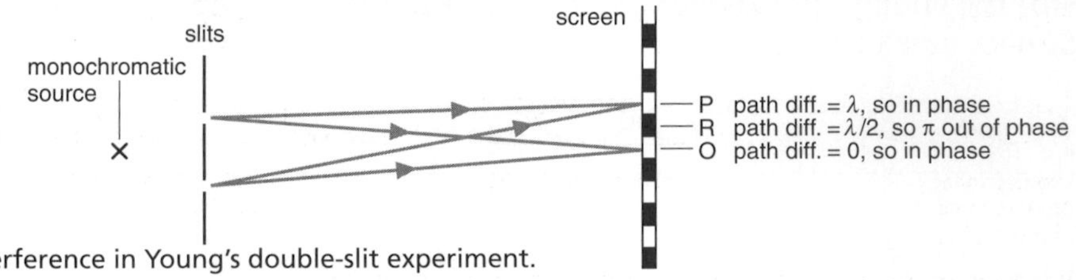

Interference in Young's double-slit experiment.

and constructive interference occurs. At point P the path difference is one wavelength, so constructive interference occurs again. At R it is half a wavelength, the waves are π out of phase, and destructive interference occurs.

● Water waves entering a harbour appear to be dragged back or slowed down by the edge of the harbour wall (see right). This bending is called **diffraction**.

Diffraction is the bending of a wave due to an edge or an obstacle; the degree of bending depends upon the wavelength and the size of the gap.

Diffraction is the reason why the two slits in the previous section could act as secondary sources.

When light from a monochromatic source passes through a small slit it produces an interference pattern of light regions and dark regions. This can be explained by considering each point across the slit as a new wave source, so the screen will have various points where constructive and destructive interference take place. More slits will produce different interference patterns, but each one fits inside the original single-slit intensity pattern (right). If there is a very large number of slits you have a **diffraction grating**. This is usually a piece of glass with many thousands of lines finely etched upon it. When light hits it, very bright interference lines are produced.

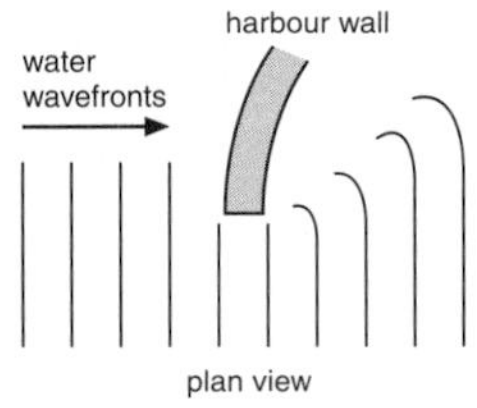

Diffraction of water waves by a harbour wall.

The positions of the maxima remain the same whatever the number of slits.

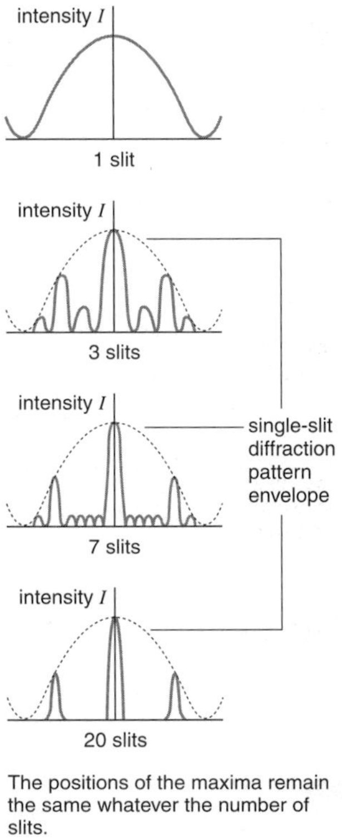

Intensity patterns produced by multiple slits.

take triangle XYZ:
$$\sin \theta = \frac{n\lambda}{d}$$
$$d \sin \theta = n\lambda$$
$$d = \frac{1}{N}$$ where N is the number of lines per metre

n is the order of diffraction

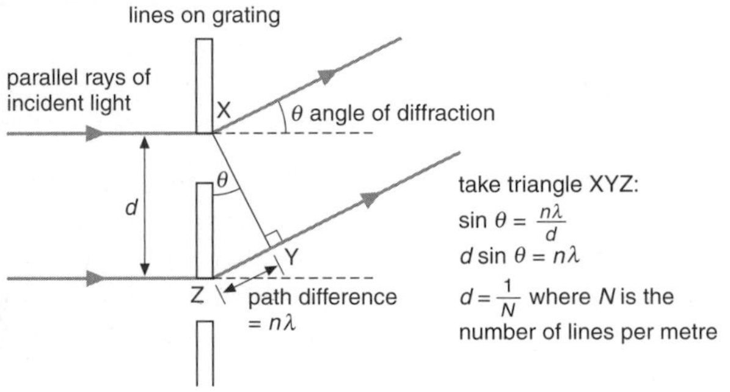

Diffraction of light at a grating.

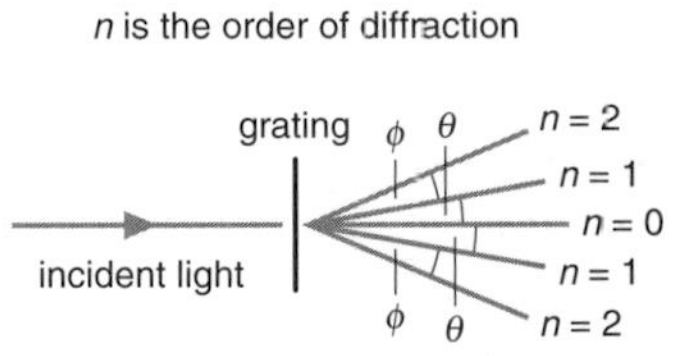

Orders of diffraction.

● If two identical waves are travelling in opposite directions, a **standing** or **stationary** wave is formed. The differences between a stationary and a progressive wave are important. Look at the figure below right.

1 A progressive wave transmits energy, a standing wave does not.

2 All points on a progressive wave have the same amplitude. A, B, C, D, and E have the same amplitude. In a standing wave, points half a wavelength apart have the same amplitude. P, R, and T have zero amplitude (**nodes**). Q and S have maximum amplitude (**antinodes**).

3 On a progressive wave points one wavelength apart are in phase with each other. On the standing wave all the points between P and R move in phase with each other. All the points between R and T move in phase with each other, but π out of phase with all the points between P and R.

Standing waves are found in strings and pipes, which are important in musical instruments. The different forms are shown below.

diffraction grating:
$d \sin \imath = n$

progressive wave

stationary wave

Progressive and stationary waves.

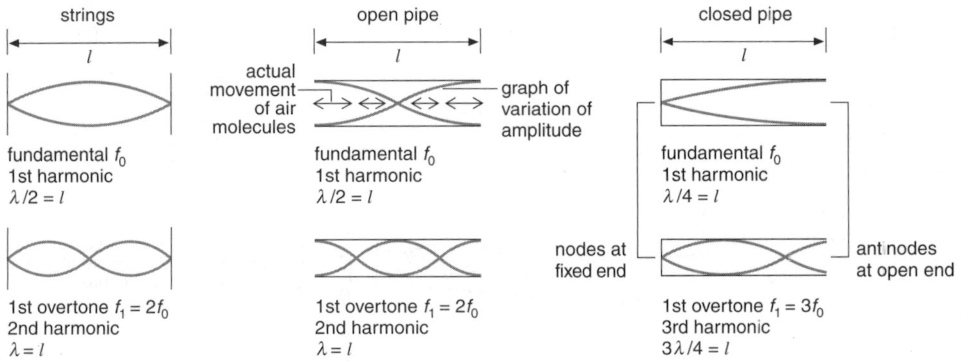

Variation of amplitude on standing waves in pipes and strings.

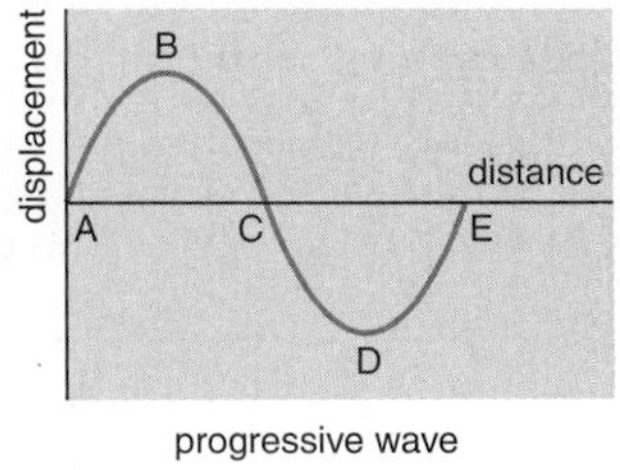

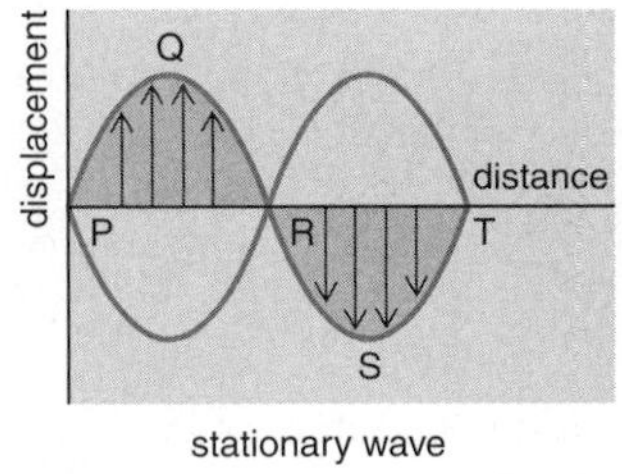

Unit 12 TESTS

RECALL TEST

1 What is 'refraction'?

(2)

2 What is 'diffraction'?

(2)

3 What happens when a wave enters a different medium and slows down?

(2)

4 What is meant by 'critical angle'?

(2)

5 What is meant by 'total internal reflection'?

(2)

6 What are the conditions for two-source interference?

(2)

7 When a wave is reflected, where is the reflected image?

(2)

8 When a wave is refracted, what changes and what stays constant?

(2)

9 What is a diffraction grating?

(2)

10 What is the main difference between a standing and a progressive wave?

(2)

(Total 20 marks)

CONCEPT TEST

Take $c = 3.0 \times 10^8$ m/s

1 What is the angle of the second-order maximum for a wave of wavelength 3.86×10^{-7} m when it passes through a diffraction grating with 5.0×10^5 lines per metre? How many orders of diffraction are possible?

(4)

the dark lines represent bright interference fringes

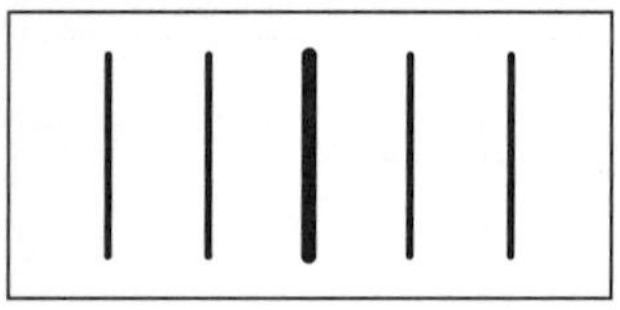

2 The figure left shows the interference pattern produced on a screen 2.0 m from a pair of slits 1 mm apart. The pattern has been magnified by a factor of four. Mark on the diagram a point where the path difference of waves from the two slits is two wavelengths. Determine the fringe spacing from the diagram, and work out the wavelength of the emitted light.

(6)

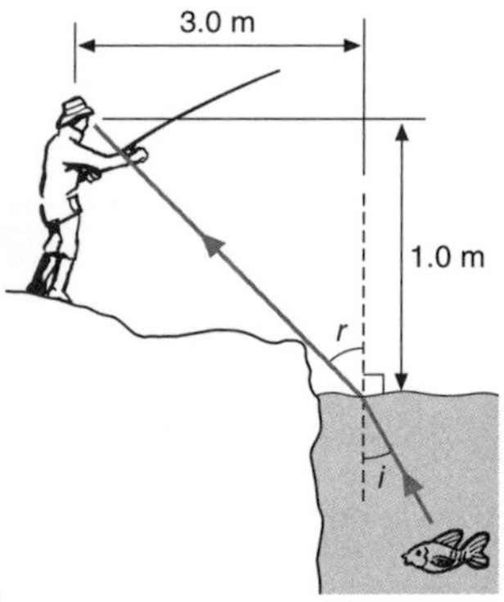

3 A fisherman is standing on a bank as shown left. Given that the absolute refractive index of water is 1.33, what must *i* be for the angler to see the fish?

(4)

4 The diagram right shows the set of apparatus needed to produce interference using Young's double-slit experiment. Explain how the interference pattern is produced and suggest appropriate values for a and D. What will be the effect on the fringe spacing if **i** the frequency is increased, and **ii** the slit separation is reduced?

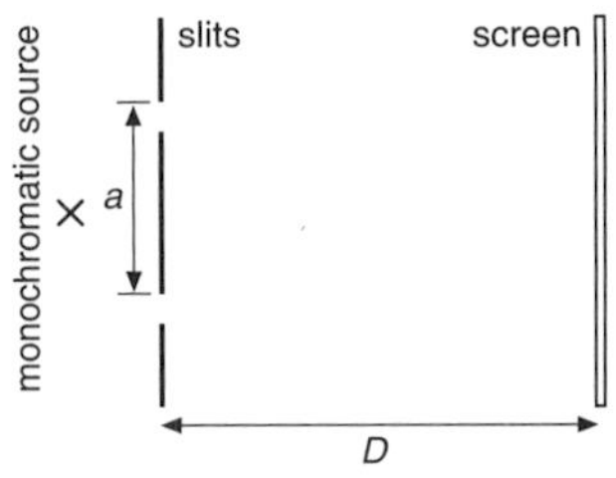

(8)

5 A washing line 2.0 m long is supported as shown in (a) below. When the wind blows, a standing wave is set up on the line. Sketch on the diagram the simplest waveform that you would expect to see. Another line 3.0 m long is propped as shown in (b) below. Draw on it the simplest waveform you would expect to see. If the lowest frequency of the wave in (a) is 6.0 Hz, calculate the wave speed. If the wave speed is the same in (b), what is the lowest frequency possible?

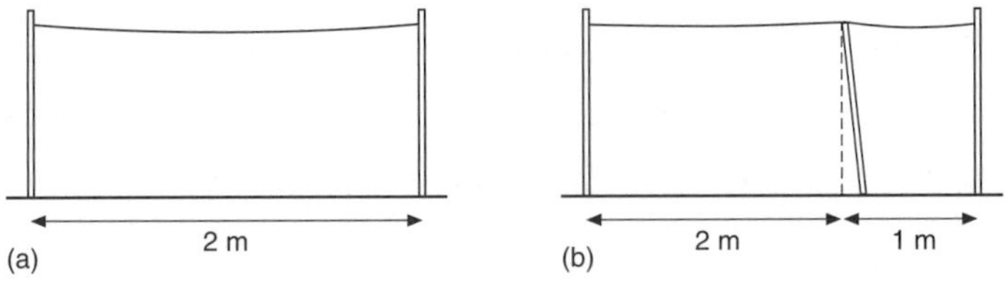

(6)

6 The figure on the right shows a ray of monochromatic light hitting a triangular prism with a refractive index of 1.4, such that the light just emerges from the face as shown. What is the speed of light in the prism? What are the values of angles i_2 and i_1? Why is i_1 the minimum angle for light to just emerge from the prism?

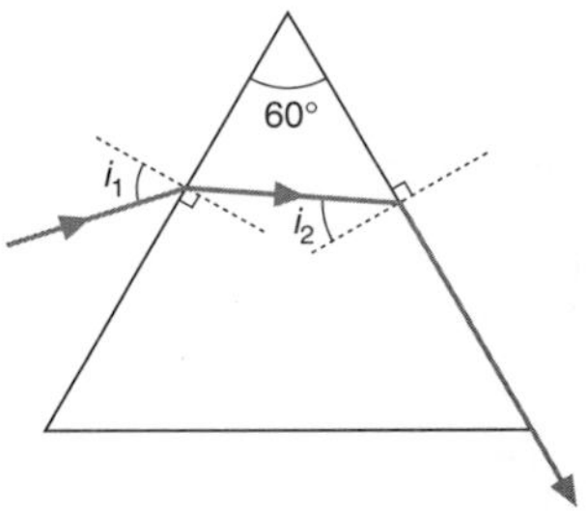

(8)

7 The diagram right shows an organ pipe open at both ends. When the pipe is sounding in its fundamental mode, how do the air molecules behave at X and Y? How is Z different from X? A normal organ has pipes ranging from 12 m to 10 mm. Take the speed of sound as 320 m/s and calculate the frequency range of the organ. If the pipes have one end blocked off, what would the new range be?

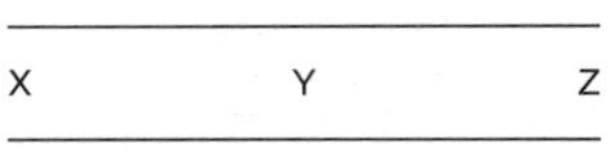

(8)

8 A glass fibre consists of a central core with absolute refractive index 1.6, surrounded by a cladding with a different refractive index. Work out the refractive index of the cladding, given that the critical angle between the cladding and the core is 12°. What is the speed of light in the core? If the glass fibre is 500 m long, what is the minimum time taken for the light to pass through it?

(6)

(Total 50 marks)

Unit 13

WAVE–PARTICLE DUALITY

- The evidence from wave experiments indicates that the nature of light is that of a waveform. One major piece of evidence contradicting this is the photoelectric effect. Like the word suggests, it deals with photons and electrons.

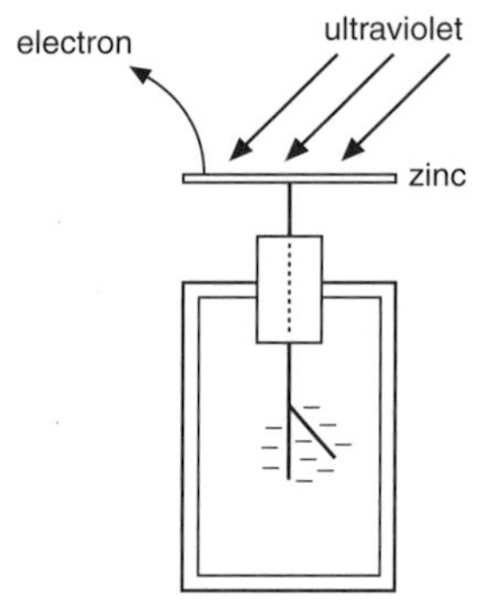

Ultraviolet radiation hitting a negatively charged electroscope.

- Ultraviolet light is incident upon (strikes) a freshly cleaned zinc surface, placed on top of a negatively charged gold-leaf **electroscope**. The surface is cleaned to remove the oxide layer; this enables electrons to be emitted more easily. An **electroscope** is a simple device used for detecting the presence of charge. It consists of a conducting plate connected to a vertical piece of conductor which is attached to a loose piece of gold leaf, hinged so that it can bend upwards (see left). The electroscope is initially negatively charged, so the gold leaf is bent upwards. When the ultraviolet light hits the surface the leaf goes down. If the experiment is repeated with a sheet of glass in the way, the leaf stays up. Glass stops ultraviolet, so it does not reach the surface. If the experiment is repeated with a positively charged electroscope, the leaf also stays up. From this we conclude that when light (electromagnetic radiation) hits the surface of some metals, electrons are emitted. This is the **photoelectric effect**.

  The **photoelectric effect** is the emission of electrons from the surface of a material when it is exposed to electromagnetic radiation.

- More detailed experimental work produces the following evidence:
  1. No electrons are emitted unless the frequency of the incident radiation is above a certain value, called the **threshold frequency**, f_0, which depends upon the type of material.
  2. The emitted electrons have a range of kinetic energy (KE) up to a specific maximum value, which changes if the frequency changes.
  3. If the intensity of the radiation is increased, more electrons are emitted but their maximum KE remains the same.

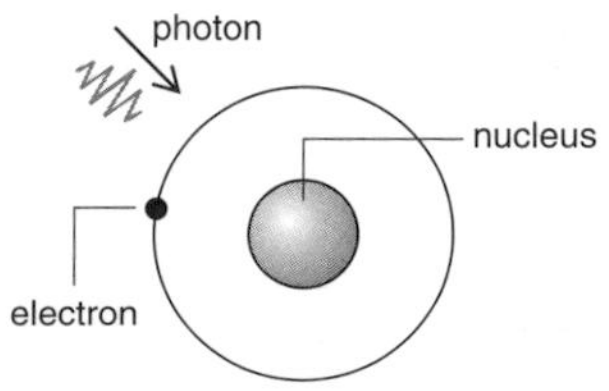

Photon absorption.

Planck's hypothesis:
$E = hf$ or $E = hc/\lambda$

- This is evidence for the particle-like nature of light, as it can only be explained if the electromagnetic radiation is arriving in little packets of energy (called **photons** or **quanta**), whose energy is directly proportional to the photon's frequency, *f*.

  This is **Planck's hypothesis**:

  $E = hf$ or $E = hc/\lambda$ where h is the **Planck constant**, 6.6×10^{-34} J s.

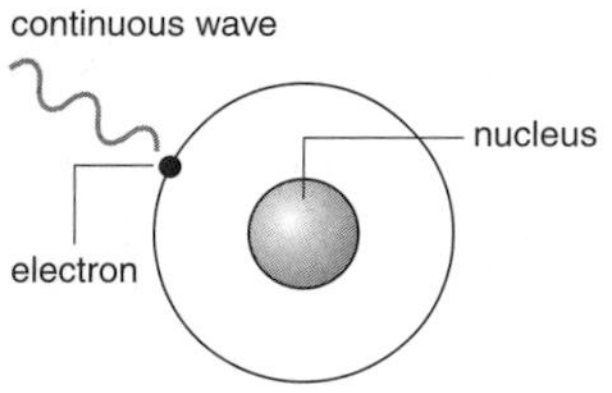

Continuous-wave absorption.

  If light had a continuous waveform, energy would be continuously pouring into the electron (see left). The electron would break free of its atom and be emitted whatever the frequency of light. The electron's KE would have no fixed limit, and an increase in intensity would increase the KE. Under this model the intensity of a wave is proportional to the square of the amplitude: $I \propto A^2$. The detailed experimental evidence mentioned before does not support this, so light is taken as being made up of photons. Under the photon model the intensity depends on the number of photons per second per square metre.

photoelectric equation:
$hf = \phi + KE_{max}$
or $hf = \phi + \frac{1}{2}mv_{max}^2$

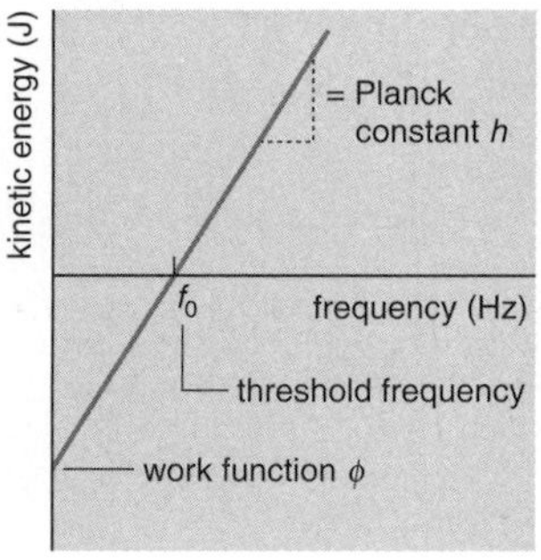

- When conservation of energy is applied to this, **Einstein's photoelectric equation** is produced (see left). The energy of the incoming photon is *hf*; some of it is used to break free of the surface (**work function**), and the rest is the electron's KE.

  The **work function**, ϕ, is defined as the minimum energy required to break free of the surface, and varies depending upon the type of material.

  If an electron is only just emitted we take the KE as 0, so $\phi = hf_0$ where f_0 is the threshold frequency. Rearrange the photoelectric equation into $KE = hf - \phi$, and plot a graph of KE against *f*. The gradient is equal to the Planck constant and is the same for any surface. The *y* intercept equals the work function, and the *x* intercept equals the threshold frequency.

Example *What is the maximum KE of an electron emitted from the surface of a metal with work function 1.2×10^{-9} J, when light of wavelength 740 nm is incident upon it?*

First determine the frequency of the light, then use Einstein's photoelectric equation:

$c = f\lambda$ so $f = c/\lambda = 3 \times 10^8/7.4 \times 10^{-7} = 4 \times 10^{14}$ Hz
$hf = \phi + KE_{max}$ so $4 \times 10^{14} \times 6.6 \times 10^{-34} = 1.2 \times 10^{-19} + KE_{max}$ and
$\boldsymbol{KE_{max} = 1.4 \times 10^{-19}}$ **J**

- To determine the value of the KE we use the idea of **stopping potential**.

  The **stopping potential** is the voltage required to stop electron emission.

  The circuit shown right contains a **photocell**, which consists of an evacuated tube containing two electrodes, one of which is made from a photo-emissive material. A p.d. opposing the electron motion is placed across the photocell and adjusted until there is no current flowing (the ammeter registers zero). The voltage across the cell at this instant is the stopping potential, V_s. The KE comes from the equation $eV_s = \frac{1}{2}mv^2$, where e is the charge on an electron (energy = charge × potential difference; see unit 9). A graph of current against potential difference will give an accurate value of V_s.

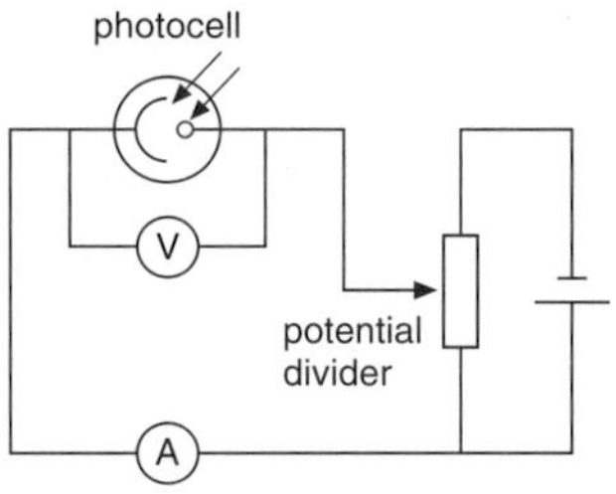

A circuit for determining stopping potential.

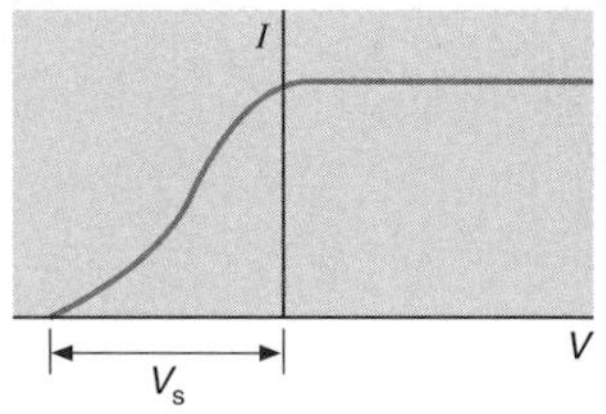

A graph of current versus voltage.

stopping potential:
$eV_s = \frac{1}{2}mv^2$

- Waves can behave like particles and vice versa. Electrons can be diffracted by thin sheets of material such as graphite (below). For diffraction to occur the atomic spacing must be close to but not smaller than the particle's wavelength (see unit 12). These are called **de Broglie waves.** A moving particle's wavelength is given by $\boldsymbol{\lambda = h/p}$ or $\boldsymbol{\lambda = h/mv}$.

de Broglie waves:
$\lambda = h/p$ or $\lambda = h/mv$

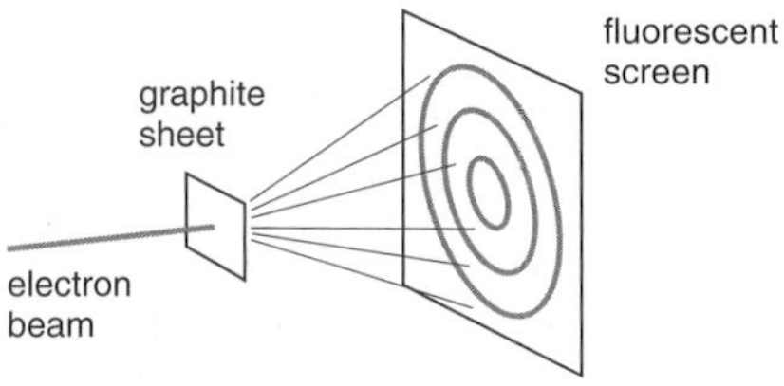

Electron diffraction.

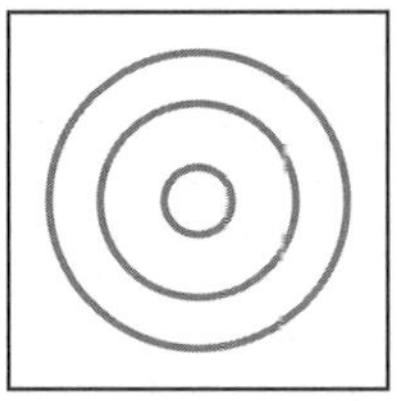
Electron diffraction pattern.

This only has any meaning on a subatomic level. It is possible to work out the wavelength of yourself, but the size of the doorway required to diffract you as you walk through it would be too small to pass through. Wavelength depends upon the speed of the particle and so it can be changed. Neutron diffraction is widely used for studying molecular and atomic structures, especially in crystals.

Example *What is the wavelength of a tennis ball, mass 0.3 kg, travelling at 5 m/s?*
Use $\lambda = h/mv$: $\lambda = 6.6 \times 10^{-34}/0.3 \times 5 = \mathbf{4.4 \times 10^{-34}}$ **m**

- The wave-like nature of particles can be used to explain why electrons around a nucleus can only have certain energy values (see unit 14). There are two models used to do this: both treat the electron as a wave applied to the Bohr model (see unit 15). In the first, a whole number of wavelengths is spread around the electron orbit (right). If there were not a complete number of wavelengths, destructive interference would occur. Only one particular wavelength will fit exactly into a specific orbit, which shows that the energy the electron can have is limited to certain values. In the second model, the electron's wavelength fitted exactly across the atom in a standing waveform (below right). The electron is confined to the atom by the attractive force of the nucleus. The standing wave can taken on various modes with different frequencies and thereby have different energies. Current theories look upon what we call electromagnetic and mechanical waves as '**probability waves**'. This means that the chance of an electron being in a particular place depends upon the intensity of the wave. (Intensity = amplitude^2.) This area of physics is called **wave mechanics**.

Amplitude^2 ∝ probability of an electron being present

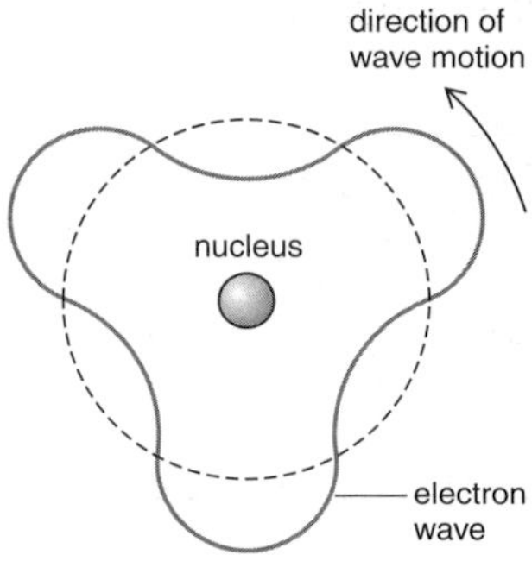

Electron wave moving around a nucleus.

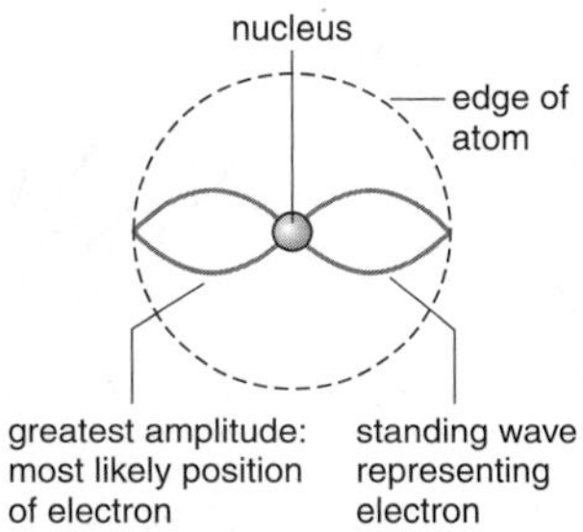

Standing wave of electron across an atom.

Unit 13 TESTS

RECALL TEST

1 What is the 'photoelectric effect'?

(2)

2 What is Planck's hypothesis?

(2)

3 What is meant by the term 'work function'?

(2)

4 What is a gold-leaf electroscope?

(2)

5 What is 'stopping potential'?

(2)

6 What is meant by 'threshold frequency'?

(2)

7 What is meant by 'de Broglie waves'?

(2)

8 In the photoelectric effect, what happens if the intensity of the incident radiation is increased?

(2)

9 What does the amplitude of an electron's wave tell you?

(2)

10 What evidence is there for the wave-like nature of particles?

(2)

(Total 20 marks)

CONCEPT TEST

Take $c = 3.0 \times 10^8$ m/s and the Planck constant $h = 6.6 \times 10^{-34}$ J s

1 What is the de Broglie wavelength of a snooker ball of mass 0.050 kg travelling at a velocity of 3.0 m/s?

(4)

2 What is a photon? What is the energy of a photon of visible light of wavelength 550 nm?

(4)

3 Light of frequency 5.80×10^{14} Hz hits a metal surface which ejects electrons with a range of kinetic energies up to a maximum of 1.20×10^{-19} J. What is the work function of the material and the threshold frequency?

(4)

4 An electron is accelerated from rest in a vacuum by a potential difference of 2000 V. Assume that all of the electric energy given to the electron is converted into KE, so that the equation $eV = \frac{1}{2}mv^2$ may be used. Calculate the electron's velocity and its de Broglie wavelength, given that the charge on an electron is 1.6×10^{-19} C and its mass is 9.1×10^{-31} kg.

(4)

5 The figure right shows how the maximum KE of electrons emitted from a metal surface varies with the frequency, *f*, of the electromagnetic radiation incident on the surface. What are the work function and the threshold frequency? How could a value for the Planck constant be found? Draw on the axes the graph you would obtain for a metal with a greater work function.

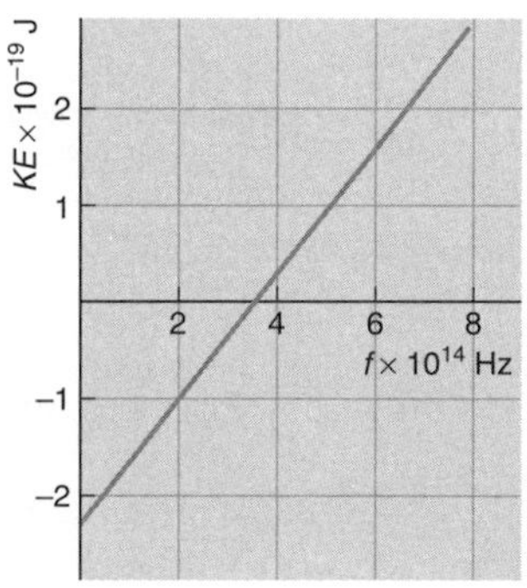

(8)

6 To get good diffraction from a slit, its size must be close to the wavelength of the wave to be diffracted. Calculate the de Broglie wavelength of an average person of mass 60 kg walking at a speed of 2.0 m/s. Explain why you do not get diffracted when you walk through a doorway.

(6)

7 What evidence is there for the particle-like nature of light and what evidence is there for the wave-like nature of light? When potassium is irradiated with light of wavelength 6.0×10^{-7} m electrons are just emitted. When it is irradiated with light of wavelength 1.6×10^{-7} m electrons leave the surface with a KE of 8.9×10^{-19} J. Determine a value for the Planck constant.

(6)

8 The figure right shows a graph of the variation of wavelength and momentum of a fundamental particle. How can we find out if this graph supports the de Broglie equation, $\lambda = h/mv$? What does this graph give as a value of the Planck constant? What information can be gained from knowing the amplitude of the de Broglie wave of an electron in an atom?

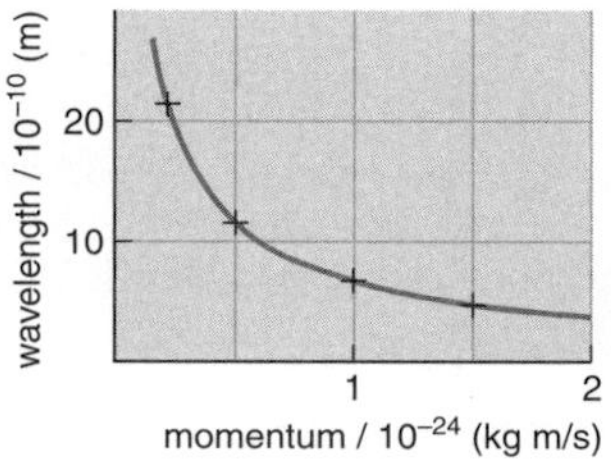

(6)

9 What potential difference would be required to stop the emission of electrons from a tungsten surface by electromagnetic radiation of wavelength 190 nm if the work function is 7.2×10^{-19} J? The graph right shows how current varies with voltage for a photocell connected to a potential divider. Draw on the axes another graph representing an increase in the frequency of the incident radiation and label it A. Draw another graph showing the effect of increasing the intensity of the radiation and label it B.

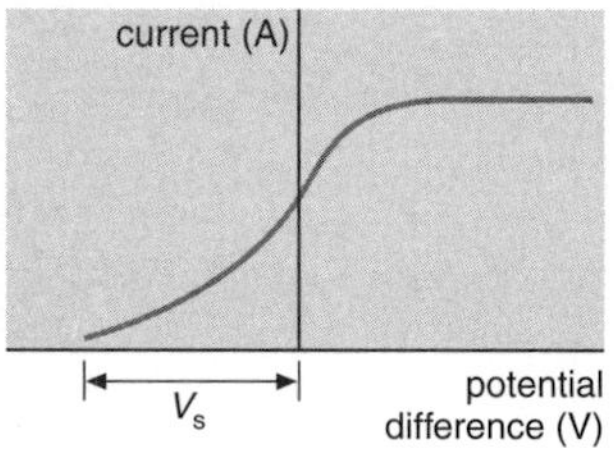

(8)

(Total 50 marks)

Unit 14

ENERGY LEVELS

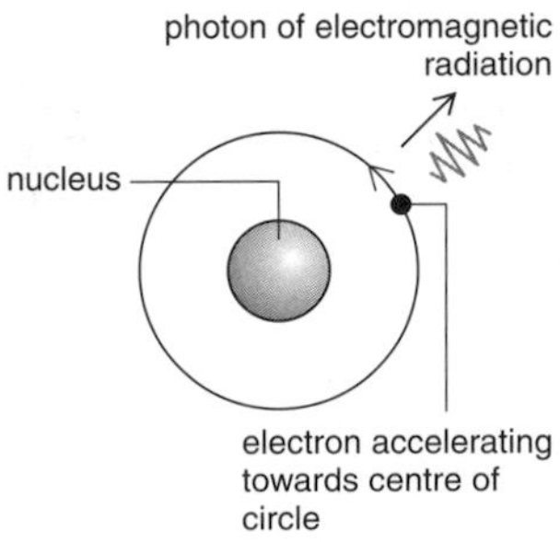

Accelerating electron.

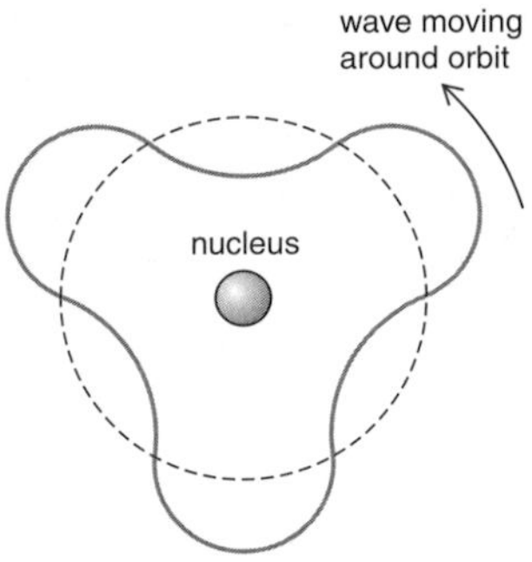

Electron waves in the atom.

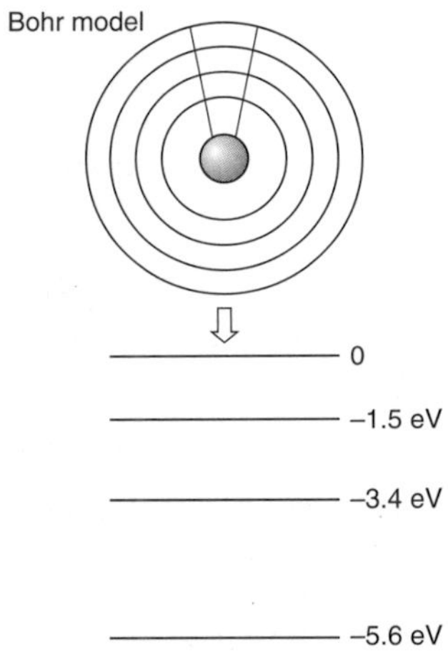

Examples of energy levels in an atom.

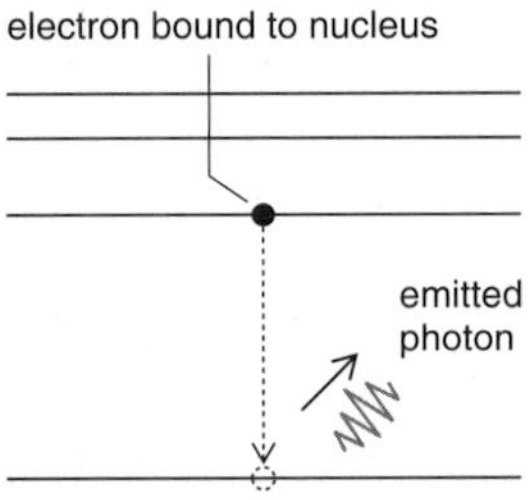

- An isolated atom consists of a central nucleus surrounded by electrons. These electrons can have certain specific energy-level values, but no value in between, and each element has a different set of energy levels. This produces a problem, because in the simple model of an electron orbiting a nucleus the electron is accelerating towards the centre of a circle. An accelerating electron should emit electromagnetic radiation, thereby losing energy, and spiral down into its atom's nucleus.

  This contradiction can explained as follows. In the previous unit we mentioned that electrons have a wavelength which can be applied to an atom in two ways: either the waveform fits exactly across the atom as a standing wave, or it fits around an electron orbit in a whole number of wavelengths. If the electron is considered to be a wave, rather than a particle, then it is no longer accelerating and the problem disappears. Wavelength is related to frequency by $v = f\lambda$, frequency is related to energy by $E = hf$, so the electrons can only have certain energy values.

- The value of energy for an electron which has left the atom is arbitrarily taken to be zero. As it approaches the atom its potential energy decreases, because the force acting on it is attractive; all energy-level values are therefore negative. They are often given in the unit **electronvolts** (eV).

  An **electronvolt** is defined as the energy given to an electron accelerated across a potential difference of one volt; $1\text{ eV} = 1.6\times10^{-19}\text{ J}$.

  This is often represented in an **energy-level diagram**, which shows the values of the different levels. If you take the simple Bohr model of the atom (see unit 15) and cut a section out of the electron orbits you get an energy-level diagram (see below left). The zero level corresponds to an electron having left the atom.

- It is possible for electrons to move between levels. They can gain the energy required to jump up in two ways:

  1 A free electron can collide with an electron bound to the nucleus and give it enough energy to jump between the levels. The free electron then moves off with the remainder of the energy. This gives us the equation $\boldsymbol{E_2 - E_1 = (\tfrac{1}{2}mv^2)_{lost}}$.

  2 An incoming photon can be absorbed by an electron which can then jump between levels. The energy of the photon must exactly equal the difference between levels. This gives us the equations $\boldsymbol{E_2 - E_1 = hf}$ and $\boldsymbol{E_2 - E_1 = hc/\lambda}$.

  In both cases the electron then drops back down and emits a photon corresponding to the difference between energy levels.

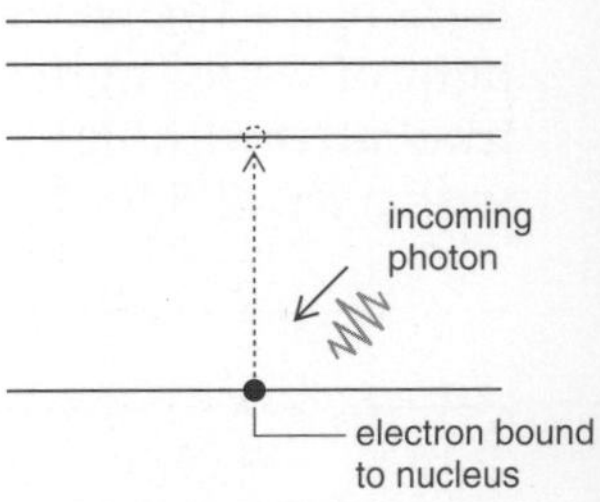

Electron transition due to photon.

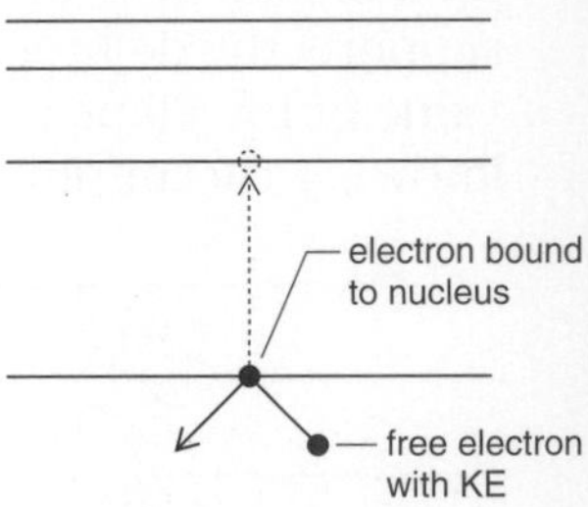

Electron transition due to collision.

Example *An electron in a hydrogen atom absorbs a photon and jumps between levels as shown right. What is the wavelength of the photon that produces this effect?*

Start by converting the units (eV) into joules:

$E_1 = -3.39\times1.6\times10^{-19} = -5.4\times10^{-19}\text{ J}$ and
$E_2 = -0.54\times1.6\times10^{-19} = -8.6\times10^{-20}\text{ J}$
$E_2 - E_1 = hc/\lambda$ so $(-8.6\times10^{-20}) - (-5.4\times10^{-19})$
$= (6.6\times10^{-34}\times3\times10^{8})/\lambda$
$\lambda = \mathbf{3.2\times10^{-7}\ m}$

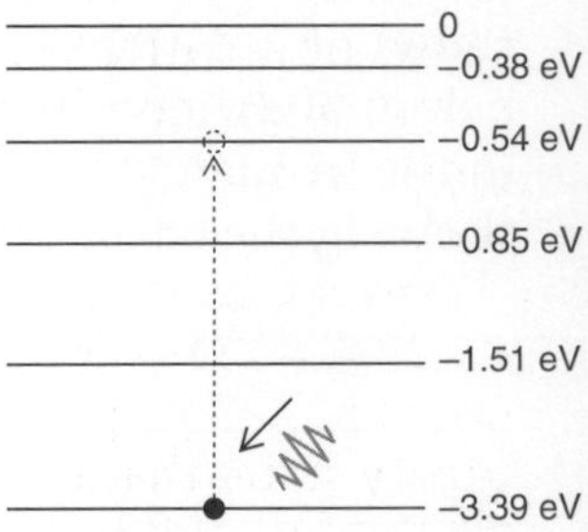

- The maximum number of electrons in each energy level is fixed. If all of an atom's electrons are in the lowest available energy levels the atom is in its **ground state.**

  When an electron has jumped to a higher level leaving an unoccupied level below it, the atom is in an excited state. The minimum energy required to raise an electron from one level to another for an atom in its ground state is called the **excitation energy**.

  When an electron is removed from an atom it is said to be ionized. The minimum energy required to remove an electron from an atom in its ground state is called the **ionization energy**.

- If a gas is heated the electrons in its atoms become excited and jump up levels. They then drop back, emitting photons in the usual way. Each element has a different set of levels and so has a different set of possible transitions, and therefore a different set of emitted frequencies. The emitted light can be analysed by breaking it up (using a diffraction grating) into a series of bright lines on a dark background; this is called an **emission line spectrum** (see right). These lines are unique to each element and are like a fingerprint which can be used to identify them. The spectrum is produced using a **spectrometer** with a diffraction grating mounted on it (see right).

- When white light passes through a gas the reverse process takes place. Some of the frequencies are absorbed by electrons jumping up levels, and then re-emitted as they fall back. They re-emit in all directions whereas the original light is only moving in one direction, so when the light's spectrum is analysed the intensity of the absorbed frequencies is less than that of the others. The spectrum appears as a normal white-light spectrum with a series of darker lines on it. This is called an **absorption line spectrum**.

  All of the above applies to isolated atoms. When atoms are combined together in molecules, in addition to the frequencies emitted due to electron transitions, frequencies are also emitted due to changes in molecular vibrational and rotational energy. This results in separate groups of lines called **bands**, which are collectively called a **band spectrum** (below right).

- Line spectra are particularly important in astronomy. When an object is moving towards you and emitting a wave, the frequency of that wave appears higher, and the wavelength shorter, than if the object was at rest (see below). This is called the **Doppler effect**. For light waves, the apparent frequency, f', is given by

  $$\boldsymbol{f'} = \left(\frac{\boldsymbol{c} + \boldsymbol{v_o}}{\boldsymbol{c} - \boldsymbol{v_s}}\right)\boldsymbol{f} \quad \text{for a source moving towards the observer}$$

  $$\boldsymbol{f'} = \left(\frac{\boldsymbol{c} + \boldsymbol{v_o}}{\boldsymbol{c} + \boldsymbol{v_s}}\right)\boldsymbol{f} \quad \text{for a source moving away from the observer}$$

  where f is the original frequency of the source, c is the speed of light, v_s is the speed of the source, and v_o is the speed of the observer.

- When the light of other galaxies is analysed their whole spectra are seen to be shifted towards lower frequencies, because they are moving away from us. This is called **redshift** and if $\boldsymbol{v \ll c}$ then $\boldsymbol{\Delta f/f = v/c}$, where v is the velocity of a distant object, c is the speed of light, f is the frequency, and Δf is the apparent change in frequency. The further the galaxies are away the greater their velocity. This is an important piece of evidence showing that the universe is expanding, and was used by Edwin Hubble to form the equation $\boldsymbol{v = H_0 d}$, where v is the velocity in km/s of an object distance d away (measured in megaparsecs), and H_0 is the **Hubble constant**. The Big Bang theory states that the universe was created at a single point by a large 'explosion'. If it is true then the age of the universe is $1/H_0$. H_0 is thought to be about 75 km s^{-1}Mpc^{-1}. (pc stands for parsec; 1 parsec = 3.258 light-years. A light-year is the distance travelled by light in one year.)

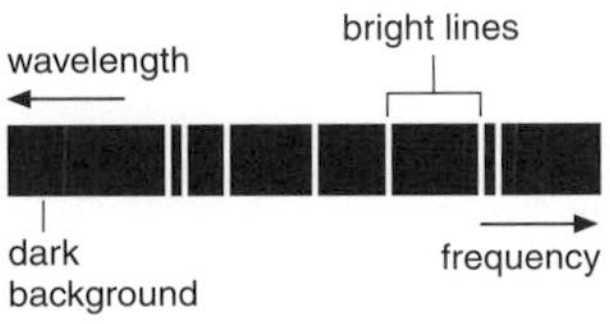

An emission spectrum.

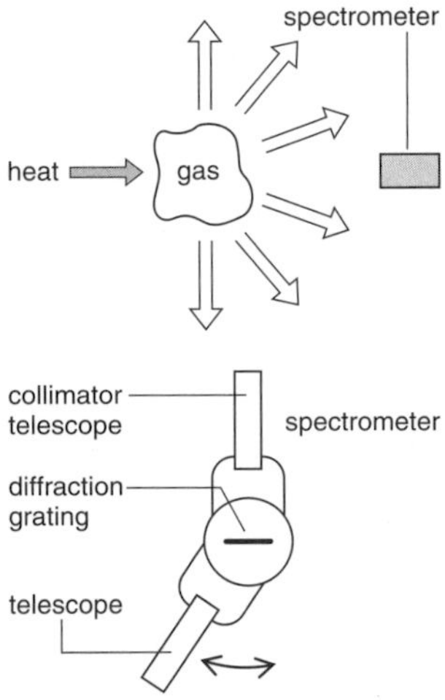

Production of an emission spectrum.

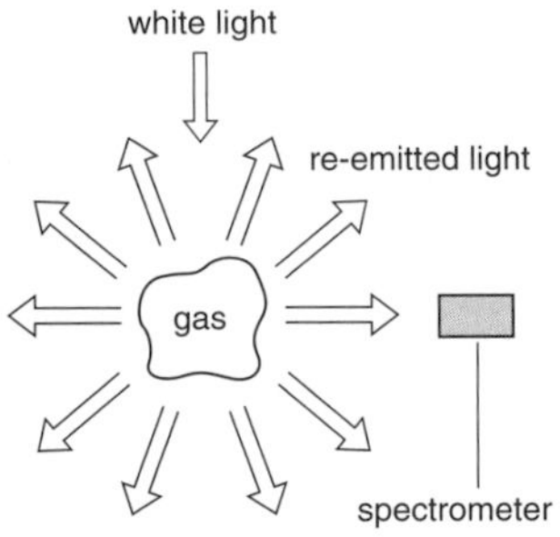

Production of an absorption line spectrum.

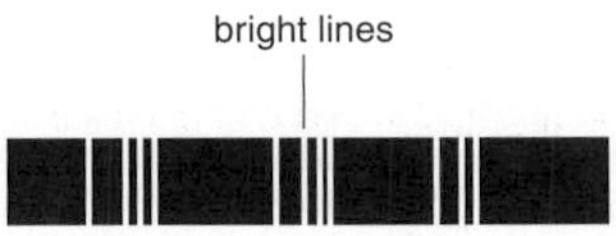

A band spectrum.

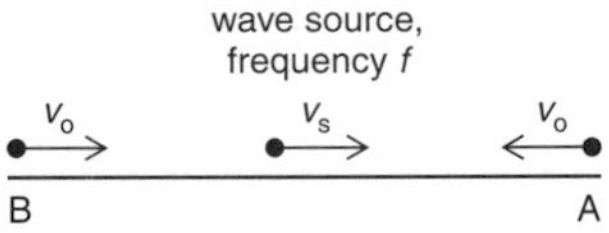

Considering the Doppler effect for a source moving towards a moving observer at A and away from a moving observer at B.

Doppler effect:

$$\boldsymbol{f'} = \left(\frac{\boldsymbol{c} + \boldsymbol{v_o}}{\boldsymbol{c} \mp \boldsymbol{v_s}}\right)\boldsymbol{f}$$

Hubble's law: $\boldsymbol{v = H_0 d}$

TESTS

RECALL TEST

1 What is an electronvolt?

______________________________ (2)

2 What is meant by 'ground state'?

______________________________ (2)

3 What is 'excitation energy'?

______________________________ (2)

4 What is 'ionization energy'?

______________________________ (2)

5 What is an 'absorption spectrum'?

____

______________________________ (2)

6 What is an 'emission spectrum'?

____

______________________________ (2)

7 What is 'redshift'?

______________________________ (2)

8 What is the 'Hubble constant'?

____

______________________________ (2)

9 What is the 'Doppler effect'?

______________________________ (2)

10 What is the Big Bang theory?

____

______________________________ (2)

(Total 20 marks)

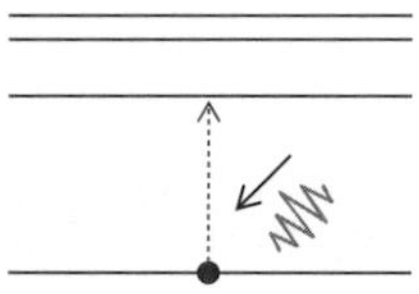

CONCEPT TEST

Take $c = 3.0 \times 10^8$ m/s and the Planck constant $h = 6.6 \times 10^{-34}$ J s

1 A photon of green light is absorbed by an electron which jumps from level A to level B in the diagram shown left. What is the wavelength of the green light?

______________________________ (4)

2 An electron with a KE of 1.2×10^{-18} J collides with an electron in orbit around an atom, causing it to jump from the –6.2 eV level to the –1.5 eV level. What is the final KE of the first electron?

____

______________________________ (4)

3 A police car is travelling towards you at 10.0 m/s, emitting a sound from its siren with a frequency of 500 Hz. What does the frequency appear to be if you are stationary and the speed of sound is 320 m/s?

______________________________ (4)

4 How is an absorption spectrum produced, given that absorbed frequencies are re-emitted? An excited electron is in the level shown right. What possible frequencies can be emitted when it drops back down to the 1.92 eV and 11.6 eV levels?

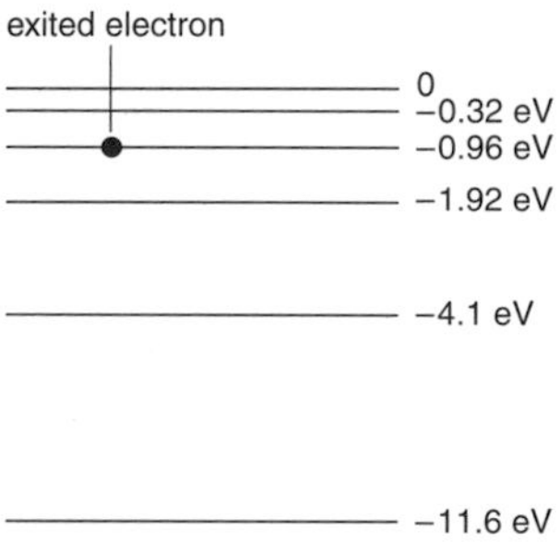

(6)

5 How fast is a galaxy moving if it is 200 Mpc away from us? If it is emitting a light of frequency 2.8×10^{14} Hz, what would the redshift change this frequency to? (Hubble constant = 75 km s^{-1} Mpc^{-1}.)

(4)

6 The first excitation energy of an atom is 4.1 eV, the second excitation energy is 5.7 eV, and the ionization energy is 6.3 eV. Complete the energy-level diagram for this atom shown right, and calculate the possible wavelength of photons that it could emit, excluding transitions from the 0 eV level. Why are the energy values in the diagram negative?

(8)

7 A star is 2.0×10^8 light years away and is moving away from you, obeying Hubble's law. What is its velocity? The star is emitting a frequency of 4.7×10^{14} Hz. What would the frequency appear to be if you left the Earth in a spaceship travelling towards the star at 3.0×10^3 km/s? (1 pc = 3.258 l.y. (light-years), Hubble constant = 75 km s^{-1} Mpc^{-1}.)

(6)

8 A 60 W monochromatic lamp is emitting light of wavelength 440 nm equally in all directions. Estimate the number of photons per second hitting a piece of paper with an area of 0.25 m^2 at a distance of 0.50 m from the lamp. (Assume that the energy from the bulb is spread over the surface area of a sphere at radius r. Surface area $A = 4\pi r^2$.)

(6)

9 Describe how an emission spectrum is produced. What is a band spectrum? The figure right shows the energy levels of a mercury atom. What does the zero level signify? What is the ionization energy of mercury if, in an unexcited state, the highest value with an electron present is the −10.4 eV level? What transition takes place when a photon of wavelength 141 nm is emitted?

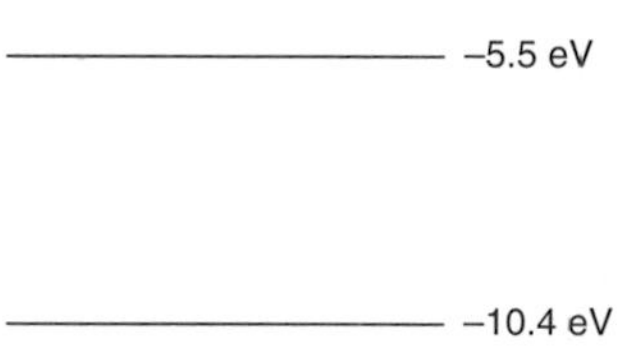

(8)

(Total 50 marks)

Unit 15 ATOMIC AND NUCLEAR

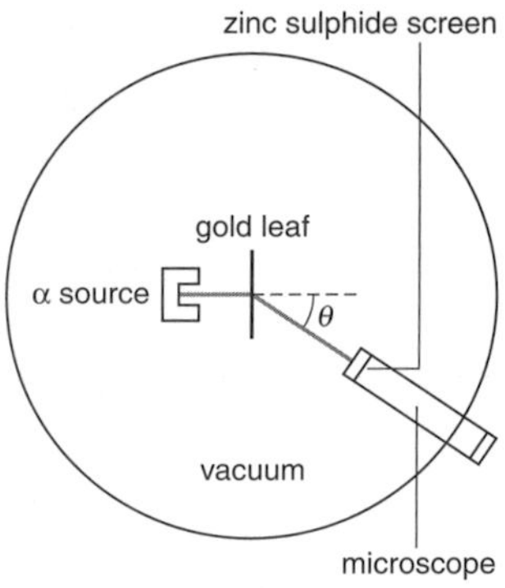

Rutherford's gold-leaf experiment.

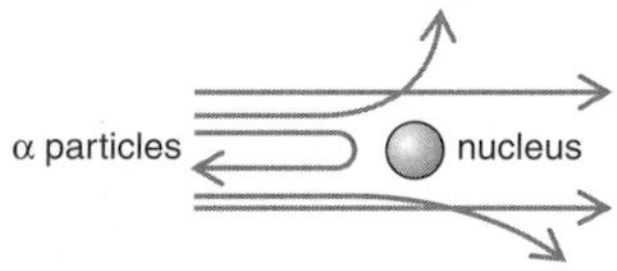

Paths of alpha particles passing close to a nucleus.

nuclear radius: $R = r_0 A^{\frac{1}{3}}$

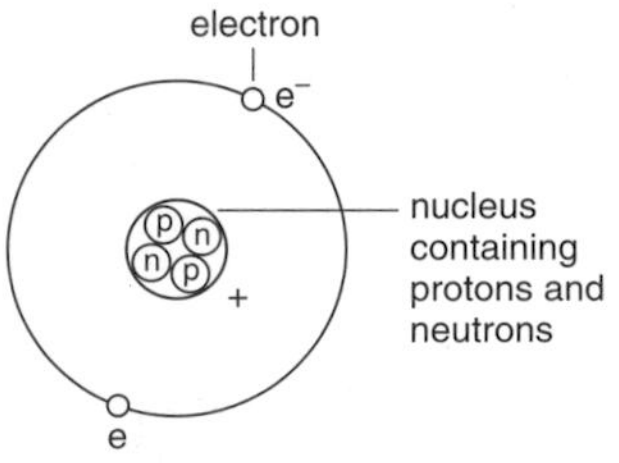

The Bohr model of the atom.

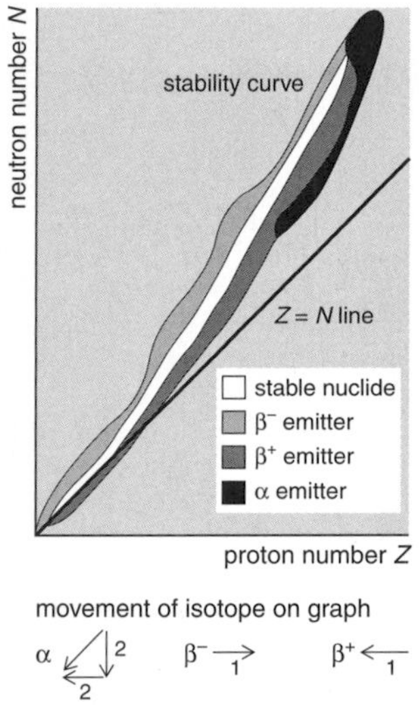

An N–Z plot for stable and unstable nuclides. Any isotopes not on the stable section of the graph will decay and try to become stable. The type of radiation emitted depends on the isotope's position on the graph. Any emission will change the isotope and its position, generally moving it towards the stable curve.

- It was known by the 1900s that matter consisted of atoms. The most important theoretical model that existed was derived from **Rutherford's alpha-particle scattering experiment**. Rutherford, with his assistants Geiger and Marsden, bombarded a thin sheet of gold with alpha particles (which are small and positively charged; see later). Most of the alpha particles passed through undeflected. This indicated that most of the gold contained empty space. Some were deflected through a small angle and some through an angle greater than 90°. This indicated that most of the mass of the atom is concentrated in a small, positively charged region. The fact that some bounce back in the original direction shows that the mass here is greater than the alpha particle's mass and has the same charge. All of this is evidence for the 'nuclear' atom.

  The central mass is called the **nucleus**. A typical nucleus has a diameter of 10^{-15} m, and a typical atom has a diameter of 10^{-10} m. This is a 10^5 difference, so if the nucleus of an atom were a 10 cm sphere, the next nucleus would be at least 10 km away. This shows how much empty space there is in matter. It is the electrostatic forces between nuclei and electrons that keep everything at this separation. The radius of a nucleus is given by

  $$R = r_0 A^{\frac{1}{3}}$$

  where R is the radius, A is the mass number (see below), and r_0 is a constant (typically 1.4 fm, where 1 fm (femtometre) = 10^{-15} m).

- The simple model used today, which is a slight refinement of Rutherford's model, is the **Bohr model**. This consists of a central nucleus containing **neutrons** (n) and positively charged **protons** (p) of roughly equal mass, with negatively charged **electrons** (e) in orbit around it. The electrons have roughly 1/2000 the mass of the neutrons and protons. In an electrically neutral atom the number of electrons equals the number of protons. If it does not, the atom is called an **ion**. The standard symbol used takes the form $^{A}_{Z}X$ where A is called the **mass** or **nucleon number**, which is the number of protons and neutrons present, and Z is called the **atomic** or **proton number**, which is the number of protons present. $(A - Z) = N$, the number of neutrons present.

- Atoms with the same proton number but different nucleon numbers are called **isotopes**. Unstable isotopes try to become stable by emitting radiation (**decaying**). This is called **radioactivity**. Radioactivity is a random process, which means that it is not possible to predict which particular nuclei will decay, but it is possible to say that, on average, a certain percentage will decay over a certain period of time. There are three types of radiation, with different properties.

	Alpha	**Beta**	**Gamma**
Nature	helium nucleus $^{4}_{2}He$	electron $^{0}_{-1}e$ or positron $^{0}_{+1}e$*	electromagnetic radiation
Charge	+2	–1 or +1	0
Mass in u**	4	1/1836	0
Penetration	1–5 cm in air	10–50 cm in air	4 cm of lead stops 90%
Velocity	up to 0.05c***	up to 0.98c	c
Deflected by magnetic and electric fields?	yes	yes – direction depends on charge	no
Affects photographic film and produces fluorescence?	yes (strong)	yes	yes (weak)
Examples of decay	$^{235}_{92}U \rightarrow ^{4}_{2}He + ^{231}_{90}Th$	$^{14}_{6}C \rightarrow ^{14}_{7}N + ^{0}_{-1}e$ $^{65}_{30}Zn \rightarrow ^{65}_{29}Cu + ^{0}_{+1}e$	

*A positron is identical to an electron except that it has positive charge (see unit 16).
**u = atomic mass unit. 1 u = 1.66×10^{-27} kg.
***c = speed of light.

Characteristics of the three types of radiation.

The type of radiation emitted depends upon the isotope present. A graph of neutron number, N, against proton number, Z, shows which type of decay is most likely (see diagram). The **activity**, A, is defined as the number of radioactive emissions per second, and is measured in becquerels (Bq). The activity is directly proportional to the number of atoms present, N. λ is the **decay constant**, which is different for each isotope.

$\boldsymbol{A \propto N}$ or $\boldsymbol{A = \lambda N}$

- The decay of nuclei is an exponential process (see diagram). It is not possible to say how long it will take a particular nucleus to decay, but we can say how long, on average, it will take half the nuclei present to decay. This length of time is called the **half-life**, $\boldsymbol{T_{\frac{1}{2}}}$.

$\boldsymbol{A = A_0 e^{-\lambda t}}$ or $\boldsymbol{N = N_0 e^{-\lambda t}}$

$\boldsymbol{T_{\frac{1}{2}} = (\ln 2)/\lambda}$ or $\boldsymbol{T_{\frac{1}{2}} = 0.693/\lambda}$

where A_0 and N_0 are initial activity and initial number of atoms respectively, e is the **transcendental number** (about 2.718), and 'ln' means 'log to the base e'.

radioactivity:
$\boldsymbol{A \propto N}$ or $\boldsymbol{A = \lambda N}$
$\boldsymbol{A = A_0 e^{-\lambda t}}$ or
$\boldsymbol{N = N_0 e^{-\lambda t}}$
$\boldsymbol{T_{\frac{1}{2}} = (\ln 2)/\lambda}$ or
$\boldsymbol{T_{\frac{1}{2}} = 0.693/\lambda}$

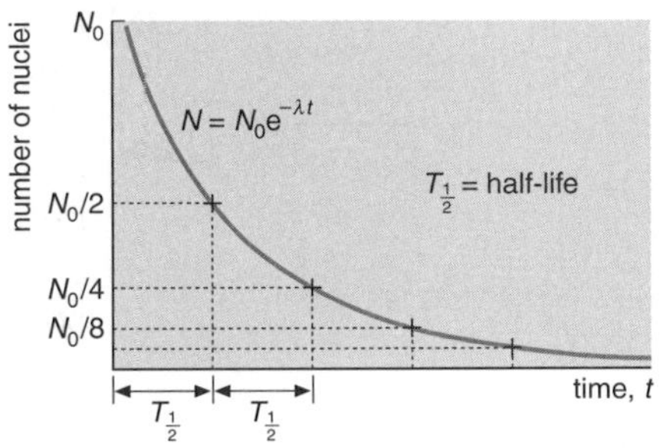

A half-life curve. An exponential curve is one in which one quantity changes by the same factor over equal changes in the other quantity. In this case the number of nuclei, N, will halve in each period of time $T_{\frac{1}{2}}$.

mass–energy equation: $\boldsymbol{E = mc^2}$

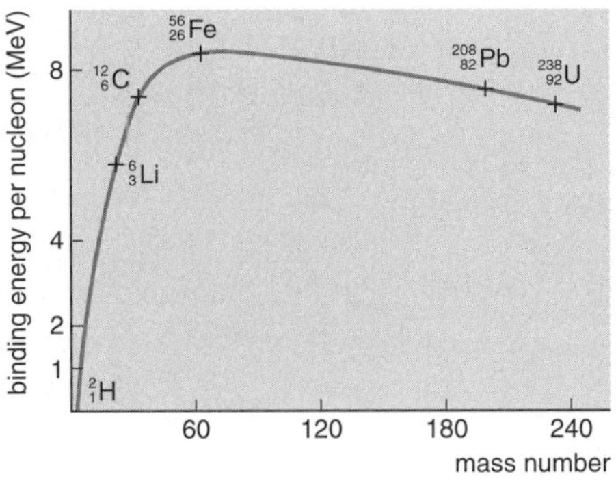

Curve for binding energy per nucleon.

- The mass of a nucleus is less than the sum of the masses of its individual protons and neutrons. The missing mass, m, is called the **mass defect** and has been converted into energy, E. This is called the **binding energy** of the nucleus, and can be determined using $E = mc^2$, where c is the speed of light.

The binding energy of a nucleus is defined as the energy required to break up a nucleus into its individual components. Energy has to be given to the nucleus so that it can be converted into the extra mass required if the nucleus is to be broken up into individual protons and neutrons.

A measure of an atom's stability is the binding energy per nucleon, which is the binding energy of the nucleus divided by the number of nucleons present. The greater the binding energy per nucleon, the more stable an atom is. A graph of the binding energy per nucleon against nucleon number is shown above right.

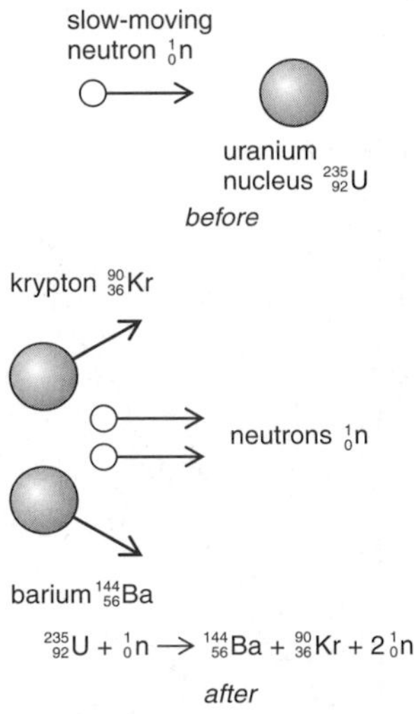

Nuclear fission.

- If a slow-moving neutron collides with a uranium nucleus, it is absorbed. The uranium nucleus becomes unstable, splits into two 'daughter' products, more neutrons, and a lot of energy. The daughter products have a greater binding energy per nucleon than the uranium, so energy is given up in the process, which is called **nuclear fission**. The neutrons released in one fission can go on and produce more fissions in a **chain reaction.** If a complete chain reaction happens then a nuclear explosion occurs. This only happens if the mass of the material present is above a certain value called the **critical mass**.

Nuclear fission is the splitting of a heavy nucleus into two lighter ones. A typical nuclear fission process is shown above right.

- It is possible to combine two light nuclei together to form a heavier one, but it is difficult because nuclei are positively charged and repel each other. To overcome this force the nuclei have to have a large amount of KE, which is done by making them very hot. The resulting product has a greater binding energy per nucleon so energy is released. The process is called **fusion**.

Nuclear fusion is the combining of two 'light' nuclei to form a heavier one. A typical nuclear fusion is shown in the diagram. In both of the above cases the products lie higher up the binding energy–nucleon curve, so they both release energy.

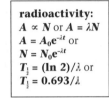

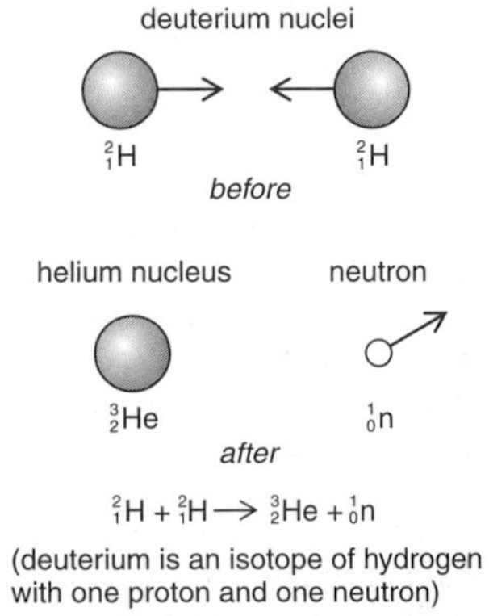

Nuclear fusion.

Unit 15

TESTS

RECALL TEST

1 What is meant by 'the nuclear atom'?

(2)

2 What is the difference in magnitude between an atomic nucleus and an atom?

(2)

3 How does Rutherford's experiment show that atoms have a small positively charged nucleus?

(2)

4 What is 'half-life'?

(2)

5 What is an alpha particle?

(2)

6 What is a beta particle?

(2)

7 Why does nuclear fusion have to take place at high temperature?

(2)

8 What is 'nuclear fission'?

(2)

9 What is the 'binding energy' of a nucleus?

(2)

10 What is a chain reaction and how does it come about?

(2)

(Total 20 marks)

CONCEPT TEST

Take $c = 3.0 \times 10^8$ m/s and the Avogadro constant = 6.0×10^{23} per mol

1 A radioactive source has a half-life of 3.6 min and an activity of 3.5×10^3 Bq. Find the radioactive decay constant and the number of nuclei present.

(4)

2 What is the mass defect and the binding energy of a $^{12}_{6}C$ nucleus? (Masses: proton = 1.0080 u, neutron =1.0087 u, carbon atom = 12.0074 u, and 1 u = 1.66×10^{-27} kg.)

(4)

3 A cardiac pacemaker containing 160 mg of plutonium-238 is fitted to a patient. The plutonium decays by emitting 5.4 MeV alpha particles and has a decay constant of 2.4×10^{-10}/s. Find the number of atoms present in the pacemaker and the initial decay rate. What is the power of a new pacemaker?

(6)

4 Deuterium ^2_1H and tritium ^3_1H fuse together in the fusion reaction $^2_1H + ^3_1H \rightarrow ^A_ZX + ^1_0n$. State the values of *A* and *Z*. Explain why energy is released in this reaction and calculate its value. (Masses: $^2_1H = 3.34250 \times 10^{-27}$ kg, $^3_1H = 5.00573 \times 10^{-27}$ kg, $^A_ZX = 6.62609 \times 10^{-27}$ kg, $^1_0n = 1.67438 \times 10^{-27}$ kg.)

(4)

5 Why is radioactive decay described as a random process? A radioactive sample has an activity of 4.8×10^4 Bq and a half-life of 2.0 days; what is the activity 13.5 days later?

(4)

6 A typical fission reaction is $^{235}_{92}U + ^1_0n \rightarrow ^{95}_{42}Mo + ^{139}_{57}La + 2^1_0n + 7^{\;0}_{-1}e$. What is the total energy released by 2.0 g of $^{235}_{92}U$ undergoing fission by this reaction? (Masses: $^1_0n = 1.0087$ u, $^{95}_{42}Mo = 94.906$ u, $^{139}_{57}La = 138.906$ u, $^{235}_{92}U = 235.044$ u, and 1 u $= 1.66 \times 10^{-27}$ kg.)

(8)

7 A uranium-bearing rock contains 9 $^{238}_{92}U$ atoms for every 8 ^4_2He atoms present. If the decay sequence that converts uranium to lead involves 8 alpha-particle emissions what is the age of the rock? (Half-life of $^{238}_{92}U = 4.5 \times 10^9$ years.)

(6)

8 A sample of iodine contains 1 atom of the unstable isotope iodine-131 for every 5.0×10^7 atoms of the stable isotope iodine-127. The figure right shows the variation of activity of an iodine-131 source with time. Use the graph to determine the decay constant, and work out the number of atoms present in the original sample and the original mass.

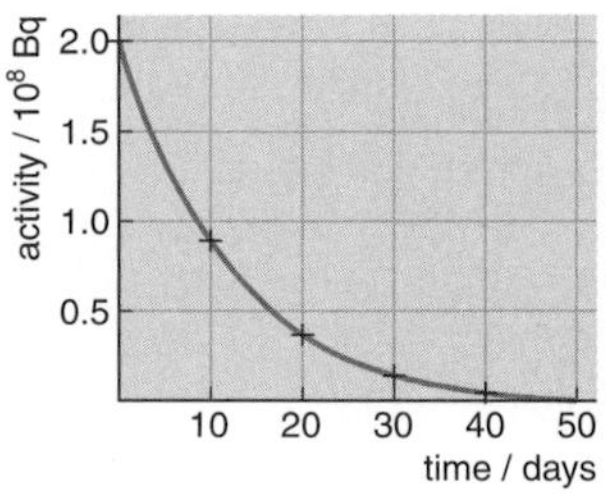

(6)

9 A small volume of a solution containing a radioactive isotope with a half-life of 15 h was used to determine the volume of blood in a patient. When the solution was injected into the patient's bloodstream the activity was 100 Bq. Calculate the total activity 20 h after the injection. What is the volume of blood if a 1.0 cm^3 sample has an activity of 0.004 Bq?

(8)

(Total 50 marks)

Unit 16 PARTICLE PHYSICS I

- This unit builds on unit 15 to look at the subatomic and **quantum mechanical** level of physics.

  Quantum mechanics is an extension of the ideas of Planck (see unit 13). Planck's idea was that the emission and absorption of electromagnetic radiation takes place in little packets of energy called quanta which have a fixed value of energy. Quantum mechanics applies this to very small subatomic (elementary) particles, assigning to them certain quantities, such as charge, which can only change by fixed amounts. **Quantum numbers** are a series of numbers, integers or half-integers (0, 1, 2, 1/2, 3/2, etc.), given to these quantities, which describe the state of a particle or system of particles. Conservation principles can be applied to quantum numbers to determine which particle interactions are possible.

- Molecules are made up of atoms and atoms are made up of smaller particles. When you can no longer break a particle up into anything smaller it is called a **fundamental particle**.

  A **fundamental particle** is a particle which cannot be broken down into any other particles.

  Leptons, **quarks**, and **gauge bosons** are fundamental particles. Particles which are not fundamental are called **hadrons**. There are two types of hadron: **mesons** and **baryons** (see unit 17).

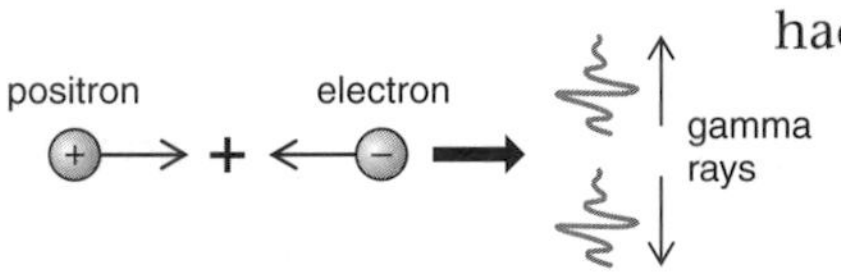

Matter–antimatter annihilation: the conversion of a positron and an electron into gamma rays.

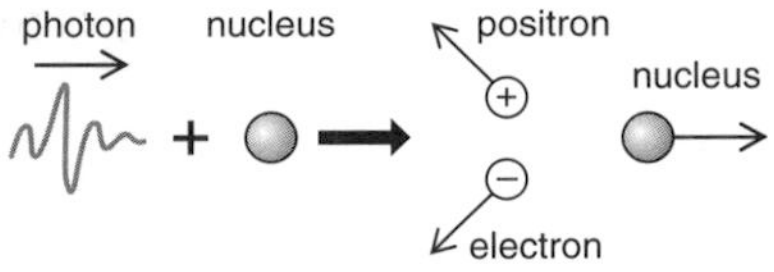

Pair production: the conversion of a gamma-ray photon into a positron and an electron.

- Every particle has its **anti-particle.** An **anti-particle** is the opposite of a **particle** and is not of ordinary matter.

  Particles and their anti-particles have the same mass and lifetime, but opposite charges. When a particle and an anti-particle meet, for example an electron and a positron, they annihilate each other (see left). Their mass is converted into energy in the form of gamma rays. The reverse process can also occur when a gamma ray spontaneously turns into an electron–positron pair, but another particle such as an atomic nucleus or electron must also be present so that energy and momentum may be conserved (see left).

beta decay:
$^{1}_{0}\mathbf{n} \rightarrow ^{1}_{1}\mathbf{p} + ^{0}_{-1}\mathbf{e} + \bar{\nu}_e$

- When a radioactive atom emits a β^- particle from its nucleus a neutron changes into a proton and an electron. Charge is obviously conserved, but if the energy released in the decay is calculated and the energy of the particles is measured there appears to be some missing. Similarly, if one applies the principle of conservation of momentum there appears to be some momentum missing. This cannot be possible, so physicists introduced the idea of **neutrinos**, ν, and **anti-neutrinos**, $\bar{\nu}$, new particles of zero **rest mass** (no mass when stationary) produced in decays and accounting for missing momentum and energy.

- All particles are said to have a **spin**, and there is a quantum number assigned to this. Think of a particle as a sphere spinning with a particular angular momentum. Angular momentum is conserved so spin must be conserved. Some particles, such as neutrinos, can only spin in one direction, but others, such as electrons, may spin in either direction. **Fermions** are particles with a half-integer spin i.e. 1/2, 3/2, 5/2, etc. **Bosons** are particles with zero or integer spin i.e. 0, 1, 2, 3, etc. All leptons have a spin of 1/2 and so are fermions.

- There are three basic types of lepton, each with an associated neutrino. If one includes the corresponding anti-particles there are 12 in total. The three basic negatively charged leptons are the electron, e^-, the muon minus, μ^-, and the tau minus, τ^-. Each of the three basic types of lepton has a **lepton number** associated with it, which must be conserved: electron (L_e), muon (L_μ), and tau (L_τ). Each lepton or its corresponding neutrino has a lepton number of +1, and each anti-lepton or its corresponding neutrino has a lepton number of –1. All other particles have a lepton number of 0.

  For any particle interaction lepton numbers must be conserved.

Particle	Relative rest mass	Symbol	Charge	Lepton number L_e	L_μ	L_τ	Spin	Anti-particle	Symbol	Charge	Lepton number L_e	L_μ	L_τ	Spin
electron	1	e^-	–1	1	0	0	$\frac{1}{2}$	positron	e^+	1	–1	0	0	$\frac{1}{2}$
muon (mu minus)	200	μ^-	–1	0	1	0	$\frac{1}{2}$	mu plus	μ^+	1	0	–1	0	$\frac{1}{2}$
tau (tau minus)	3500	τ^-	–1	0	0	1	$\frac{1}{2}$	tau plus	τ^+	1	0	0	–1	$\frac{1}{2}$
electron neutrino	0	ν_e	0	1	0	0	$\frac{1}{2}$	electron anti-neutrino	$\bar{\nu}_e$	0	–1	0	0	$\frac{1}{2}$
muon neutrino	0	ν_μ	0	0	1	0	$\frac{1}{2}$	muon anti-neutrino	$\bar{\nu}_\mu$	0	0	–1	0	$\frac{1}{2}$
tau neutrino	0	ν_τ	0	0	0	1	$\frac{1}{2}$	tau anti-neutrino	$\bar{\nu}_\tau$	0	0	0	–1	$\frac{1}{2}$

The leptons and their anti-particles.

- Unit 4 mentioned the four fundamental forces: strong and weak nuclear, electromagnetic, and gravitational. When dealing with subatomic particle physics the first three are more important, as the gravitational force is insignificant at this level. We also refer to them as 'interactions'. The strong interaction only has a very short range (10^{-15} m), but is about a hundred times stronger than the electromagnetic interaction that produces a repulsion between protons, so it can hold the nucleus together. The weak interaction causes beta decay and the decay of certain unstable particles. Weak interaction processes take much longer than strong interactions.

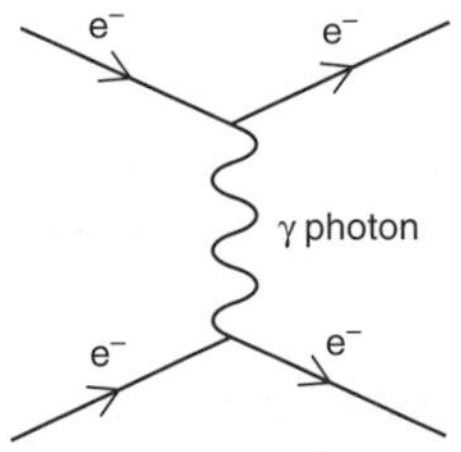

A Feynman diagram showing two electrons repelling each other.

Interaction/ force	electromagnetic	gravitational	weak nuclear	strong nuclear
Range (m)	∞	∞	10^{-18}	10^{-15}
Relative strength	10^{-2}	10^{-38}	10^{-5}	1
Exchange particles	photon (γ)	graviton (G)	W^+, W^-, Z^0	gluon (g)
Affects	all charged particles	all particles with mass	all particles, and changes in quark type	quarks, not leptons

Quarks are explained in unit 17. Gravitons are only predicted to exist and have not yet been observed, and gluons (see unit 17) are not directly observable as they cannot exist independently of fundamental particles.

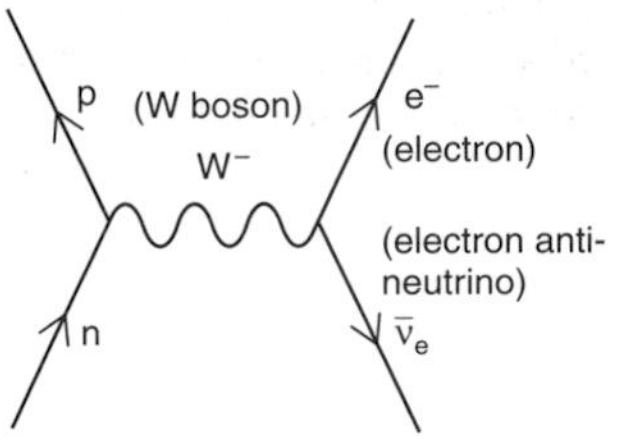

A Feynman diagram showing β^- decay. A neutron decays into a proton and a W^- gauge boson. The gauge boson then decays into an electron and an electron anti-neutrino.

Hadrons can take part in all interactions. Leptons can take part in all interactions except the strong nuclear interaction. All leptons interact by the weak interaction, but only charged leptons interact by the electromagnetic.

Generally a force is thought to act on an object because of its presence in a field, for example gravitational or electromagnetic, produced by another object (see unit 18). The quantum mechanical way of describing this is to say that the force is due to an exchange of particles which are emitted and absorbed by the objects. These particles which carry the force between objects (mediate), causing the interactions, are called **gauge bosons**. They can only exist for a short time and their range is inversely proportional to their mass. Electromagnetic and gravitational forces have infinite range because their exchange particles have no mass. These interactions are illustrated using **Feynman diagrams**.

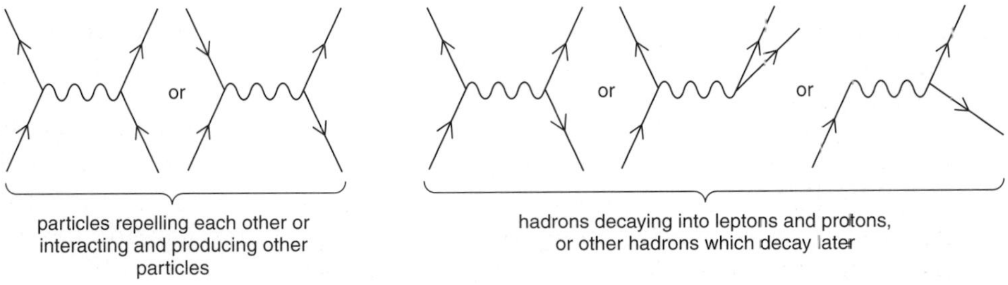

Types of Feynman diagram. One type shows particles exerting forces on or interacting with each other. The other shows particles decaying into other particles. The directions of the arrows indicate which is which.

TESTS

RECALL TEST

1 What is 'quantum mechanics'?

(2)

2 What are the four fundamental interactions?

(2)

3 What are 'exchange particles'?

(2)

4 What is a fundamental particle?

(2)

5 What interactions can leptons 'feel'?

(2)

6 What are the lepton numbers of a lepton, an anti-lepton, a hadron, and a baryon?

(2)

7 What are 'fermions' and 'bosons'?

(2)

8 What is a 'neutrino'?

(2)

9 What does 'conservation of lepton numbers' mean?

(2)

10 Why is the range of gravitational and electromagnetic interactions infinite?

(2)

(Total 20 marks)

CONCEPT TEST

1 Which interaction (force) is involved in each of the following situations? **a** A satellite orbiting the Earth; **b** a gas atom being ionized; **c** beta decay.

(3)

2 Which of the interactions (forces) would act between the following pairs of particles? **a** Two positrons; **b** two anti-neutrinos; **c** two protons; **d** an electron and a proton.

(4)

3 Describe in words what happens in each of these particle interactions: **a** ${}_{0}n \rightarrow {}_{1}^{1}p + {}_{-1}^{0}e + \bar{\nu}_e$; **b** $\mu^- \rightarrow e^- + \nu_\mu + \bar{\nu}_e$; **c** $\pi^+ \rightarrow \mu^+ + \nu_\mu$. ($\pi^+$ is a particle called a pion plus. It is a meson and is explained in unit 17.)

(6)

4 What is meant by 'pair production'? Why is another particle required for it to take place?

____

____

____ (4)

5 Describe what is happening in each of the Feynman diagrams right.

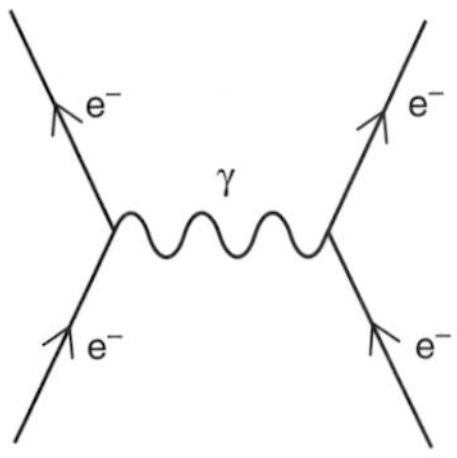

a

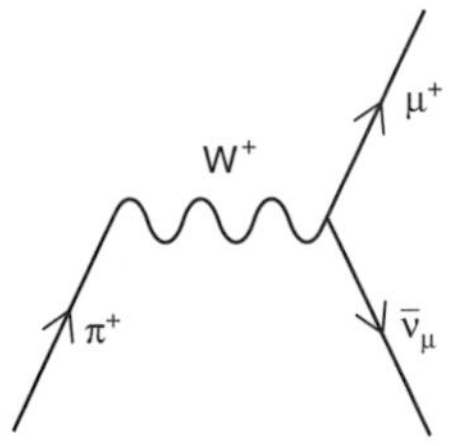

b

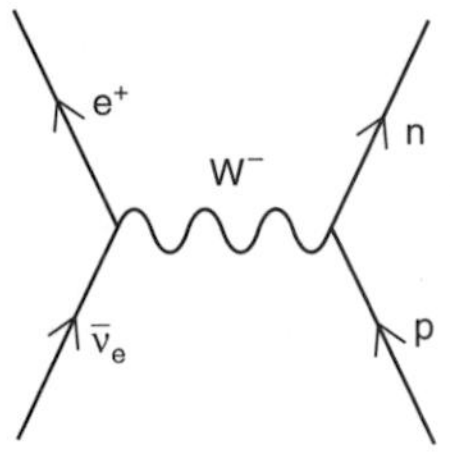

c

____

____

____

____

____ (6)

6 Draw Feynman diagrams for these particle interactions:
a β^- decay: ${}^1_0n \rightarrow {}^1_1p + {}^0_{-1}e + \bar{\nu}_e$; **b** repulsion between two positrons;
c electron–positron annihilation: $e^+ + e^- \rightarrow \mu^- + \mu^+$ (Z^0 boson interaction);
d electron anti-neutrino–proton collision: $p + \bar{\nu}_e \rightarrow n + e^+$ (W^+ boson interaction).

a **b**

c **d**

(8)

7 Lepton numbers are conserved. By applying this principle and using the table of leptons, determine which of the following interactions are possible:
a $\mu^- = e^- + \bar{\nu}_e + \nu_\mu$; **b** $\mu^+ \rightarrow e^+ + \nu_e$; **c** $\nu_e + n \rightarrow p + e^-$.

____

____

____

____

____

____ (9)

8 Use the lepton table to determine the unknown particle X in each of these interactions: **a** $\mu^+ \rightarrow e^+ + X + \nu_e$; **b** $\tau^- + \mu^+ + p \rightarrow e^+ + \nu_e + \bar{\nu}_\mu + X$.

____

____

____

____

____ (10)

(Total 50 marks)

Unit 17 PARTICLE PHYSICS II

- This unit looks at **hadrons**, **quarks**, and **particle detectors**. There are a large number of hadrons, of which there are two types: **baryons** and **mesons**. Each has a **baryon number**, which must be conserved in interactions involving hadrons. Baryons have a baryon number of 1, and anti-baryons –1. Mesons and **anti-mesons** have a baryon number of 0. From their spin all baryons are fermions and all mesons are bosons. Some of the different types of baryons and mesons are shown in the table below. Particles produced by the strong interaction should decay by the strong interaction. Some particles which are produced by the strong interaction decay more slowly than expected, by the weak interaction, and are always produced in pairs. To describe this strange behaviour, we use a quantum number which is conserved in strong interactions but not necessarily in weak ones; this is **strangeness**. It may change by +1 or –1. Particles and their anti-particles have equal and opposite strangeness values.

Particle	Relative rest mass	Average lifetime (s)	Symbol	*Q*	*B*	*S*	Spin	Anti-particle	Symbol	*Q*	*B*	*S*	Spin
Baryons													
proton	1	stable	p	1	1	0	$\frac{1}{2}$	anti-proton	$\bar{p}$	–1	–1	0	$\frac{1}{2}$
neutron	1	898	n	0	1	0	$\frac{1}{2}$	anti-neutron	$\bar{n}$	0	–1	0	$\frac{1}{2}$
lambda	1.2	2.6×10^{-10}	Λ	0	1	–1	$\frac{1}{2}$	anti-lambda	$\bar{\Lambda}$	0	–1	1	$\frac{1}{2}$
omega	1.8	0.8×10^{-10}	Ω^-	–1	1	–3	$\frac{3}{2}$	anti-omega	Ω^+	1	–1	3	$\frac{3}{2}$
Mesons													
pion zero	0.14	0.8×10^{-16}	π^0	0	0	0	0	self	π^0	0	0	0	0
pion plus	0.14	2.6×10^{-8}	π^+	1	0	0	0	pion minus	π^-	–1	0	0	0
kaon plus	0.52	1.2×10^{-8}	K^+	1	0	1	0	kaon minus	K^-	–1	0	–1	0
kaon zero	0.53	8.9×10^{-11}	K^0	0	0	1	0	anti-kaon zero	$\bar{K}^0$	0	0	–1	0

Some common hadrons and their anti-particles. *Q*, charge; *B*, baryon number; *S*, strangeness.

All hadrons are unstable except protons, and all other baryons will eventually decay into protons and leptons. A good example of this is the decay of the free neutron, $n \rightarrow p + e^- + \bar{\nu}_e$ (baryon numbers: 1 = 1 + 0 + 0). Mesons eventually decay into photons or leptons. Neutrons are stable in a nucleus if the energy released by a neutron decay is less than the energy binding them to the nucleus. For other examples of hadron interactions see the figure left.

(a) $\pi^+ \rightarrow \mu^+ + \nu_\mu$

charge Q $1 \rightarrow 1 + 0$ ✓

baryon number B $0 \rightarrow 0 + 0$ ✓

strangeness S $0 \rightarrow 0 + 0$ ✓

(b) $p + \pi^- \rightarrow \Lambda + K^0$

Q $1 - 1 \rightarrow 0 + 0$ ✓

B $1 + 0 \rightarrow 1 + 0$ ✓

S $0 + 0 \rightarrow -1 + 1$ ✓

(c) $K^+ + K^- \rightarrow \pi^0$

Q $1 - 1 \rightarrow 0$ ✓

B $0 + 0 \rightarrow 0$ ✓

S $1 - 1 \rightarrow 0$ ✓

(d) $K^+ \rightarrow \mu^+ + \nu_\mu$

Q $1 \rightarrow 1 + 0$ ✓

B $0 \rightarrow 0 + 0$ ✓

S $1 \rightarrow 0 + 0$ ✗

∴ weak decay

(a) and (d) show some typical decay reactions and (b) and (c) show some typical production reactions. Conservation principles are applied.

- Hadrons are not fundamental: they can be broken down into smaller particles called **quarks**, which cannot exist on their own. The exchange particle that binds quarks together is called a **gluon**. There are six types or **flavours** of quark, each with its own anti-quark (see table opposite). This is the same as the number of leptons; in fact, physicists used the idea of this symmetry to predict the existence of the top and bottom quarks. Mesons consist of a quark and an anti-quark ($q\bar{q}$), baryons consists of three quarks (qqq), and anti-baryons consist of three anti-quarks ($\bar{q}\bar{q}\bar{q}$). Quarks and anti-quarks are fermions (spin = $\frac{1}{2}$). The quark structure of a proton is (uud), and a neutron is (udd). Quarks can also have '**colour charge**' (red, blue, or green) or '**anti-colour charge**' (anti-red, anti-blue, or anti-green). Any quark can have any one of the three colour charges, and any anti-quark any one of the three anti-colour charges. Colour determines how a quark interacts through the strong interaction. In a baryon there must be three different-coloured quarks, red, blue, and green, so it is white or colourless. A meson is colourless because it contains a colour–anti-colour combination, such as red and anti-red. Quarks cannot be observed on their own because the forces between them are very large (of the order of 100 000 N) and so the energy required to split them up would be enormous.

Quarks can take part in all four interactions. Quark numbers (*S*, *C*, *T*, *B*) are only conserved in strong and electromagnetic interactions, because weak interactions can change the nature of one of the quarks, as in β^- decay. The strong interaction binds quarks together.

Particle	Symbol	Charge	Baryon number	*S*	*C*	*T*	*B*	Spin	Anti-particle	Symbol	Charge	Baryon number	*S*	*C*	*T*	*B*	Spin
up	u	$\frac{2}{3}$	$\frac{1}{3}$	0	0	0	0	$\frac{1}{2}$	anti-up	$\bar{u}$	$-\frac{2}{3}$	$-\frac{1}{3}$	0	0	0	0	$\frac{1}{2}$
down	d	$-\frac{1}{3}$	$\frac{1}{3}$	0	0	0	0	$\frac{1}{2}$	anti-down	$\bar{d}$	$\frac{1}{3}$	$-\frac{1}{3}$	0	0	0	0	$\frac{1}{2}$
strange	s	$-\frac{1}{3}$	$\frac{1}{3}$	–1	0	0	0	$\frac{1}{2}$	anti-strange	$\bar{s}$	$\frac{1}{3}$	$-\frac{1}{3}$	1	0	0	0	$\frac{1}{2}$
charm	c	$\frac{2}{3}$	$\frac{1}{3}$	0	1	0	0	$\frac{1}{2}$	anti-charm	$\bar{c}$	$-\frac{2}{3}$	$-\frac{1}{3}$	0	–1	0	0	$\frac{1}{2}$
top	t	$\frac{2}{3}$	$\frac{1}{3}$	0	0	1	0	$\frac{1}{2}$	anti-top	$\bar{t}$	$-\frac{2}{3}$	$-\frac{1}{3}$	0	0	–1	0	$\frac{1}{2}$
bottom	b	$-\frac{1}{3}$	$\frac{1}{3}$	0	0	0	–1	$\frac{1}{2}$	anti-bottom	$\bar{b}$	$\frac{1}{3}$	$-\frac{1}{3}$	0	0	0	1	$\frac{1}{2}$

All the quarks and their anti-particles. *S*, strangeness; *C*, charm; *T*, topness; *B*, bottomness. In most A-level questions you will only need to concern yourself with up, down, and strange quarks.

- There are two methods used to investigate subatomic structure:
  1. **Diffraction** Electrons with energy of 10 GeV have a de Broglie wavelength (see unit 13) of 10^{-16} m, which is smaller than the size of a proton. When these electron de Broglie waves hit a nucleus, diffraction takes place. They are smaller than the nucleus and so can only be diffracted by particles smaller than the nucleus. This is often referred to as **deep inelastic scattering**, because it takes place deep in the nucleus and the kinetic energy of the electrons is not always conserved (energy is available to create other particles). Such electron diffraction was the first evidence of quarks.
  2. **Collision of particles** High-energy particles can smash into each other in a variety of different ways, and the products are analysed with a detector, such as a **cloud chamber** or a **bubble chamber**, that uses the ionizing effect of the particles to show their paths. For the types of tracks produced and their interpretation, see the figure below.

Both methods need a **particle accelerator**. A particle accelerator uses electric and magnetic fields to exert forces on particles, making them accelerate to reach very high speeds and energies. Either linear or circular accelerators can be used (see units 19 and 22 respectively).

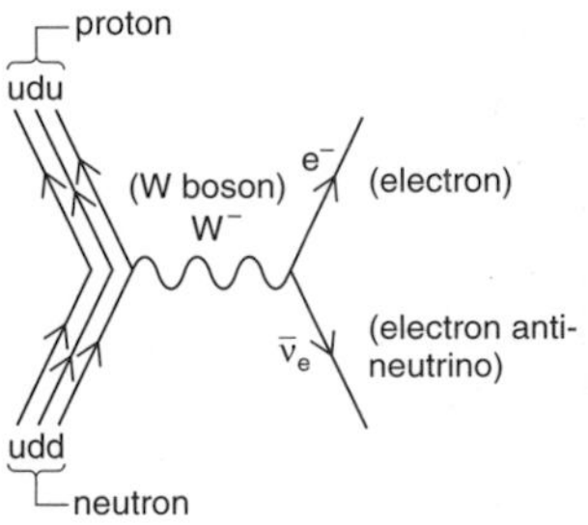

A Feynman diagram showing β⁻ decay with quarks. A down quark changes into an up quark by emitting a W⁻ boson.

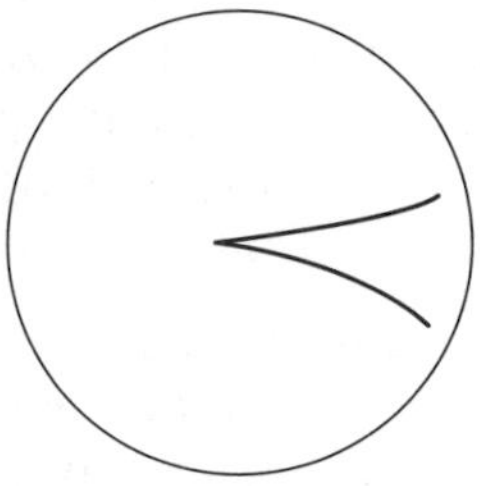

This is an uncharged particle, which has no track, decaying into two charged particles.

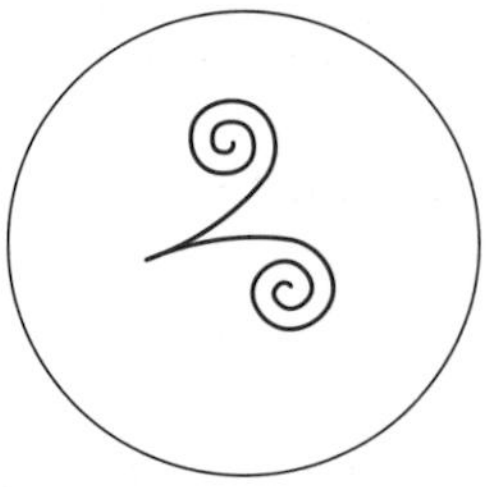

Pair production: a gamma ray producing an electron–positron pair. They spiral because as they ionize they slow down, and as they slow down the radius of curvature is reduced.

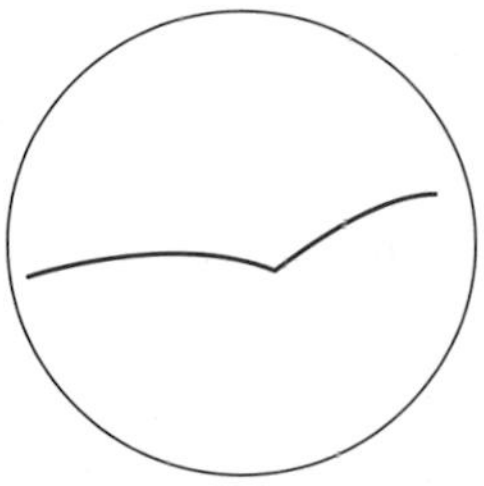

This is a charged particle emitting one or more uncharged particles.

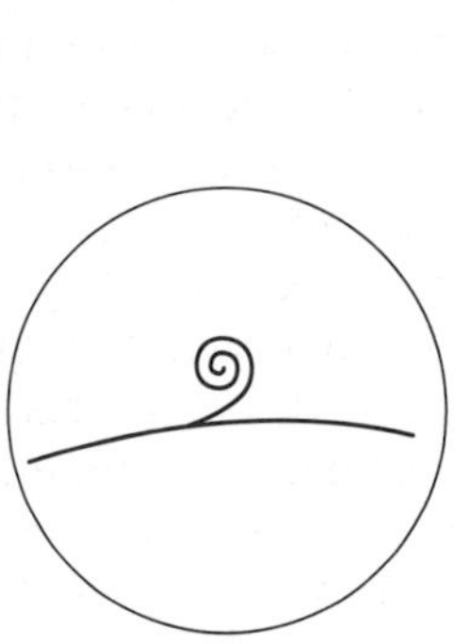

This an electron produced during ionization as a particle interacts with a gas atom.

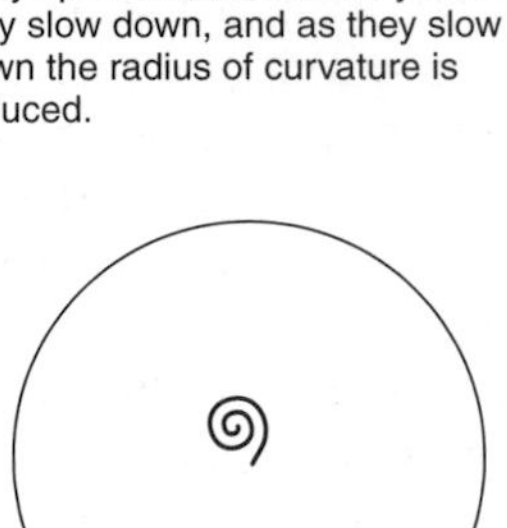

This is an isolated electron produced by **Compton scattering**. An electron is emitted by an atom as a gamma ray rebounds off it.

A particle slows down as it uses up its energy ionizing the gas. It produces a 'thicker' trail as it gets slower because slower-moving particles produce more ionization.

Examples and interpretation of cloud- and bubble-chamber tracks. Charged particles moving through a gas produce ions when they collide with the gas atoms. These ions act as points around which bubbles or droplets of vapour can form, in bubble or cloud chambers respectively. In this way they can show the path the charged particle has taken. They cannot show the paths of uncharged particles. A magnetic field is normally acting directly downwards or upwards so curved paths are produced (see unit 21).

Unit 17

TESTS

RECALL TEST

1 What is a hadron?

(2)

2 What is the difference between a baryon and a meson?

(2)

3 What is the exchange particle that binds quarks together?

(2)

4 What is 'strangeness'?

(2)

5 What is the difference between a lepton and a hadron?

(2)

6 What are 'baryon numbers' and why do we use them?

(2)

7 What is a quark?

(2)

8 What evidence is there for the existence of quarks?

(2)

9 What particles are made from quarks?

(2)

10 What are the two methods of investigating subatomic structure?

(2)

(Total 20 marks)

CONCEPT TEST

1 What change must take place in the quark structure of a neutron when it decays into a proton, and which of the interactions is responsible for this decay?

(4)

2 Tritium decays into helium and emits an a beta particle (electron) and an anti-neutrino. Which of the interactions produces **a** the stability of the electrons in their orbits; **b** the decay of tritium into helium?

(4)

3 Draw a Feynman diagram, including quarks, for the decay $\Sigma^+ \rightarrow p + \pi^0$ given that it is produced by the weak nuclear interaction and mediated by a Z^0 gauge boson. Quark structures are Σ^+ = uus, p = uud, and $\pi^0 = u\bar{u}$. (6)

4 Write down two differences between mesons and baryons. Baryon numbers are conserved; by applying this principle and using the table of baryons and mesons, determine which of the following interactions are possible:
a $\Lambda \rightarrow \pi^- + p$; **b** $K^+ + K^- \rightarrow \pi^0$; **c** $n \rightarrow p + e^- + \bar{\nu}_e$.

(8)

5 Using the table of baryons and mesons determine the unknown particle X in each of these interactions: **a** $\pi^+ + n \rightarrow \Lambda + X$; **b** $p + p \rightarrow p + X + \pi^0$.

(6)

6 What are the quark structures of a proton and a neutron? Deduce the baryon number and charge of a neutron from its quark structure. Why are quark numbers (S, C, T, and B) not conserved in weak interactions? Use the tables of baryons and mesons and of quarks to determine the quark content of **a** each of the mesons K^0, K^-, and π^-, and **b** each of the baryons Λ, Ω^-, and Ω^+, given that they consist only of up, down, and strange quarks.

(12)

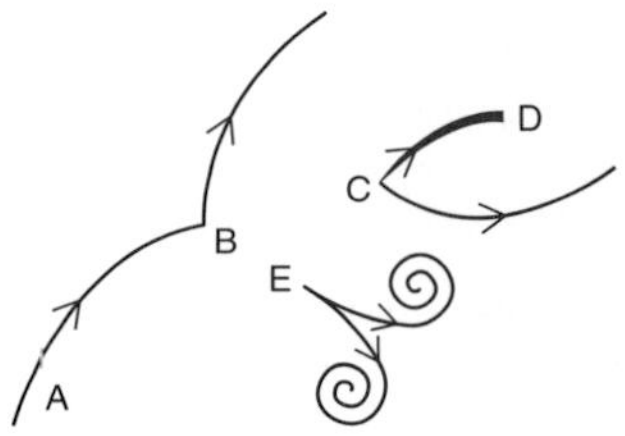

7 In the figure right a positively charged particle enters a cloud chamber at A. Explain the other tracks by referring to what is happening at each point.

(10)

(Total 50 marks)

Unit 18

ELECTROSTATIC AND GRAVITATIONAL FIELDS

- Forces such as tension, upthrust, and normal reaction are due to the **electrostatic** force between charges. Weight is due to the **gravitational** force produced between masses. A region of space where a force can be felt is called a **field**. This unit deals with the fundamental forces of gravity and electrostatics, and their fields. They are similar to each other so it is useful to look at them in a comparative way.

Gravitational

A gravitational field is a region of space where a mass will experience a force due to the presence of another mass.

Electrostatic

An electrostatic field is a region of space where a charge will experience a force due to the presence of another charge.

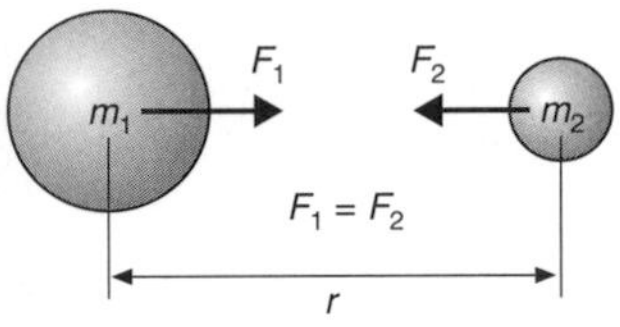

Forces between two masses.

- It is very useful to be able to calculate the actual force between two charges or two masses. In both cases the force decreases the further they are apart (both obey inverse-square laws i.e. they are proportional to $1/r^2$) and increases with the values of charge or mass. The behaviour of each is governed by a law:

Newton's law of gravitation

The force between two masses (m_1 and m_2) is directly proportional to the product of the two masses, and inversely proportional to the square of the distance between their centres, r (see above left).

Coulomb's law

The force between two charges (Q_1 and Q_2) is directly proportional to the product of the two charges, and inversely proportional to the square of the distance between their centres, r (see below left).

Newton's law:

$$F = \frac{-Gm_1m_2}{r^2}$$

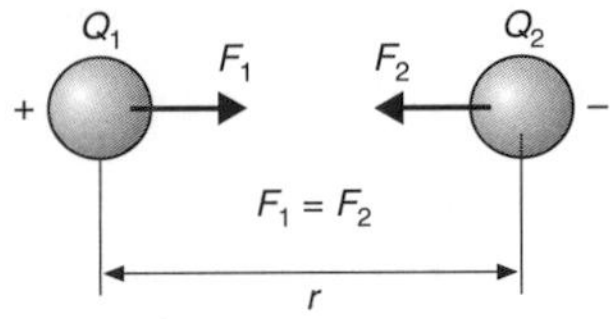

Forces between two charges.

G is the **universal gravitational constant**, and ε is a constant, called the **permittivity**, of the material in which the charges lie. G is always constant, but ε varies depending upon the material. G is always 6.7×10^{-11} and is measured in $\text{N m}^2/\text{kg}^2$. ε is measured in farads per metre (F/m) and has a value of about 8.85×10^{-12} F/m in a vacuum (ε_0, the **permittivity of free space**). Its value in air is usually taken to be the same as in a vacuum.

Coulomb's law:

$$F = \frac{Q_1Q_2}{4\pi\varepsilon r^2}$$

Example *Calculate and compare the electrostatic and gravitational forces acting between the proton and electron in a hydrogen atom. (Masses: electron = 9.1×10^{-31} kg, proton = 1.7×10^{-27} kg. Charges: electron = -1.6×10^{-19} C; proton = 1.6×10^{-19} C. Atomic radius = 1.0×10^{-10} m, $G = 6.7 \times 10^{-11}$ Nm^2/kg^2, $\varepsilon_0 = 8.8 \times 10^{-12}$ F/m.)*

$$\text{gravitational: } F = \frac{-Gm_1m_2}{r^2} = \frac{6.7 \times 10^{-11} \times 9.1 \times 10^{-31} \times 1.7 \times 10^{-27}}{(1.0 \times 10^{-10})^2} = \mathbf{-1.0 \times 10^{-47}\ N}$$

$$\text{electrostatic: } F = \frac{Q_1Q_2}{4\pi\varepsilon_0 r^2} = \frac{-1.6 \times 10^{-19} \times 1.6 \times 10^{-19}}{4\pi \times 8.8 \times 10^{-12} \times (1.0 \times 10^{-10})^2} = \mathbf{-2.3 \times 10^{-8}\ N}$$

(The negative values for the forces show they are attractive.)

The gravitational force is negligible compared with the electrostatic force.

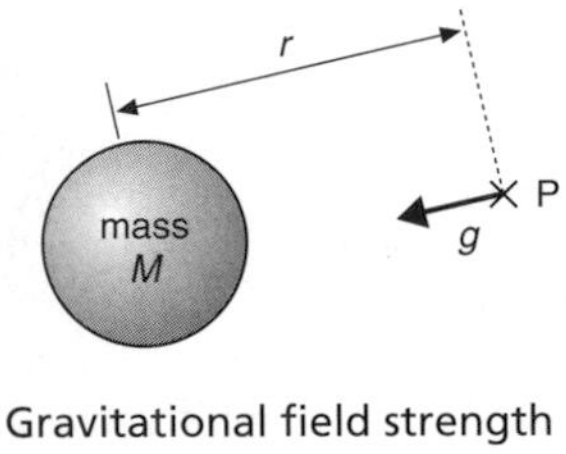

Gravitational field strength at a point P.

gravitational field strength:

$$g = \frac{F}{m} \text{ or } g = \frac{-GM}{r^2}$$

- In order to compare the strength of these fields at different places we determine the force acting on a 1 kg mass or a 1 C charge at each point. This is called **field strength**, and is a vector because it is based on force.

Gravitational field strength, g, is the force acting on unit mass in a field (**$g = F/m$**; units N/kg). For a point mass in the field of a mass M

$$g = \frac{-GM}{r^2} \text{ (see above left)}$$

Electric field strength, E, is the force acting on unit positive charge in a field (**$E = F/q$**; units N/C). For a point charge in the field of a charge Q

$$E = \frac{Q}{4\pi\varepsilon r^2} \text{ (see below left)}$$

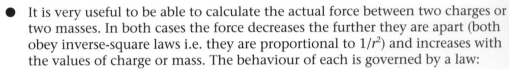

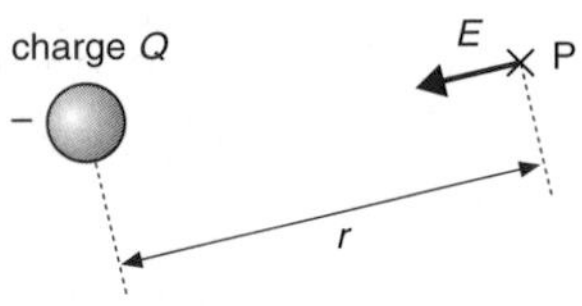

Electric field strength at a point P.

electric field strength:

$$E = \frac{F}{q} \text{ or } E = \frac{Q}{4\pi\varepsilon r^2}$$

- When an object has a force acting on it and it moves in the direction of that force, work is done and there is a corresponding change in its potential energy. The amount of potential energy that a 1 kg mass or a 1 C charge has at a particular point is called the **field potential**. Infinity is taken as the starting point where any object has zero potential energy. Field potential is a scalar quantity because it is based on energy.

Gravitational field potential, V_g, is the work done, W, in moving unit mass from infinity to its position in space ($\boldsymbol{V_g = W/m}$; units J/kg). For a point mass in the field of a mass M

$$V_g = \frac{-GM}{r} \text{ (see above right)}$$

V_g is always negative because gravity is only attractive.

Electric field potential, V_E, is the work done, W, in moving unit positive charge from infinity to its position in space ($\boldsymbol{V_E = W/q}$; units J/C or V). For a positive point charge in the field of a charge Q

$$V_E = \frac{Q}{4\pi\varepsilon r} \text{ (see below right)}$$

V_E can be negative or positive depending on the type of charge.

Example *What are the field strength and potential at a point 400 km above the Earth's surface? ($G = 6.7 \times 10^{-11}\,\text{N m}^2/\text{kg}$, radius of Earth $= 6.4 \times 10^6\,\text{m}$, mass of Earth $= 6 \times 10^{24}\,\text{kg}$.)*

When calculating the distance r always do it from the Earth's centre.

$$g = \frac{-GM}{r^2} = \frac{6.7 \times 10^{-11} \times 6.0 \times 10^{24}}{(6.4 \times 10^6 + 4.0 \times 10^5)^2} = \mathbf{-8.7\,N/kg}$$

$$V_g = \frac{-GM}{r} = \frac{-6.7 \times 10^{-11} \times 6.0 \times 10^{24}}{6.4 \times 10^6 + 4.0 \times 10^5} = \mathbf{-5.9 \times 10^7\,J/kg}$$

- Field strength and potential are connected. Unit 5 used $W = Fs$ to show that in a graph of F against s the area under the graph is the work done, and in a graph of W against s the gradient is the force. Field strength and potential are linked similarly. In a graph of potential against distance the gradient equals the field strength. So a graph of field strength against distance will show the variation of gradient in the potential graph. In a similar way the area under a graph of field strength against distance will show the change of field potential (see below right).

  Field strength is equal to the potential gradient.

  The – sign in the equations makes them mathematically accurate when using calculus. Do not worry about this, but remember it is why the field-strength graphs have the opposite sign to the potential graphs. All the above works for both the inverse-square law (radial) fields used up to now, and for uniform fields (unit 19).

- It is often necessary to represent visually a field's strength and shape. We do this using **field lines** and **equipotential lines.**

  A **field line** is a line which shows the direction of the force acting on unit mass or charge in a electric field. In both cases it can be thought of as the path taken by either unit mass or unit charge in the field. Thus the direction of field lines will always be towards a mass, and from positive to negative charges.

  An **equipotential** is a line joining up points of equal potential in either type of field. It is always perpendicular to the field lines. Equipotentials are usually drawn at regular intervals of potential (e.g. 10 V, 20 V, 30 V, etc.), so if they are closer together the potential gradient is greater.

  Below are typical examples of field lines and equipotentials. Field lines and equipotentials are equally spaced in a uniform field.

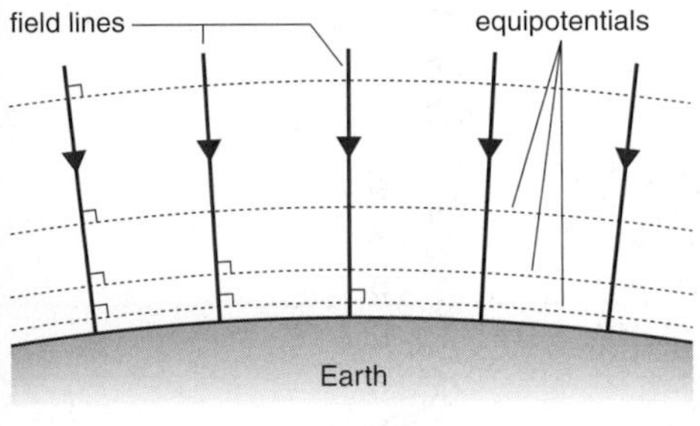

Field lines and equipotentials in a typical gravitational field.

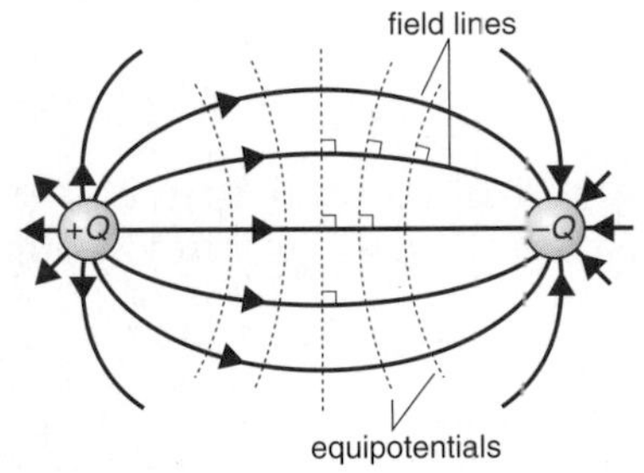

Field and equipotential lines in a typical electric field.

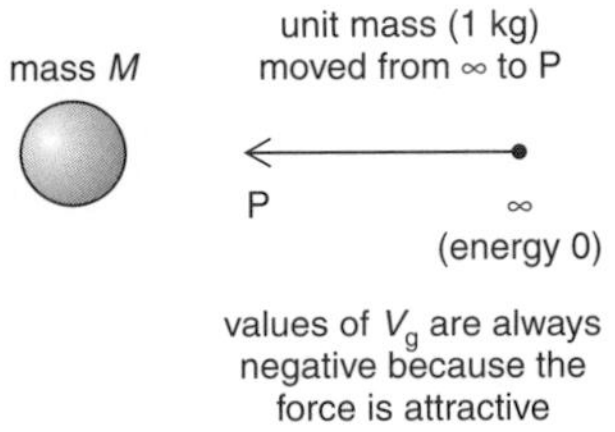

Gravitational field potential at a point P.

gravitational field potential:

$$\boldsymbol{V_g = \frac{W}{m}} \text{ or } \boldsymbol{V_g = \frac{-GM}{r}}$$

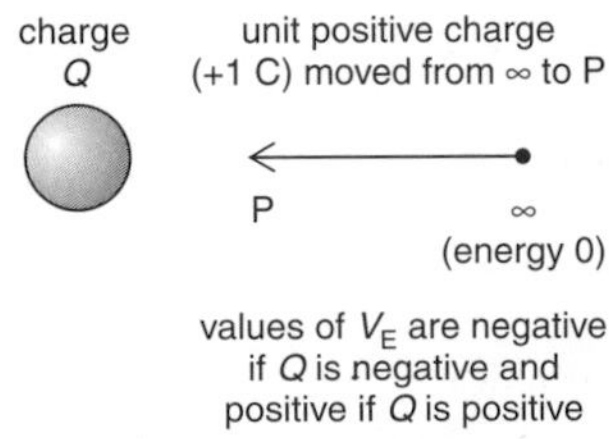

Electric field potential at a point P.

electrical field potential:

$$\boldsymbol{V_E = \frac{W}{q}} \text{ or } \boldsymbol{V_E = \frac{Q}{4\pi\varepsilon r}}$$

potential gradients:

$$\boldsymbol{E = -\frac{\Delta V_E}{\Delta r}} \text{ and } \boldsymbol{g = -\frac{\Delta V_g}{\Delta r}}$$

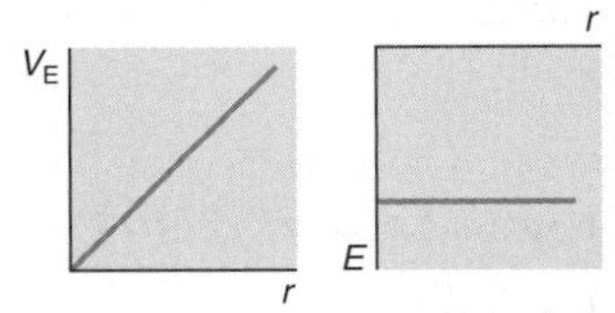

Uniform positive gradient gives constant negative field strength.

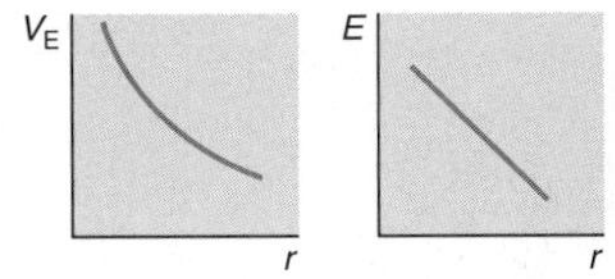

Negative gradient with uniformly decreasing value gives positive field strength with decreasing value.

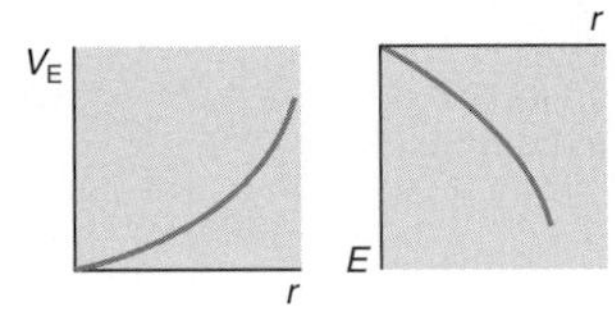

Positive gradient increasing at an increasing rate gives negative field strength with increasing gradient.

Graphs of potential against distance and their corresponding field-strength graphs. To produce a field-strength graph from a potential graph, consider the gradient and change its sign.

Unit 18

TESTS

RECALL TEST

1 What is a gravitational field?

(2)

2 What is an electric field?

(2)

3 What is Newton's law of gravitation?

(2)

4 What is Coulomb's law?

(2)

5 Define 'electric field strength'.

(2)

6 Define 'gravitational field strength'.

(2)

7 Define 'gravitational field potential'.

(2)

8 Define 'electric field potential'.

(2)

9 What is a gravitational field line?

(2)

10 What is an equipotential line?

(2)

(Total 20 marks)

CONCEPT TEST

Take $G = 6.67 \times 10^{-11}\,\text{N m}^2/\text{kg}^2$ and $\varepsilon_0 = 8.8 \times 10^{-12}\,\text{F/m}$

1 What is the force between the Earth and the Moon? From what points can this force be assumed to act and in what direction? (Distance from centre of Earth to centre of the Moon = 3.8×10^5 km. Masses: Moon = 7.4×10^{22} kg, Earth = 6.0×10^{24} kg.)

(4)

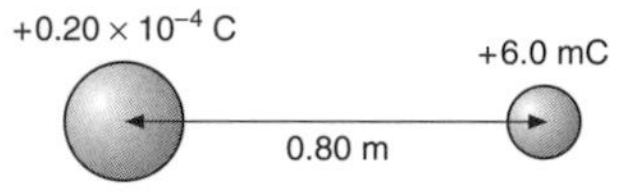

2 What is the force between the two charged spheres shown left?

(4)

3 A sphere of radius 0.20 m is charged with a 10 mC positive charge. What are the field strength and field potential 0.50 m from the surface of the sphere?

(4)

4 How are electric fields lines and equipotential lines related? Sketch in the electric field lines and equipotential lines on the figure right. Remember that field lines are always perpendicular to lines of equipotential and that a conducting surface is all at the same potential.

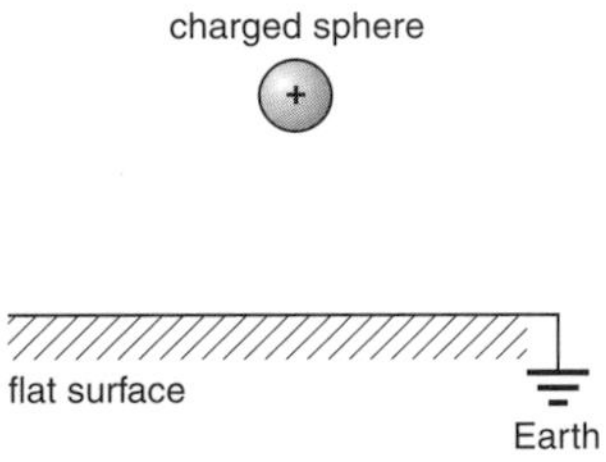

(4)

5 The planet Zargon has a gravitational field strength of 16.7 N/kg at its surface. If the radius of the planet is 8.20×10^{6} m, what is its mass and what would be the value of the field potential 40.0 km above its surface?

(4)

6 How are electric field strength and potential related? The graph right shows how the electric field potential varies with distance in a region of space. Sketch the corresponding graph that shows how the field strength varies. What is the value of field strength between A and B?

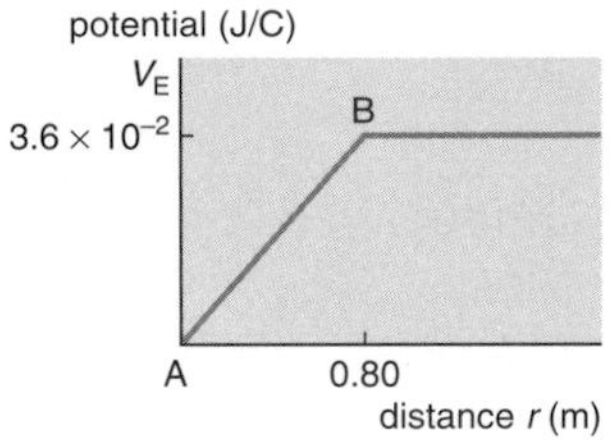

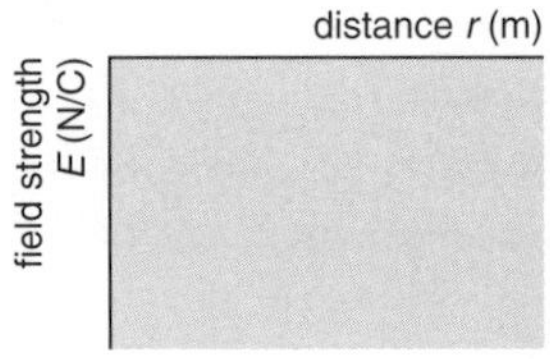

(6)

7 What are the electric field strength and potential at a point 1.0×10^{-8} m from the centre of a uranium nucleus, $^{235}_{92}U$? (Charge on a proton = 1.6×10^{-19} C.)

(4)

8 Two asteroids are floating in space as shown right. A spaceship approaches them and is in the position shown. What are the values of gravitational field strength and potential where the spaceship is?

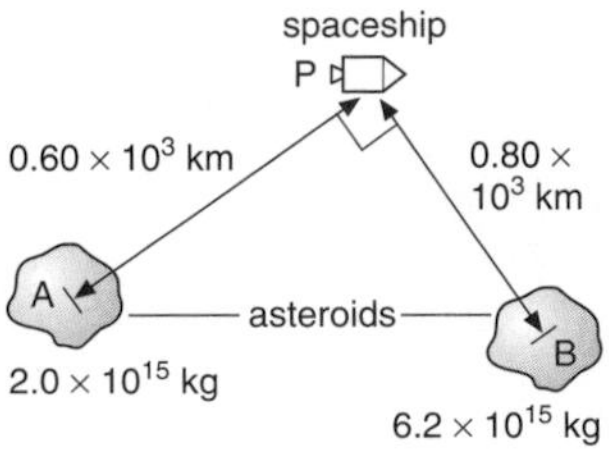

(6)

9 The Sun and the Moon combine together to produce the tides on the Earth. Determine the resultant force and its direction on 1 kg of seawater when the Earth, Moon, and Sun are in the position shown here. What do you conclude from these calculations? (Masses: Moon = 7.4×10^{22} kg, Sun = 2.0×10^{30} kg.)

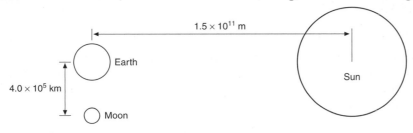

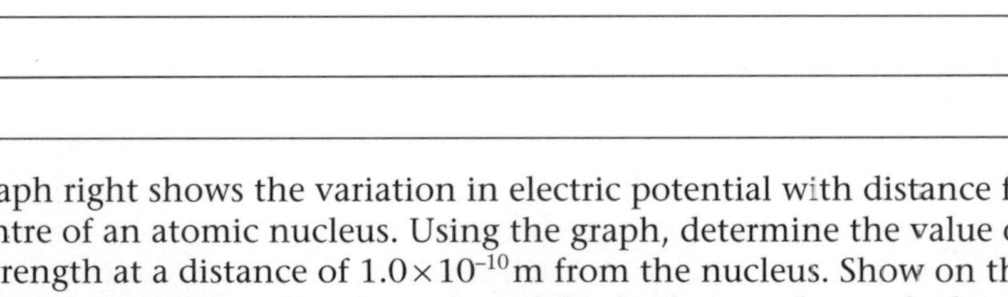

(6)

10 The graph right shows the variation in electric potential with distance from the centre of an atomic nucleus. Using the graph, determine the value of the field strength at a distance of 1.0×10^{-10} m from the nucleus. Show on the graph how you did this. Use this value of the field strength to calculate the force on an electron at a distance of 1.0×10^{-10} m from the nucleus.

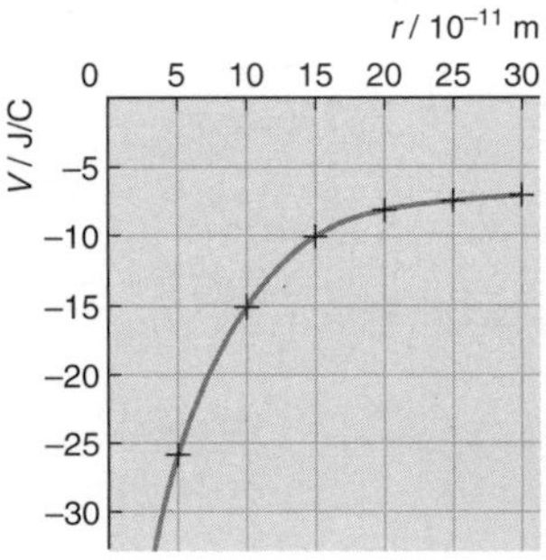

(8)

(Total 50 marks)

Unit 19

APPLIED FIELD THEORY

change in potential energy:
gravitational: $\Delta PE = \Delta V_g m$

electric: $\Delta PE = \Delta V_E Q$

- The previous unit dealt with the basic theory behind fields. This unit will look at the application of these ideas to different situations. First consider the energy changes that an object undergoes when it moves through a field. The field potential that an object has is the potential energy that unit mass or unit charge has, so to find the change in potential energy (PE) of a complete object, multiply the change in field potential, ΔV_g or ΔV_E, by the mass, m, or charge, Q, of the object accordingly.

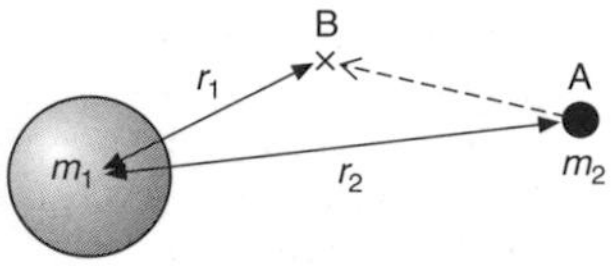

gravitational: $\Delta PE = \Delta V \times m$

For an object of mass m_2 in the field of mass m_1, moving from A to B:

$$\Delta PE = \left(\frac{-Gm_1}{r_2} - \frac{-Gm_1}{r_1}\right)m_2$$

$$\Delta PE = Gm_1m_2\left(\frac{1}{r_1} - \frac{1}{r_2}\right)$$

electric: $\Delta PE = \Delta V \times Q$

For an object with charge Q_2 in the field of charge Q_1, moving from A to B:

$$\Delta PE = \left(\frac{Q_1}{4\pi\varepsilon r_2} - \frac{Q_1}{4\pi\varepsilon r_1}\right)Q_2$$

$$\Delta PE = \frac{Q_1Q_2}{4\pi\varepsilon}\left(\frac{1}{r_2} - \frac{1}{r_1}\right)$$

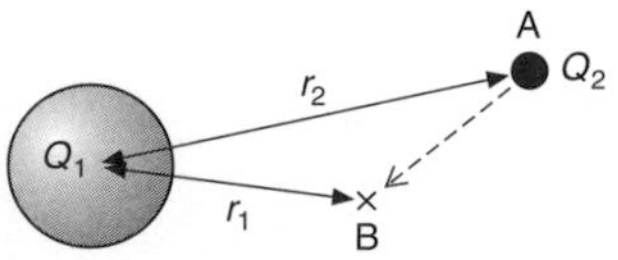

For an object of mass m to break free of the Earth's gravitational field it must be given KE

$$\tfrac{1}{2}mv^2 = GMm\left(\frac{1}{R} - \frac{1}{\infty}\right)$$

escape velocity:
$v_{esc}^2 = \frac{2GM}{R}$

where M and R are the mass and radius of the Earth respectively.
m cancels and $1/\infty = 0$, so

$$v_{esc}^2 = \frac{2GM}{R}$$

v_{esc} is **escape velocity** (about 11km/s on Earth). The mass of the object does not matter, so escape velocity is the same for all objects.

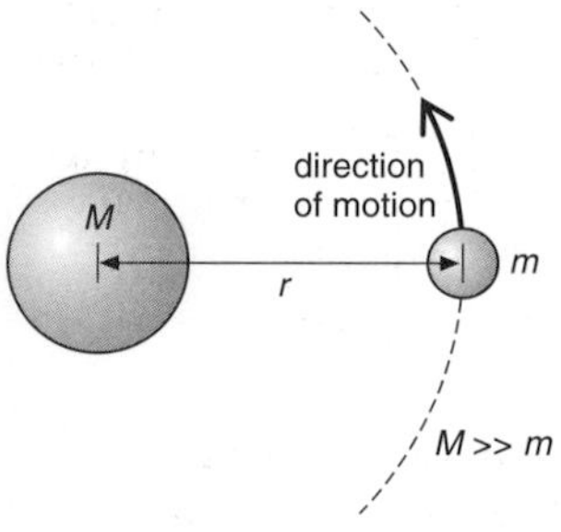

Orbit in a gravitational field.

- One of the most important applications of field theory is in analysing the orbits of planets and satellites in gravitational fields, and the motion of charged particles in circles. In both cases the force of attraction provides the centripetal force (see unit 8) required for the circular motion.

The force between two masses provides the centripetal force:

$$\frac{GMm}{r^2} = \frac{mv^2}{r} \text{ so } \boldsymbol{v^2 = \frac{GM}{r}}$$

$$\frac{GMm}{r^2} = mr\omega^2 \text{ so } \omega^2 = \frac{GM}{r^3}$$

$$\text{Now } \omega = \frac{2\pi}{T} \text{ so } \boldsymbol{T^2 = \frac{4\pi^2r^3}{GM}}$$

The force between two charges provides the centripetal force:

$$\frac{Q_1Q_2}{4\pi\varepsilon r^2} = \frac{mv^2}{r} \text{ so } \boldsymbol{v^2 = \frac{Q_1Q_2}{4\pi\varepsilon mr}}$$

$$\frac{Q_1Q_2}{4\pi\varepsilon r^2} = mr\omega^2 \text{ so } \omega^2 = \frac{Q_1Q_2}{4\pi\varepsilon mr^3}$$

$$\text{Now } \omega = \frac{2\pi}{T} \text{ so } \boldsymbol{T^2 = 16\pi^3\varepsilon mr^3}$$

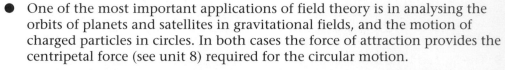

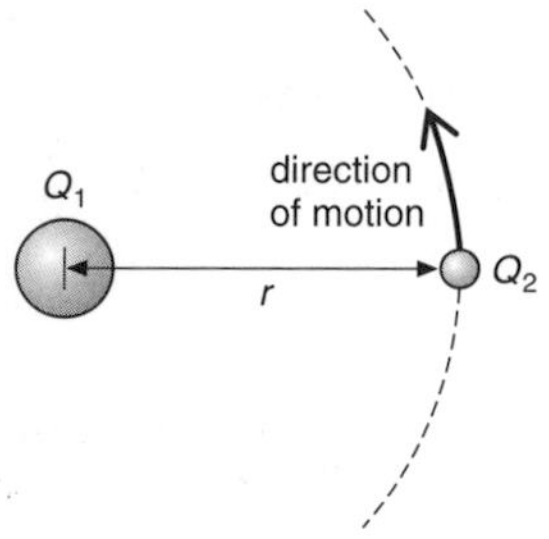

Orbit in an electric field.

Kepler's third law: $T^2 \propto r^3$.

In each type of field both objects rotate about their common centre of mass. The gravitation equations only work if $M \gg m$.

orbital speed:
g field: $v^2 = \frac{GM}{r}$

E field: $v^2 = \frac{Q_1Q_2}{4\pi\varepsilon mr}$

orbital time period:
g field: $T^2 = \frac{4\pi^2r^3}{GM}$

E field: $T^2 = \frac{16\pi^3\varepsilon mr^3}{Q_1Q_2}$

Example *Calculate the velocity and time period for the electron in orbit around a proton in a hydrogen atom. (Mass of electron = 9.1×10^{-31} kg. Charges: electron = -1.6×10^{-19} C, proton = $+1.6 \times 10^{-19}$ C. Atomic radius = 1.0×10^{-10} m.)*

Ignore the signs on the charges because you cannot root a negative number. A negative sign just means that the force between the charges is attractive, which is necessary for circular motion anyway. Take $\varepsilon = \varepsilon_0 = 8.8 \times 10^{-12}$ F/m.

$$v^2 = \frac{Q_1Q_2}{4\pi\varepsilon mr} = \frac{1.6\times10^{-19}\times1.6\times10^{-19}}{4\pi\times8.8\times10^{-12}\times9.1\times10^{-31}\times1\times10^{-10}} \text{ so } \boldsymbol{v = 1.6 \times 10^6\ \text{m/s}}$$

$$T^2 = \frac{16\pi^3\varepsilon mr^3}{Q_1Q_2} = \frac{16\times\pi^3\times8.8\times10^{-12}\times9.1\times10^{-31}\times(1.0\times10^{-10})^3}{1.6\times10^{-19}\times1.6\times10^{-19}} \text{ so } \boldsymbol{T = 4.0 \times 10^{-16}\text{s}}$$

- The fields dealt with up till now emanate out from a single point, and are called **radial fields**. There are also **uniform fields** (see figures right). Their equations come from the definitions of field strength and potential in unit 18.

From $g = \frac{F}{m}$, $\boldsymbol{F = mg}$. From $E = \frac{F}{Q}$, $\boldsymbol{F = EQ}$. From $E = \frac{-\Delta V_E}{\Delta r}$, $\boldsymbol{E = \frac{-V}{d}}$ (units V/m).

$\boldsymbol{F = \frac{-VQ}{d}}$ where d is the separation between two parallel charged plates.

This equation is used to determine the force present in a linear particle accelerator.

- The variation of g inside and outside the Earth is shown below.
The uniform increase of g inside the Earth can be shown by considering the Earth as a sphere with uniform density.

mass (m) = volume (v) × density (ρ)

$v = \frac{4}{3}\pi r^3$ and $g = \frac{-Gm}{r^2}$

so $\boldsymbol{g = \frac{-4}{3}G\pi\rho r}$, or $\boldsymbol{g \propto -r}$

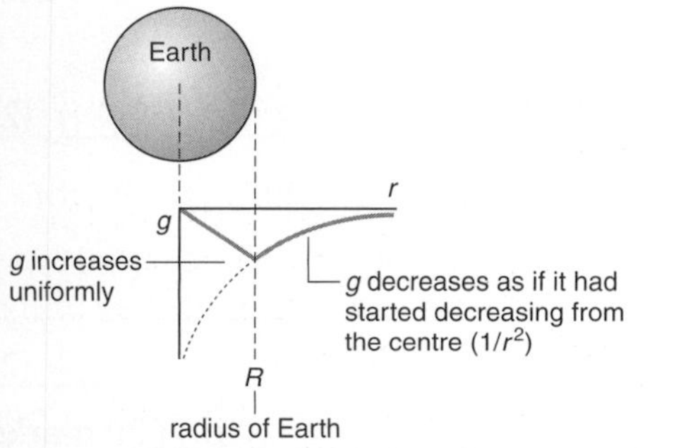

Variation of g inside and outside the Earth. The values of g and V_g are always calculated using the distance to the centre of the Earth, which is why their values decrease above the surface as if they had started decreasing from the centre, and g can be taken as constant for small changes of height above the surface.

The variation of E and V_E inside and outside a hollow conductor is shown below right. The field strength inside a hollow conductor is always zero.

On a conductor, charge concentrates where the surface is most sharply curved, e.g. at a point (see below right).

There are two quantities used to describe the distribution of charges on objects: **charge density**, ρ (units C/m³), and **surface charge density**, σ (units C/m²).

$$\sigma = \frac{\text{charge}}{\text{area}} = \frac{Q}{A} \qquad \rho = \frac{\text{charge}}{\text{volume}} = \frac{Q}{V}$$

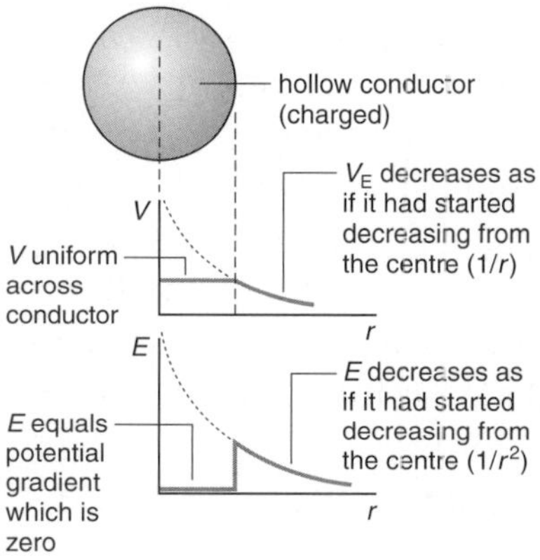

Variation of E and V_E inside and outside a hollow charged conductor.

- Similarities between E and g fields:
  1. Both fields produce forces that decrease with distance according to an inverse-square law ($1/r^2$).
  2. Both produce forces that are directly proportional to the product of masses or charges.

- Differences between E and g fields:
  1. Electric fields can be attractive and repulsive, whereas gravitational fields are only attractive.
  2. Electric fields depend upon the material in which they lie, gravitational ones do not (G is constant but ε varies from medium to medium).
  3. It is possible to shield things from electric fields, but it is not possible to shield things from a gravitational field. A metal cage (**Faraday cage**) will shield any object inside it from an electric field.

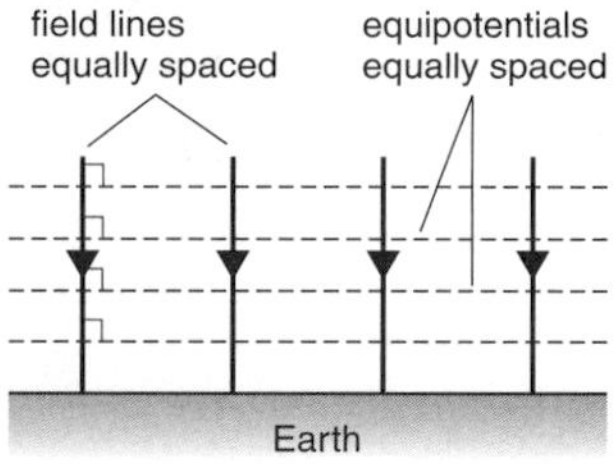

Uniform gravitational field close to Earth's surface. Equally spaced field lines and equipotentials show a field is uniform. Field strength is the same at all points in a uniform field.

uniform g fields:

$\boldsymbol{g = \frac{F}{m}}$ or $\boldsymbol{F = mg}$

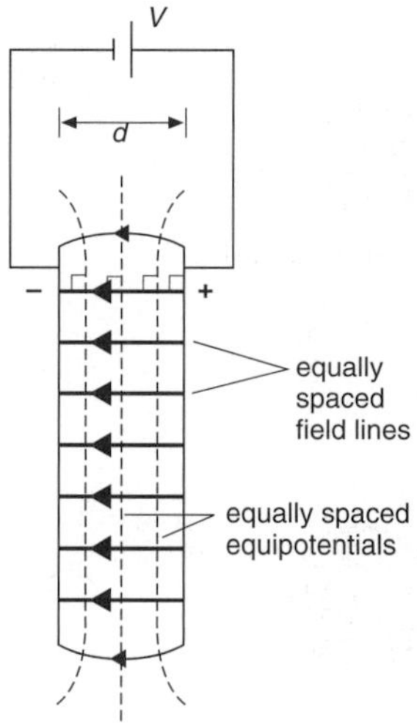

Uniform electric field between two parallel charged plates.

uniform E fields:

$\boldsymbol{E = \frac{F}{Q}}$ or $\boldsymbol{F = EQ}$ $\qquad \boldsymbol{E = \frac{V}{d}}$

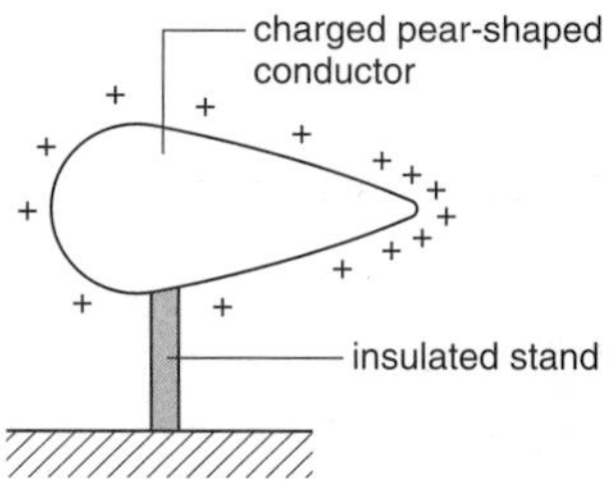

Charge on a conductor's surface concentrates at sharply curved points.

Unit 19

TESTS

RECALL TEST

1 How do you find out the potential energy of a mass in a gravitational field?

(2)

2 What is meant by 'escape velocity'?

(2)

3 What provides the centripetal force for a satellite in orbit around the Earth?

(2)

4 What is the difference between 'surface charge density' and 'charge density'?

(2)

5 Why can electric fields be both attractive and repulsive while gravitational fields are only attractive?

(2)

6 What is a uniform field?

(2)

7 What is the difference between G and ε?

(2)

8 Where does charge concentrate on an object?

(2)

9 How is g related to r inside the Earth?

(2)

10 Why does the fact that the potential is uniform across a hollow conductor mean that the field strength is zero?

(2)

(Total 20 marks)

CONCEPT TEST

Take $G = 6.67 \times 10^{-11}\,\text{N m}^2/\text{kg}^2$, $\varepsilon_0 = 8.8 \times 10^{-12}\,\text{F/m}$, and $g = 9.8\,\text{m/s}^2$

1 A satellite is in orbit 300 km above the Earth's surface. What is the gravitational field strength at this point, and the time period of the satellite's orbit? (Radius of Earth = 6.4×10^6 m, mass of Earth = 6.0×10^{24} kg.)

(4)

2 An electron in a hydrogen atom is 1.0×10^{-10} m away from the proton. How much energy is needed to remove it from the atom? Ignore the effects of gravity. (Charge on an electron or proton = 1.6×10^{-19} C.)

(4)

3 What is the escape velocity on Saturn? How much energy must be given to a 2.0 kg mass for it to escape the influence of the planet? (Mass of Saturn = 6.0×10^{26} kg, radius of Saturn = 5.0×10^4 km.)

(4)

4 Kepler's law states that $T^2 = kr^3$, where T is the time period of a planet's orbit, r is the average radius of orbit, and k is a constant. The radius of the Earth's orbit is 1.50×10^{11} m and the radius of Jupiter's orbit is 8.00×10^{11} m. What is the time period of Jupiter's orbit?

(4)

5 If a woman can jump 1.2 m on the Earth, how high can she jump on a planet where the density is half that of the Earth and the radius is three-quarters that of the Earth?

(6)

6 Draw a graph of the variation of field strength with distance from the Earth's centre on the axes right. Estimate the potential energy needed to raise a satellite of mass 600 kg from the surface to a height of 19.2×10^3 km.

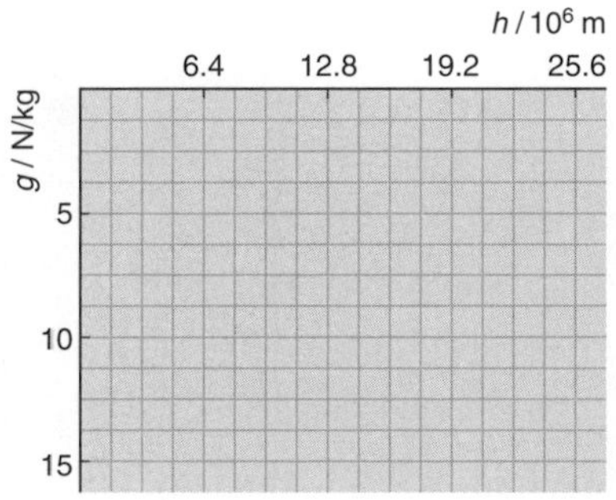

(6)

7 In Rutherford's gold-leaf experiment an alpha particle approaches a gold nucleus. If the kinetic energy of the alpha particle is 2.4×10^{-13} J, what is the closest the alpha particle can get to the nucleus? (The atomic number of gold is 79, and the charge on an electron is -1.6×10^{-19} C.)

(6)

8 How can you tell if a field is uniform? A charged oil drop is suspended between two parallel plates 0.50 cm apart with a potential difference of 400 V across them (see right). What is the charge on the oil drop if its mass is 4.0×10^{-15} kg? How many electrons are present? (Charge on electron = -1.6×10^{-19} C.)

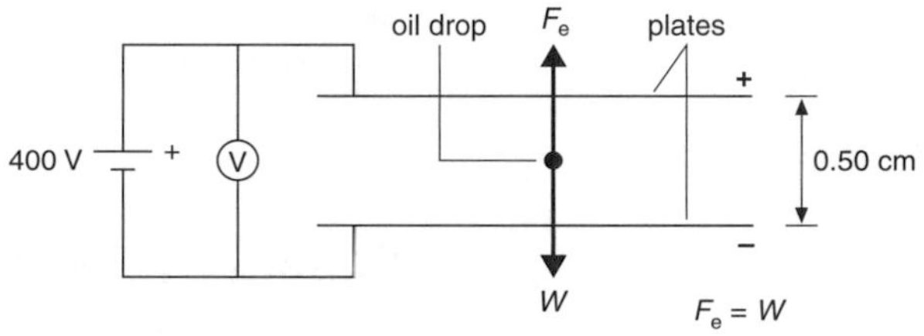

(8)

9 In a binary star system the two stars rotate about their common centre of mass. If each star has a mass of 2.0×10^{32} kg and their separation is 1.0×10^{12} m what is the time period of each orbit, and what is the angular velocity?

(8)

(Total 50 marks)

CAPACITANCE

capacitance: $C = \frac{Q}{V}$

parallel-plate capacitor:

$C = \frac{\varepsilon_0\varepsilon_r A}{d}$

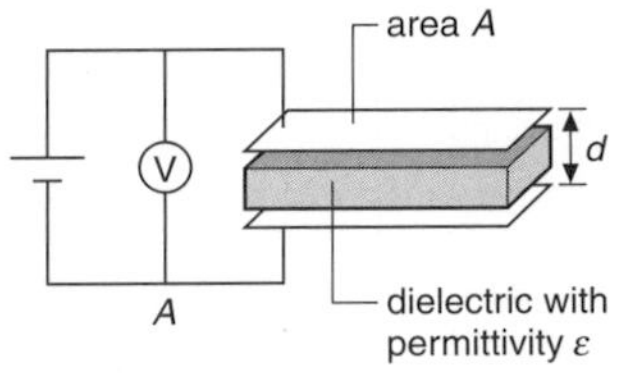

A parallel-plate capacitor.

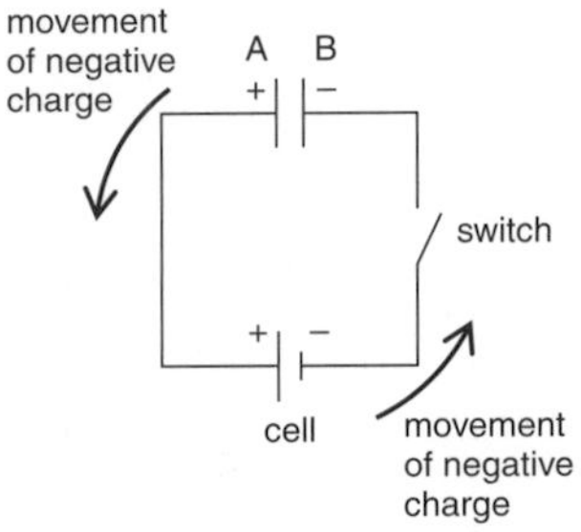

A simple capacitor circuit.

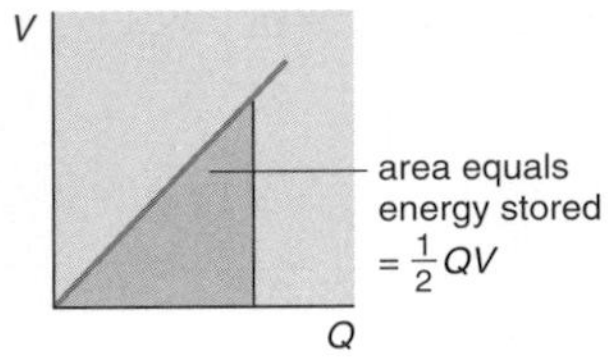

Graph of p.d. against charge for a charging capacitor.

capacitor's energy:

$E = \frac{1}{2}QV$

$E = \frac{1}{2}CV^2$

$E = \frac{Q^2}{2C}$

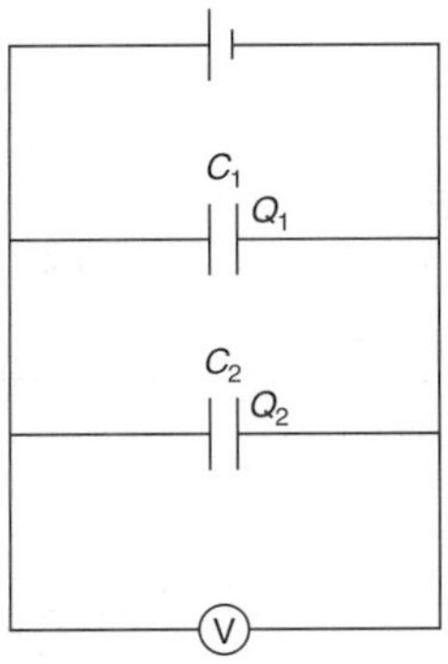

Capacitors in parallel.

parallel capacitors:

$C_T = C_1 + C_2$

series capacitors:

$\frac{1}{C_T} = \frac{1}{C_1} + \frac{1}{C_2}$

- A capacitor consists of two conducting plates, with a large surface area, separated by an insulator or **dielectric**. Its main function is to store electric charge. The amount of charge it can store for each volt of potential difference placed across it is its **capacitance**.

  Capacitance is the ability to store charge, and is equal to the charge stored per unit p.d. placed across it. The unit of capacitance is the farad (F) or more usually pF (10^{-12}), μF (10^{-6}), or nF (10^{-9}). Capacitance is given by the equation $\boldsymbol{C = Q/V}$.

- The capacitance of a practical parallel-plate capacitor (see left) is given by the equation

$$C = \frac{\varepsilon_0\varepsilon_r A}{d}$$

where A is the area of overlap of the plates, d is the plate separation, ε_0 is the permittivity of free space, and ε_r is the **relative permittivity** of the dielectric. (If ε is the permittivity of the material then $\varepsilon_r = \varepsilon/\varepsilon_0$.)

- A simple circuit containing a capacitor, a cell, and a switch is shown left. When the switch is closed, negative charge flows around the circuit onto plate B. An equal amount of negative charge moves off plate A around to the cell. The capacitor charges up uniformly as shown left, until the potential difference across the capacitor is the same as the cell. As the charge builds up on the plate it exerts a repulsive force on any further charge approaching the plate. Consequently work must be done against this repulsive force by the cell, and so a capacitor not only stores charge, but also stores energy. We can use the graph left to formulate an expression for this energy. The definition of potential difference is $V = E/Q$. In a graph of V against Q the area under the graph is the energy. In the graph left the area of the triangle would give us

$$E = \tfrac{1}{2}QV$$

substituting in $Q = CV$ gives $\boldsymbol{E = \tfrac{1}{2}CV^2}$ or $\boldsymbol{E = \frac{Q^2}{2C}}$.

- Capacitors are used a lot in electrical circuits. In a parallel circuit such as that shown below left, the total charge stored is the sum of the charges on each capacitor:

$$Q_T = Q_1 + Q_2$$

From $Q = CV$, $C_T V = C_1 V + C_2 V$.

V is constant (Kirchhoff's second law) so $\boldsymbol{C_T = C_1 + C_2}$.

In a series circuit, such as that shown below right, the charge on each capacitor is the same whatever the value of capacitance, and is equal to the total charge stored in the circuit. This is because as electrons move around the circuit onto plate A of the first capacitor, they charge it negatively and induce an equal and opposite positive charge on plate B, by repelling electrons from it. The electrons repelled from plate B then go onto plate C, making it negatively charged, and so on.

The sum of the p.d.s is the same as the e.m.f. of the cell (from Kirchhoff's second law): $V_T = V_1 + V_2$.

From $V = Q/C$:

$$\frac{Q}{C_T} = \frac{Q}{C_1} + \frac{Q}{C_2}$$

As Q is the same in each term:

$$\frac{1}{C_T} = \frac{1}{C_1} + \frac{1}{C_2}$$

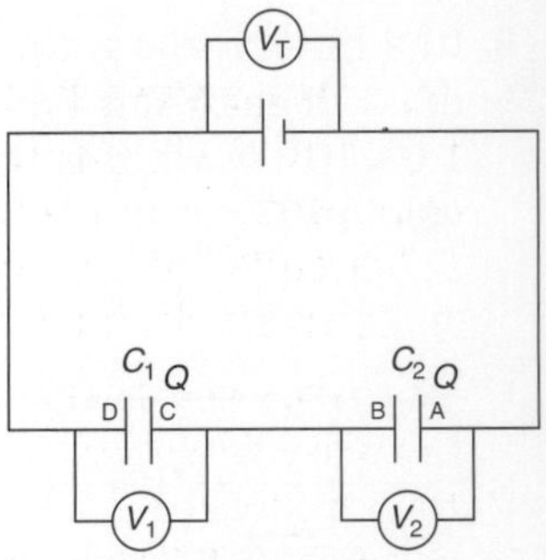

Capacitors in series.

Example *Work out the total capacitance of the circuit right. Calculate the total charge stored in the circuit, and the energy stored on capacitor Y. If Y is effectively a parallel-plate capacitor of area 20 m² and a dielectric permittivity of 8.8 × 10⁻¹² F/m, what is the separation of its plates?*

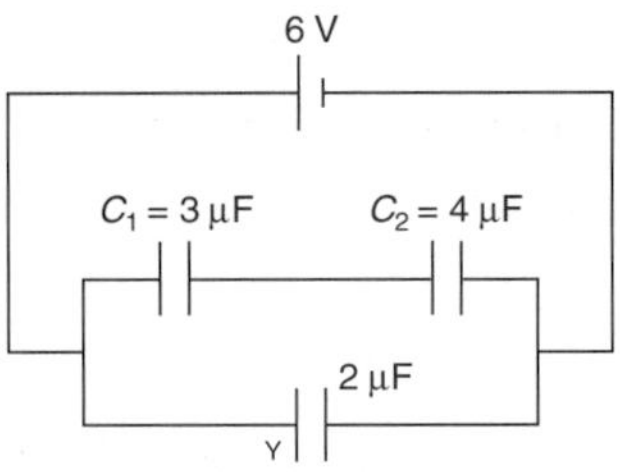

Let C_T be the total capacitance and $C_{(1+2)}$ the capacitance of C_1 and C_2:

$$\frac{1}{C_{(1+2)}} = \frac{1}{C_1} + \frac{1}{C_2} = \frac{1}{3} + \frac{1}{4} \text{ so } C_{(1+2)} = 1.7\,\mu F$$

$$C_T = 2\,\mu F + 1.7\,\mu F = 3.7\,\mu F$$

$$\text{Total charge} = C_T V = 3.7 \times 10^{-6} \times 6 = \mathbf{22\,\mu C}$$

$$\text{Energy on Y} = \tfrac{1}{2} C V^2 = 0.5 \times 2 \times 10^{-6} \times 6^2 = \mathbf{36\,\mu J}$$

$$d = \frac{\varepsilon A}{C} = \frac{8.8 \times 10^{-12} \times 20}{2 \times 10^{-6}} = \mathbf{8.8 \times 10^{-5}\,m}$$

- Capacitors are often used in conjunction with a resistor to make timer circuits, such as a car indicator flasher unit. The resistor slows down the build-up of charge. When the switch in the circuit shown right is in position A the capacitor charges up, and when it is in position B it discharges. Graphs of this charge and discharge are shown below with their relevant equations.

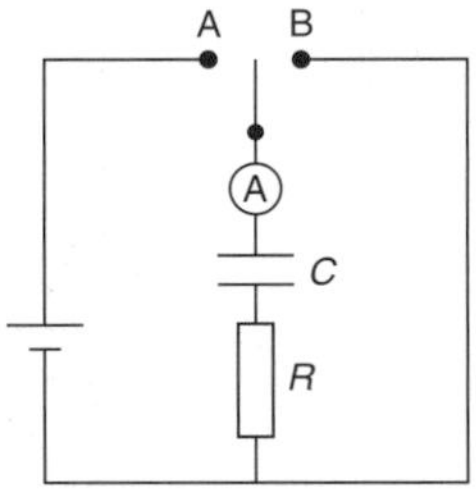

Charge–discharge circuit for a capacitor and resistor.

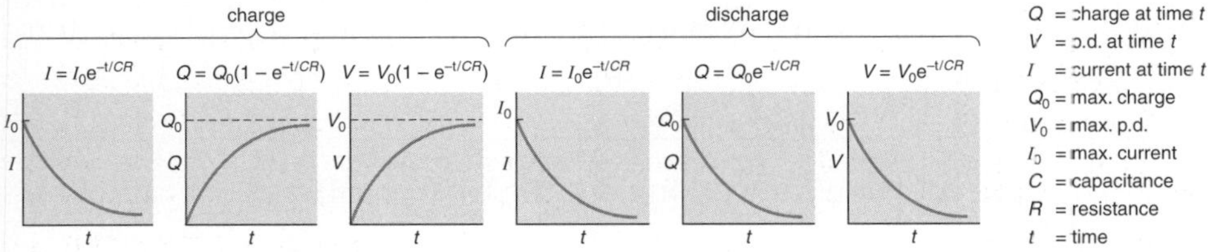

The current graph is decreasing in both cases. All the curves are **exponential**.

An **exponential** curve is one in which one of the quantities changes by the same factor over equal changes in the other quantity. This is confusing and is best understood from the half-life curve in unit 15, where the activity halves in equal periods of time.

Exponential changes usually take place with respect to time. To see whether a quantity is changing in an exponential manner, take two values over a fixed period of time, and determine the ratio between them. Repeat the process with two more pairs of points. If the ratios are the same, it is an exponential change.

Example *A 2.0 μF capacitor is connected in series with a 5.0 MΩ resistor, a 12 V cell, and a switch. What is the initial current when the switch is closed, and what is the current 3.0 s later?*

$$I_0 = \frac{V}{R} = \frac{12}{5 \times 10^6} = \mathbf{2.4 \times 10^{-6} A} \quad \text{so at } t = 3.0\text{:}$$

$$I = I_0 e^{-t/CR} = 2.4 \times 10^{-6} \times e^{-3.0/(2.0 \times 10^{-6} \times 5.0 \times 10^6)} = \mathbf{1.8 \times 10^{-6} A}$$

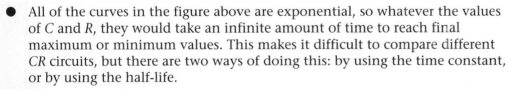

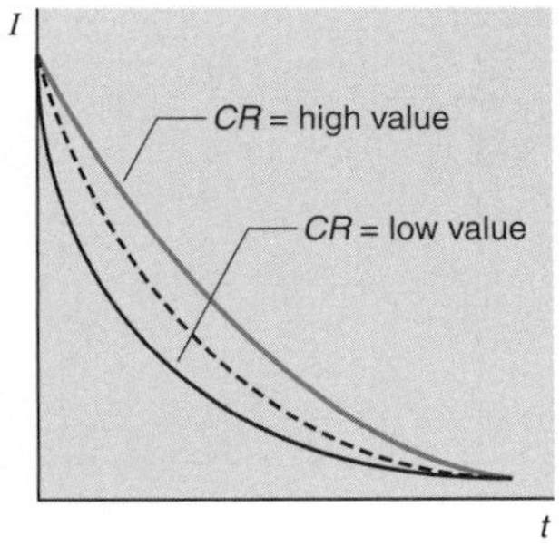

Exponential curves for different values of *CR*.

- All of the curves in the figure above are exponential, so whatever the values of *C* and *R*, they would take an infinite amount of time to reach final maximum or minimum values. This makes it difficult to compare different *CR* circuits, but there are two ways of doing this: by using the time constant, or by using the half-life.

The **time constant**, τ_c, is the time taken for the variables *Q*, *V*, or *I* to decrease to 1/e (0.37) of their original value; $\boldsymbol{\tau_c = CR}$. The **half-life**, $t_{\frac{1}{2}}$, is the time taken for the variables *Q*, *V*, or *I* to decrease to half their original value; $\boldsymbol{t_{\frac{1}{2}} = CR \ln 2}$. These equations can be verified as follows:

time constant: $\tau_c = CR$

half-life: $t_{\frac{1}{2}} = CR \ln 2$

$$I = I_0 e^{-t/CR} \text{ so if } t = CR \text{ then } I = I_0 e^{-1} \text{ or } I = \frac{I_0}{e}$$

$$I = I_0 e^{-t/CR} \text{ so at } t = t_{\frac{1}{2}},\ \frac{I_0}{2} = I_0 e^{-t_{\frac{1}{2}}/CR} \text{ so } 2 = e^{t_{\frac{1}{2}}/CR} \text{ and } \ln 2 = t_{\frac{1}{2}}/CR$$

TESTS

RECALL TEST

1 Define 'capacitance'.

(2)

2 What is 'relative permittivity'?

(2)

3 In a series circuit, what will be the same for each capacitor?

(2)

4 What is meant by an 'exponential change'?

(2)

5 What is meant by 'time constant'?

(2)

6 What is meant by 'half-life'?

(2)

7 What is the equation for working out the total capacitance of capacitors in series?

(2)

8 What is the equation for working out the total capacitance of capacitors in parallel?

(2)

9 Explain how energy is stored in a capacitor.

(2)

10 What factors determine the capacitance of a parallel-plate capacitor?

(2)

(Total 20 marks)

CONCEPT TEST

Take $\varepsilon_0 = 8.8 \times 10^{-12}$ F/m

1 A capacitor consists of two parallel plates 0.0040 m apart, with an area of overlap of 0.4 m^2. What is its capacitance? How much charge is stored if there is a potential difference of 4.0 V across the capacitor?

(4)

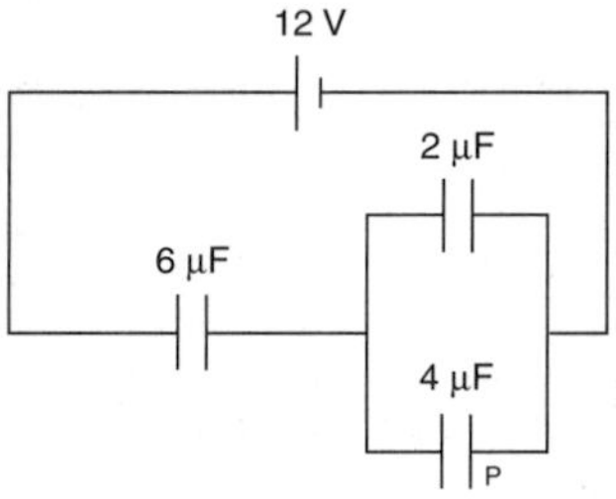

2 What is the total capacitance of the circuit left? What are the charge and energy stored on capacitor P?

(4)

3 A sphere of radius 4.0 cm has a negative charge of 4.8 nC. Calculate the potential at the surface of the sphere and determine its capacitance.

(4)

4 Treat a thundercloud as a capacitor with an area of 10 km^2, which is 500 m above the ground (see right). What is the capacitance and the energy stored if the p.d. is 1.0×10^4 V? If the cloud rises to a height of 1.0 km, assuming the charge remains constant, what is the new energy stored? Explain the difference.

(6)

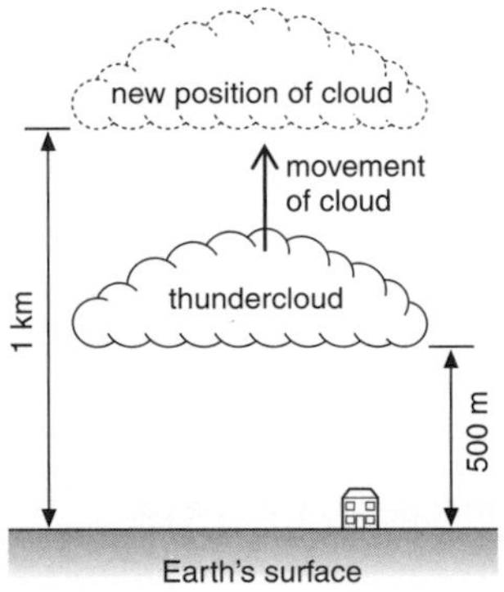

Thundercloud above the Earth's surface.

5 A 1.2 μF capacitor is charged by a 12 V cell. What is the charge stored? It is then disconnected from the cell, retaining its charge, before being connected to another capacitor of 4 μF. What is the final p.d. across the two capacitors and the energy lost in the process?

(6)

6 A car indicator circuit is required to flash once every half second. It switches off when the p.d. is reduced to a certain value. The capacitor of 4.0 nF in the circuit is connected to a 12 V car battery and a resistor of 200 MΩ. What is the p.d. across the capacitor when the indicator switches off?

(6)

7 The figure right shows the decrease in current of a capacitor as it is charged by a 6.0 V cell through a resistor. What is the total charge stored and the capacitance of the capacitor?

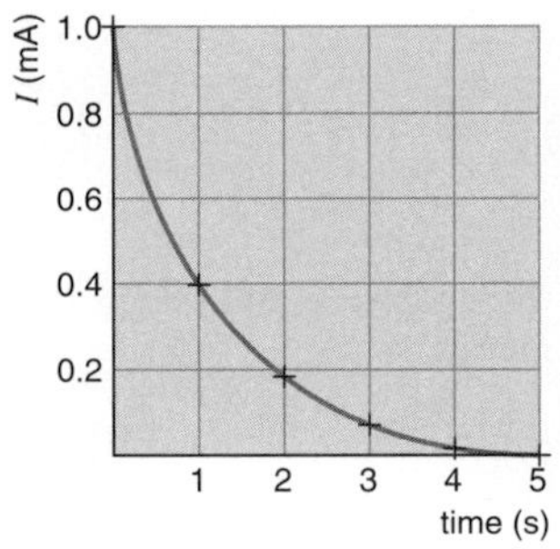

(6)

8 A capacitor of 6.0 μF is connected to a vibrating reed switch, a 12 V cell, and a 4.0 MΩ resistor (see right). If the reed switch oscillates 50 times a second, how much charge is stored each time it is in position A and what is the average current flowing in the ammeter?

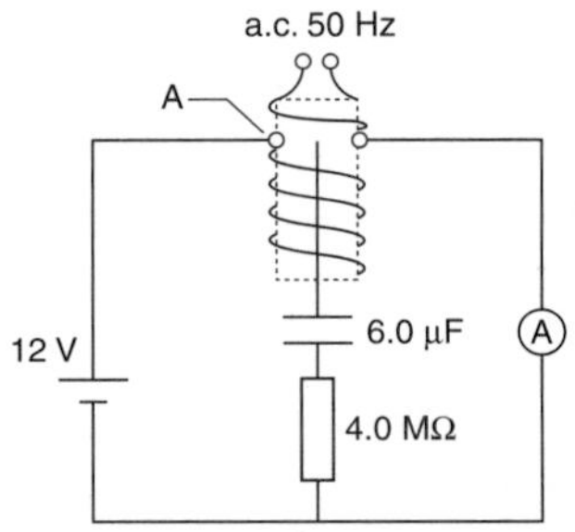

(6)

9 The circuit shown right has a combination of resistors and capacitors. What is the p.d. across each capacitor? If the switch S_1 is closed what are the new p.d.s? When the switch is first closed in which direction does the current flow?

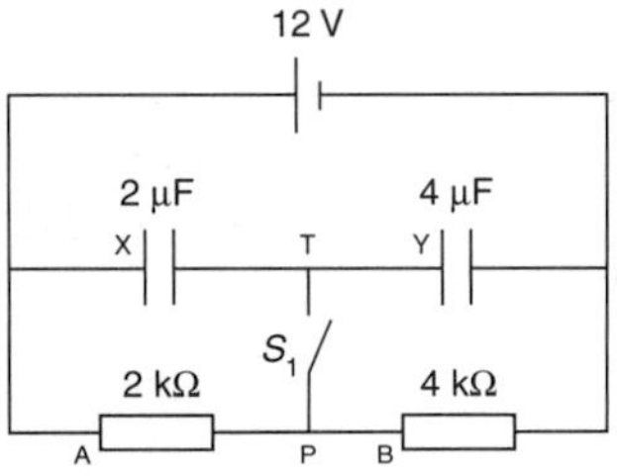

(8)

(Total 50 marks)

Unit 21

MAGNETIC FIELD THEORY

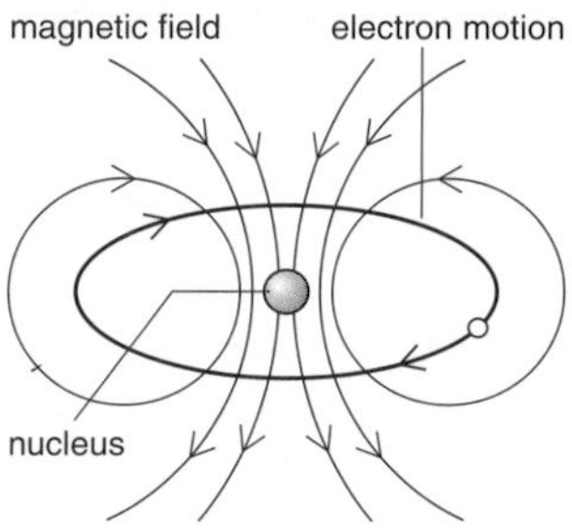

Magnetic field due to electron motion.

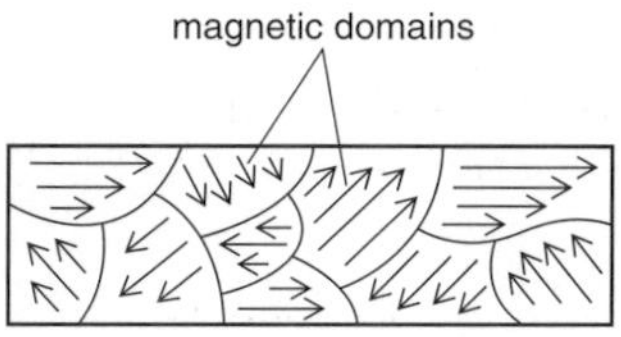

Magnetic domains inside an iron bar.

magnetic flux: ϕ = *BA*

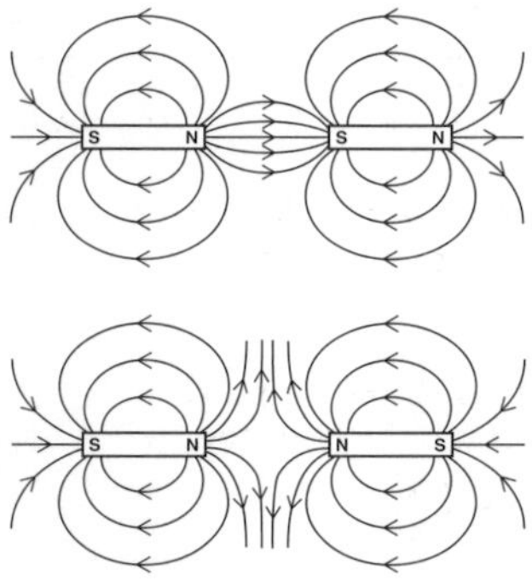

Field patterns due to bar magnets.

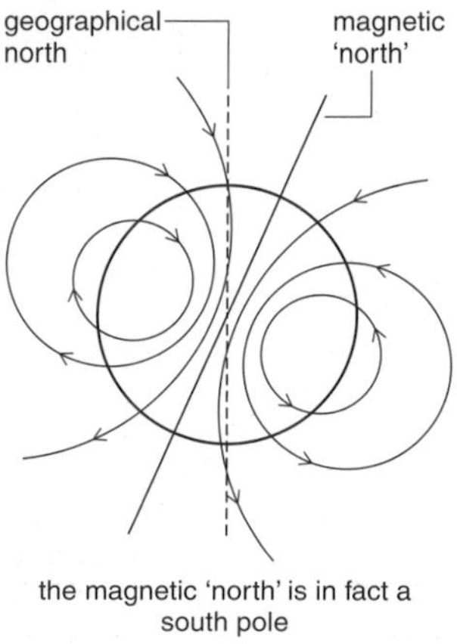

The Earth's magnetic field.

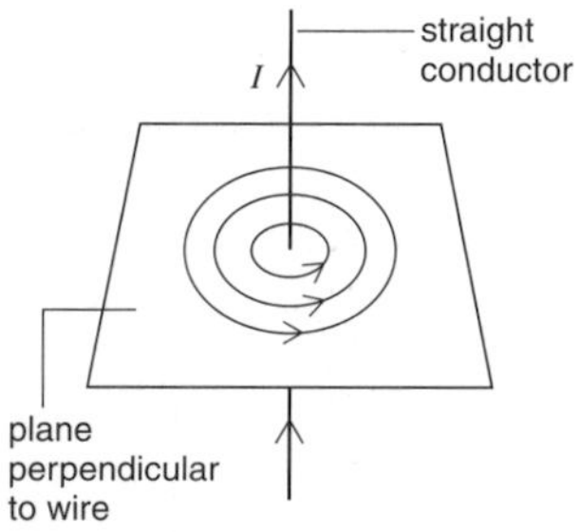

Field pattern produced by straight conducting wire.

- Units 18 and 19 covered inverse-square law fields. This unit deals with another type of field, called a magnetic field. Fields are linked with forces; where there is one of the fundamental types of force, there is an associated type of field. Magnetic forces are directly linked with electric forces, and so are classified as one force, namely the **electromagnetic force**.

  When a current flows in a wire, a circular magnetic field is produced around that wire. In the simple Bohr model of the atom (see unit 15) an electron is in orbit around the nucleus of an atom. The movement of this electron will produce a magnetic field around the atom. In most materials these fields are all pointing in different directions and so effectively cancel each other out. In some materials, such as iron, the fields of atoms in a particular region are more likely to point in the same direction. This region is called a **magnetic domain**. In an unmagnetized piece of iron the domains are all pointing in different directions, and so effectively cancel each other out. When a piece of iron is magnetized all of these domains become oriented in the same direction.

- Magnetic fields are quantified using two concepts: **magnetic flux density** (B) and **magnetic flux** (ϕ).

  Magnetic flux density is defined as the force per unit length per unit current on a current-carrying conductor lying at right angles to the direction of a magnetic field passing through it ($B = F/Il$). It can be more simply thought of as a measure of how much magnetic field is passing through unit area ($1\ m^2$) at right angles to it. The unit is the tesla (T).

  Magnetic flux is the product of the magnetic flux density and the area through which the field is passing at right angles to the direction of the field. It can be more simply thought of as a measure of how much magnetic field is passing through a given area. The unit is the weber (W).

  The two are linked by the equation $\phi = BA$ where A is area. B is a vector and its direction is always at a tangent to the magnetic field lines.

- A simple magnet has two poles, a north and a south. Like poles repel each other and unlike poles attract. **Magnetic fields** and **magnetic field lines** are defined in a similar way to gravitational and electric fields.

  A **magnetic field** is a region of space where a magnetic object will experience a force due to the presence of another magnetic object.

  A **magnetic field line** shows the direction of the magnetic flux. It can be thought of as the direction of the force acting on, or the path taken by, an isolated north pole in a magnetic field. This is a purely theoretical concept as there is no such thing as an isolated north pole, but if there were it would move towards a south pole. The closer the lines the stronger the field.

- The field patterns produced between two bar magnets show the attractive or repulsive nature of the fields. There are two simple tools for investigating magnetic fields: iron filings or plotting compasses. **Iron filings** show the overall pattern but not the direction of the field, and **plotting compasses** show the direction. It is easy to become confused about plotting compasses and normal compasses. A plotting compass shows the direction of a magnetic field from north to south, whereas a normal compass points towards the north pole of the Earth.

- The Earth has a magnetic field thought to be due to electric currents in its liquid iron-based outer core. The strength of the Earth's magnetic field is typically 10^{-4}T. Its angle of inclination will vary depending upon your position on the Earth (see above left).

- As mentioned before, any moving charge produces a magnetic field. The shape of the field depends upon the shape of the current-carrying conductor.

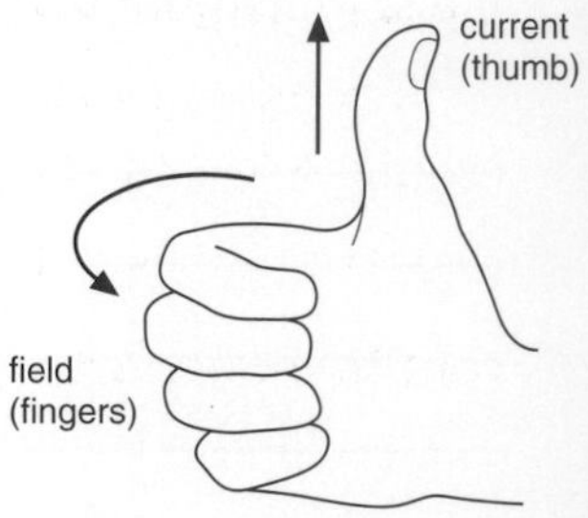

TThe right-hand grip rule.

- A straight wire produces a circular field around itself. The direction of the field in relation to the direction of the current may be determined using the **right-hand grip rule** (see diagram).

  The magnetic flux density at a point near the wire is given by the equation

  $$B = \frac{\mu I}{2\pi a}$$

  where I is the current, a is the perpendicular distance from the wire, and μ is a constant called the **permeability** of the medium in which the field lies. This is usually μ_0, the **permeability of free space** (or of a vacuum). **Relative permeability** $\mu_r = \mu/\mu_0$. The flux density decreases inversely with distance from the wire (see right).

flux density due to straight wire:

$$B = \frac{\mu I}{2\pi a}$$

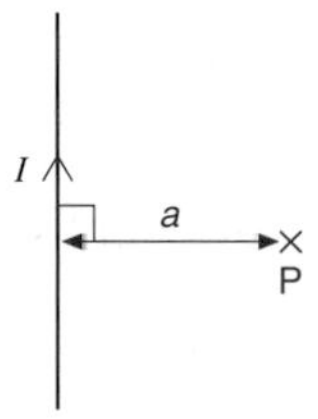

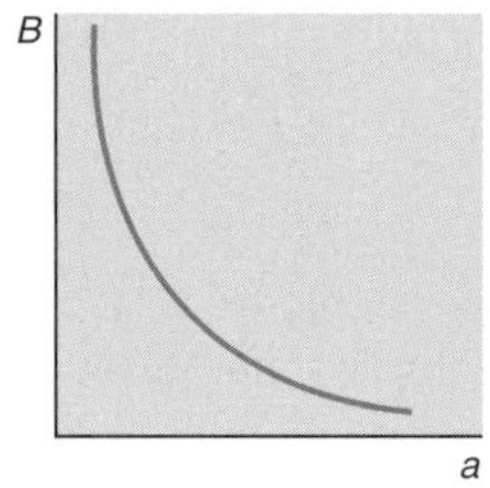

Flux density at point P depends upon, among other things, the distance a; B decreases inversely with distanc from the wire.

- A **coil** or a **solenoid** produces a field pattern very similar to a bar magnet. The polarity of the solenoid may be determined by looking at the ends. If the current flow is in a clockwise direction, it is a south pole; if it is in an anticlockwise direction, it is a north pole. It is important to remember that the field goes from north to south outside the solenoid, and from south to north inside. The magnetic field well inside the solenoid is uniform and given by the equation

  $$B = \frac{\mu NI}{l} \text{ or } B = \mu nI$$

  where I is current, N is the number of turns, l is the length, n is the number of turns per unit length, and μ is the permeability of the medium in which the field lies. The flux density at a point at the end of the coil is half that in the centre, simply because it only has contributions of field from turns of coil on one side of it and not both.

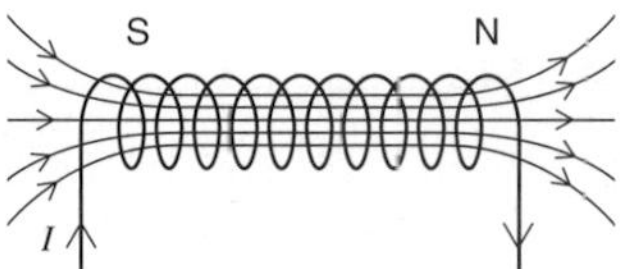

flux density due to solenoid:

$B = \mu nI$

Field pattern produced by a solenoid.

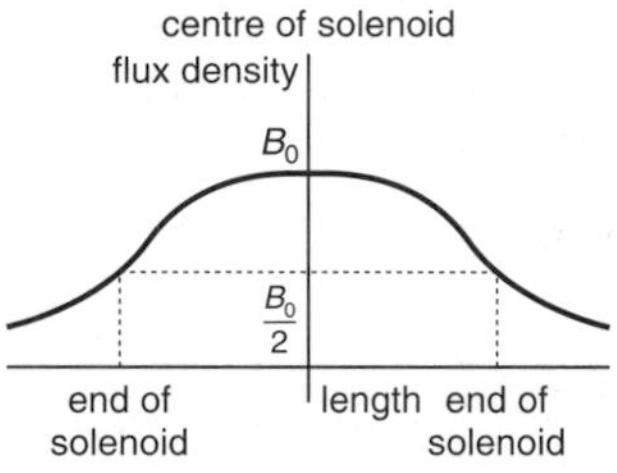

Variation of flux density along a solenoid.

flux density due to a flat circular coil:

$$B = \frac{\mu NI}{2r}$$

- The third type of conductor is a flat circular coil. This produces a uniform field at the centre of the coil, which obeys the equation

  $$B = \frac{\mu NI}{2r}$$

  where r is the radius of the coil, N is the number of turns of coil, I is the current, and μ is the permeability of the medium.

  Two of these coils placed parallel to each other can be used to produce a uniform field (see right). These are sometimes referred to as **Helmholtz coils**. The spacing between the coils should equal the radius of the coils.

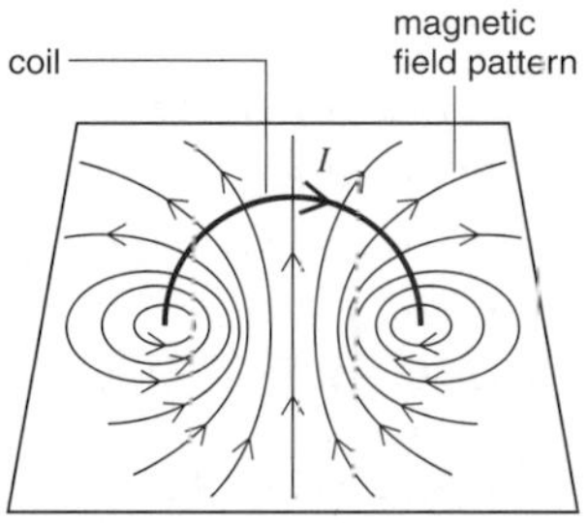

Magnetic field pattern produced by a flat circular coil.

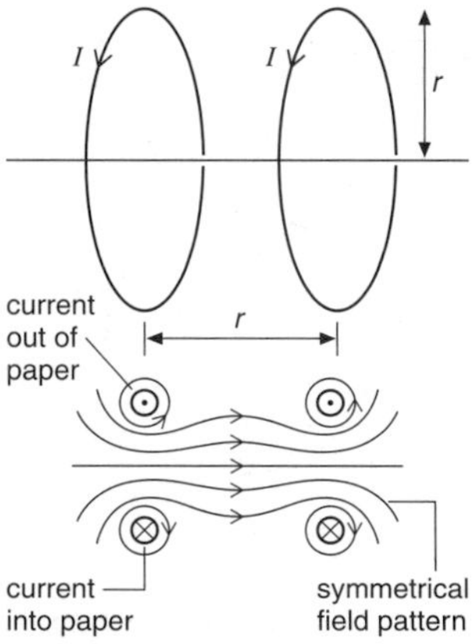

Helmholtz coil arrangement, with the field pattern between the coils.

Example *At point P the Earth's magnetic field has a downward component of 4.0×10^{-4}T. What is the current required in a solenoid of 20 turns per cm, positioned as shown in the diagram, that would produce a resultant field of 3.0×10^{-4}T acting upwards? If a straight wire carrying a current of 10 A produces the same resultant field, what is its distance from point P? ($\mu_0 = 4\pi \times 10^{-7}$ N/A^2.)*

Field required to act upwards = $(3.0 + 4.0) \times 10^{-4}$T.

For the solenoid: $B = \mu_0 nI$ and $n = 20 \times 100 = 2000$.

$$7.0 \times 10^{-4} = 4\pi \times 10^{-7} \times 2000 \times I \text{ so } \mathbf{I = 0.28A}$$

For the straight wire: $B = \dfrac{\mu_0 I}{2\pi a}$ so $a = \dfrac{4\pi \times 10^{-7} \times 10}{7.0 \times 10^{-4} \times 2\pi} = \mathbf{2.9\,mm}$

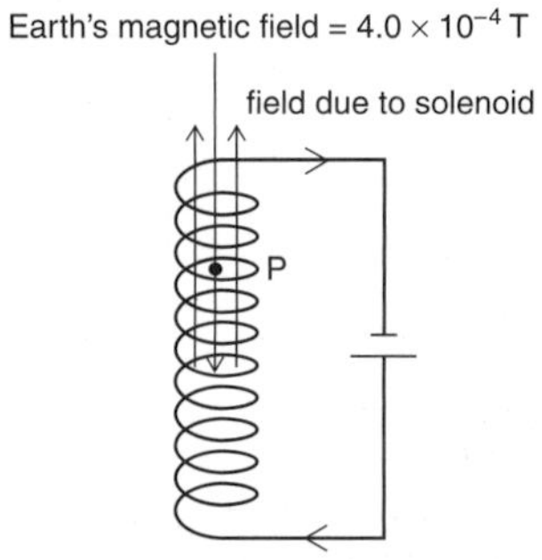

Unit 21

TESTS

RECALL TEST

1 What is a magnetic domain?

(2)

2 Define 'magnetic flux'.

(2)

3 Define 'magnetic flux density'.

(2)

4 What is a magnetic field?

(2)

5 What is a magnetic field line?

(2)

6 What are the two simplest methods of investigating magnetic fields?

(2)

7 How do you work out the direction of the magnetic field around a straight conductor?

(2)

8 What is a solenoid?

(2)

9 What is meant by 'permeability'?

(2)

10 How do you determine the polarity of a solenoid?

(2)

(Total 20 marks)

CONCEPT TEST

Take $\mu_0 = 4\pi \times 10^{-7}\,\mathrm{N/A^2}$

1 What is the magnetic flux density at a point 0.80 m from a straight wire carrying a current of 2.0 A in a medium with a relative permeability of 2.5?

(4)

2 What is the magnetic flux density inside and at the end of a solenoid of 1000 turns, 16 cm in length, carrying a current of 4.0 A? If the coil has a radius of 2.0 cm, how much magnetic flux is passing through the coil well inside the solenoid?

(6)

3 A large flat circular coil of 200 turns is being used to produce a magnetic field of flux density $2.0 \times 10^{-3}\,\mathrm{T}$ at its centre. If the maximum current available is 12 A, how big can the coil be?

(4)

4 At a point on the Earth's surface the Earth's magnetic field is inclined at 40° to the horizontal and has a value of 6.0×10^{-4} T. A solenoid of 4000 turns per metre is positioned as shown right. What current is required to produce a resultant vertical field of zero?

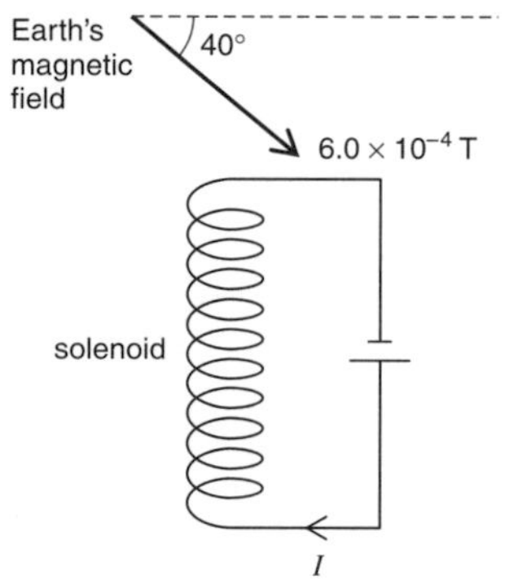

(6)

5 Two solenoids are positioned as shown right. Solenoid A has 500 turns per cm and is carrying a current of 4.0 A. Solenoid B has 1000 turns per cm and is carrying a current of 3.0 A. What is the resultant magnetic flux density and its direction at P?

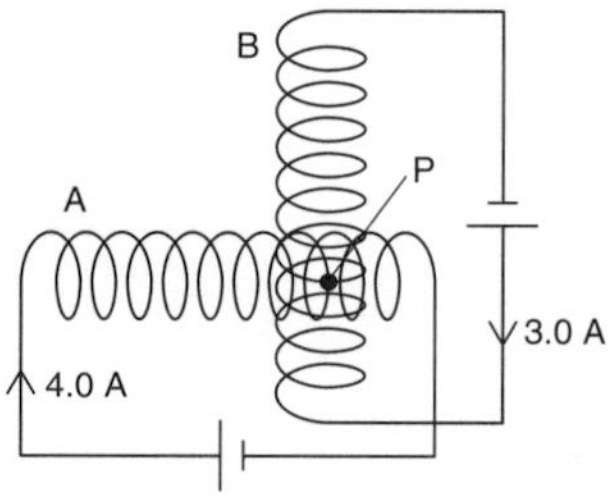

(6)

6 A solenoid and a straight wire are in the position shown right. What is the resultant magnetic flux density at point X and its direction?

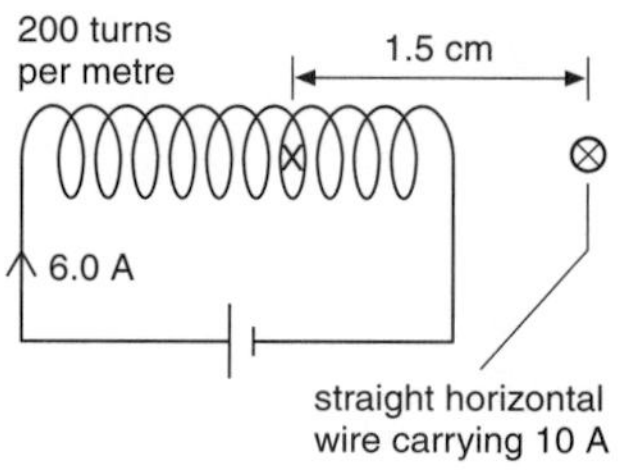

(6)

7 Two Helmholtz coils are to be used to produce a uniform field, as shown right. Given that the flux density of this arrangement at the centre of the coil is given by the equation $B = (8\mu_0NI)/(125^{0.5}r)$ determine the flux passing through the uniform field region if the diameter of each coil is 10 cm.

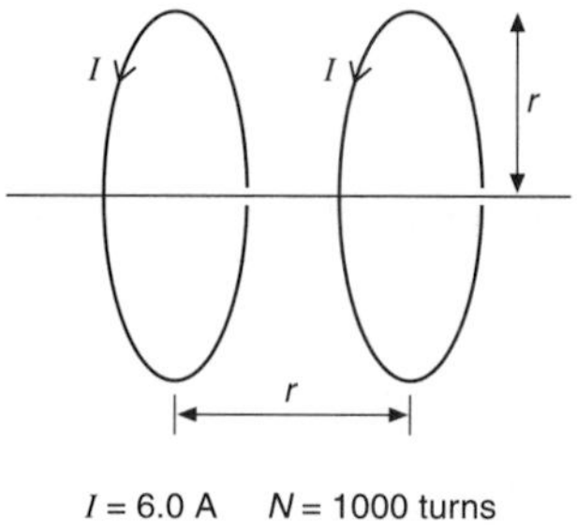

(8)

8 If you take a circular coil as shown right, the magnetic flux density at any point on the axis of that coil is given by the equation $B = (\mu_0NIa^2)/(2(a^2 + x^2)^{1.5})$, where a is the coil's radius and x is the distance from the coil's centre to the point where the flux density is being determined. Determine the flux density 0.50 m from the centre of a coil of 50 turns, carrying a current of 6.0 A and with a diameter of 0.40 m. Sketch a graph of the variation of flux density with radius on the axes right.

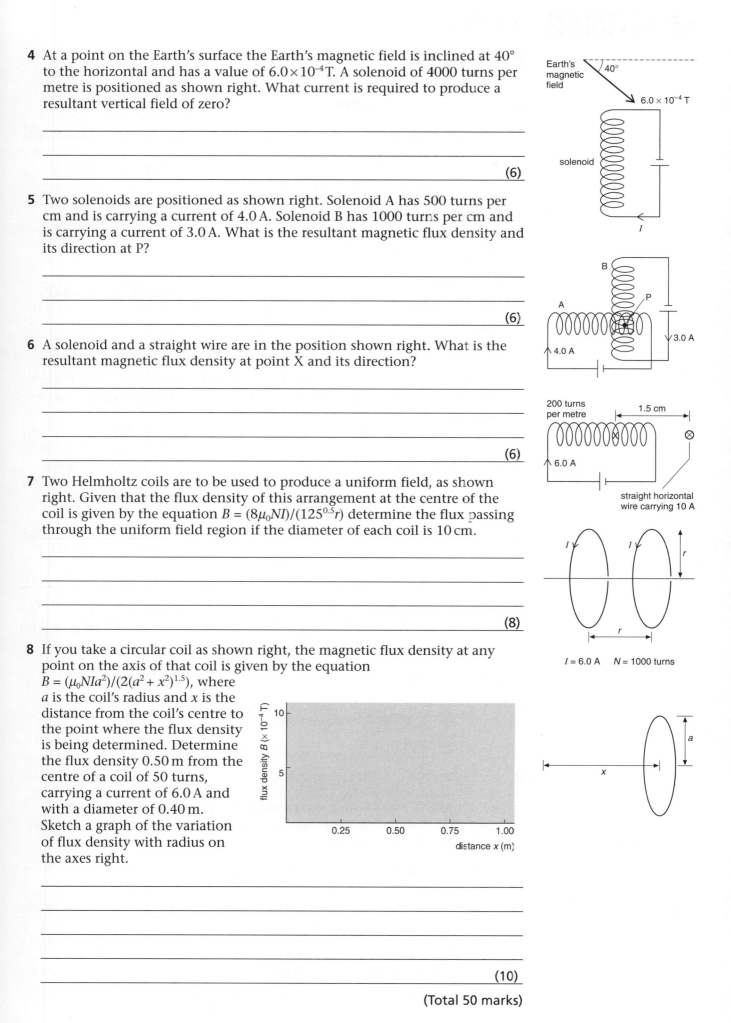

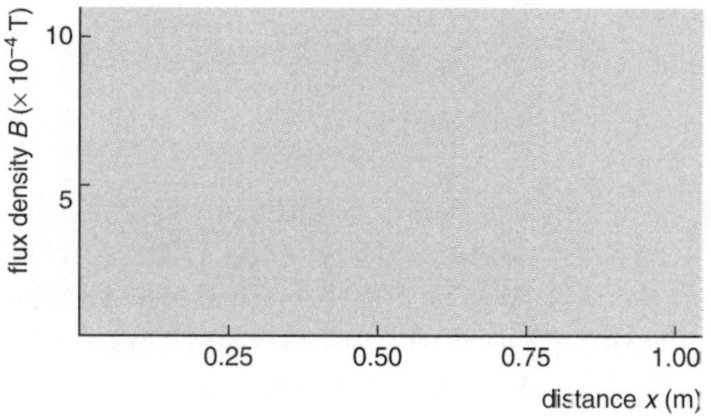

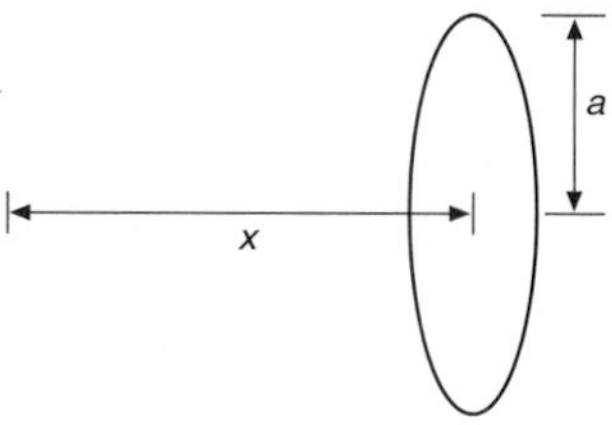

(10)

(Total 50 marks)

Unit 22

APPLIED MAGNETISM

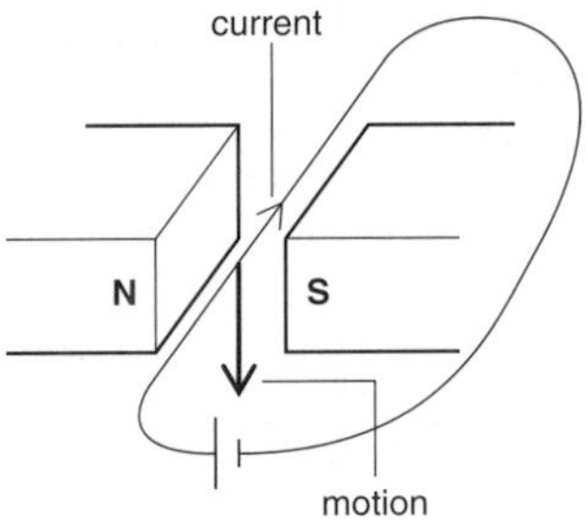

Force on a current-carrying conductor.

- If an object made of a magnetic material lies in a magnetic field, it will have a force acting on it. The object does not have to be magnetized beforehand. A **magnetic** material is one that is naturally magnetic or can be magnetized. A current flowing in a wire produces a magnetic field around it, so it is possible for a wire carrying a current to experience a force if it lies in an external magnetic field. Essentially, what happens is that the field produced by the wire combines with the external field to produce a force on the wire. In the arrangement shown left, the fields add together on top of the wire and cancel out on the bottom.

  From experimental evidence the size of this force can be shown to be directly proportional to the size of the current, I, the strength of the field, B, the length of the conductor, l, and $\sin\theta$ where θ is the angle between the field and the conductor. This is summarized in the equation

  $$\boldsymbol{F = BIl\sin\theta} \text{ and when } \theta \text{ is } 90°, \boldsymbol{F = BIl}$$

  The direction of the force may be determined using **Fleming's left-hand rule** (see below left). This equation is used to define the tesla: one tesla is the magnetic flux density that produces a force of 1 N on a 1 m length of conductor carrying a current of 1 A.

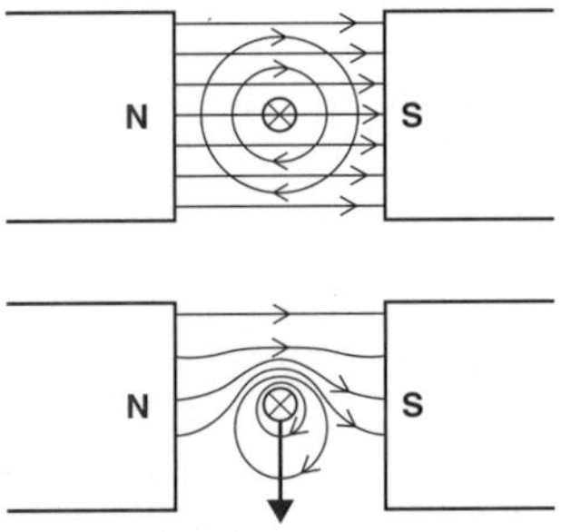

force on a current-carrying conductor:
$\boldsymbol{F = BIl\sin\theta}$

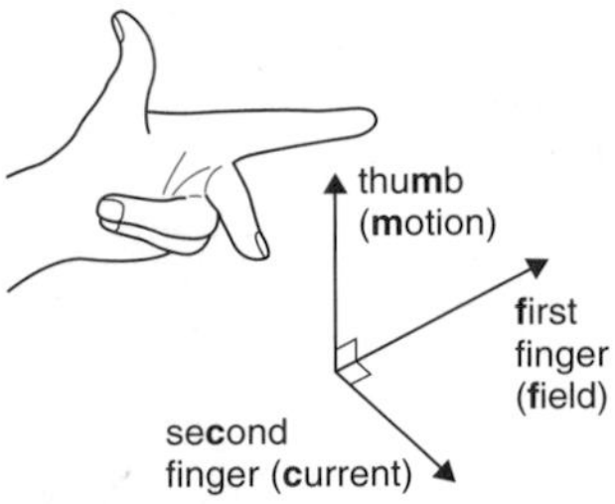

Fleming's left-hand rule.

force on a charged particle:
$\boldsymbol{F = BQv}$

- In the same way that a wire carrying a current in a magnetic field experiences a force, a charged particle with charge Q travelling with velocity v in a magnetic field also experiences a force. The equation used for determining this force is easily derived. Take the equation $F = BIl$ and substitute in $I = Q/t$ to get $F = BQl/t$. $v = l/t$ so this gives $\boldsymbol{F = BQv}$.

  To work out the direction of the force, remember that Fleming's left-hand rule uses the direction of conventional current, from positive to negative. Electrons flow the other way, so for a negatively charged particle you would have to have your current finger pointing in the opposite direction to the particle's motion.

- If the field is perpendicular to the direction of motion, the charged particle will experience a force perpendicular to its path (see right). As the particle changes direction the force continues to act at 90° to its direction of motion, so the charged particle moves into a circular path. The magnetic force of deflection provides the centripetal force. This gives

  $$\frac{\boldsymbol{mv^2}}{\boldsymbol{r}} = \boldsymbol{BQv} \text{ or } \boldsymbol{mr\omega^2 = BQv} \text{ or } \boldsymbol{m\omega = BQ}$$

  $$\omega = \frac{2\pi}{T} \text{ so } \boldsymbol{T = \frac{2\pi m}{BQ}} \text{ and } f = \frac{1}{T} \text{ so } \boldsymbol{f = \frac{BQ}{2\pi m}}$$

  These equations are used in particle accelerators. (see unit 17).

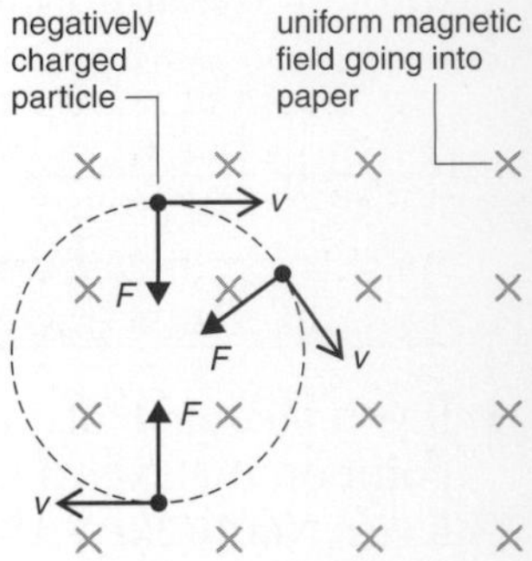

A charged particle moving in a circular path in a field.

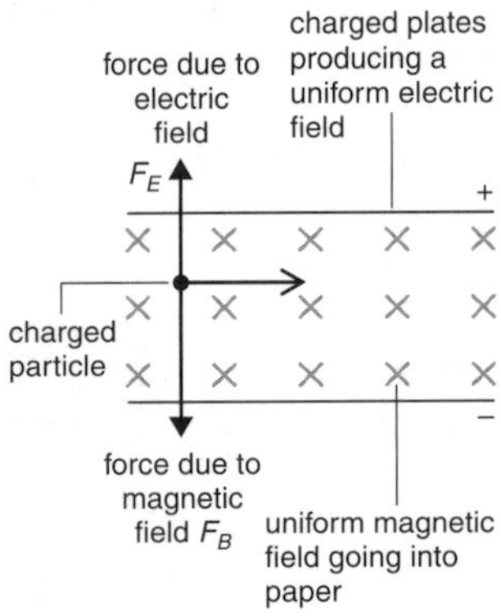

A velocity selector.

velocity selector: $\boldsymbol{v = \frac{E}{B}}$

- A magnetic field and an electric field producing forces in opposite directions create a **velocity selector**. This only allows particles with a certain velocity to pass through it (i.e. those for which the magnetic and electric forces cancel each other out). Other particles are deflected to one side or the other. For a particle to pass through:

  $$\boldsymbol{F_B = F_E}, \text{ so } \boldsymbol{BQv = EQ} \text{ and } \boldsymbol{v = \frac{E}{B}}$$

- Most things in physics repel each other if they are the same, for example like charges and like magnetic poles. So when a current flows down two wires next to each other, in the same direction, you would expect them to follow the same pattern and repel each other. However, they actually do the opposite and attract each other (see left). The reason for this is quite clear if you consider the field patterns of each wire in detail.

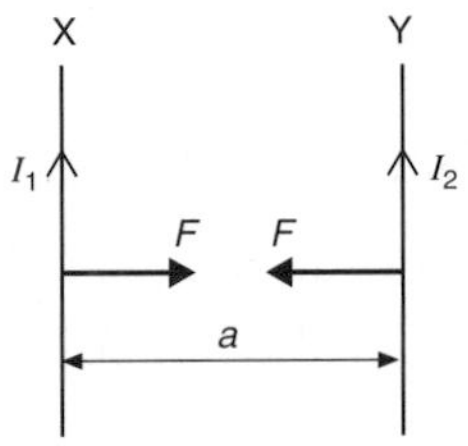

Force between two parallel conductors with current flow in the same direction.

In the region between the two wires, the fields produced by them act in opposite directions, thereby reducing the overall field strength, and on the 'outside' of each wire the fields reinforce each other, producing a stronger field. The resulting field pattern is shown right. This imbalance in the field strengths produces a force that pushes the wires together and looks like an attractive force between the two wires. If the currents in the wires flow in opposite directions, the force is repulsive, with the field pattern shown far right. To derive an equation for this force, take wire X carrying current I_1, producing a field of flux density B at wire Y, which is at a distance of a from it.

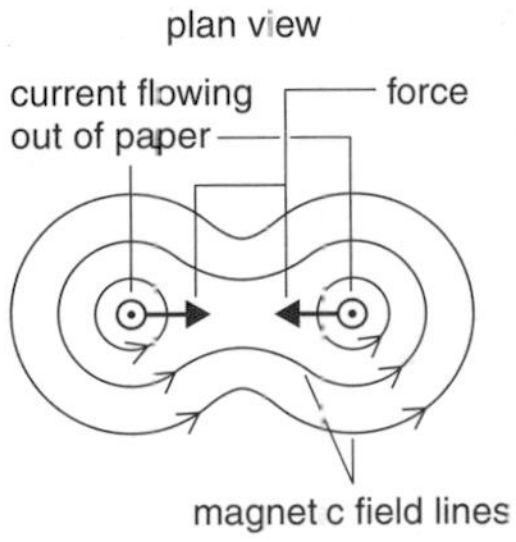

Field pattern for current flowing in the same direction.

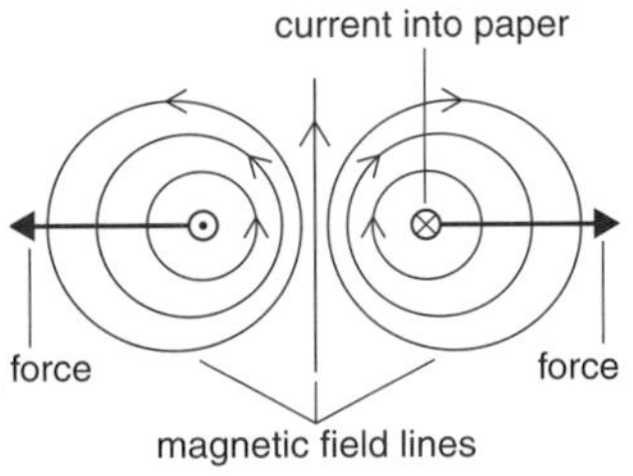

Field pattern for current flow in opposite directions.

$B = \frac{\mu_0 I_1}{2\pi a}$ and the force on Y due to B is $F = BI_2 l$

so $F = \frac{\mu_0 I_1 I_2 l}{2\pi a}$ or $\frac{F}{l} = \frac{\mu_0 I_1 I_2}{2\pi a}$

force between parallel wires:
$\frac{f}{l} = \mu_0 I_1 I_2$

This unusual equation is used to define the ampere.

1 ampere is the current flowing in each of two infinitely long parallel wires of negligible cross-sectional area separated by 1 m in a vacuum that produces a force per unit length of 2×10^{-7} N/m. To get this, just use 1 for each quantity in the equation, except μ_0, which is $4\pi \times 10^{-7}$ N/A^2 for a vacuum.

Example *Estimate the force per unit length between the live and the neutral wire in an electric kettle lead. Take $\mu_0 = 4\pi \times 10^{-7}$ N/A^2.*

A typical kettle has a 13 A fuse. Wires would be about 0.50 cm apart.

$$\frac{F}{l} = \frac{\mu_0 I_1 I_2}{2\pi a} = \frac{4\pi \times 10^{-7} \times 13 \times 13}{2\pi \times 0.005} = \mathbf{6.8 \times 10^{-3}\ N/m}$$

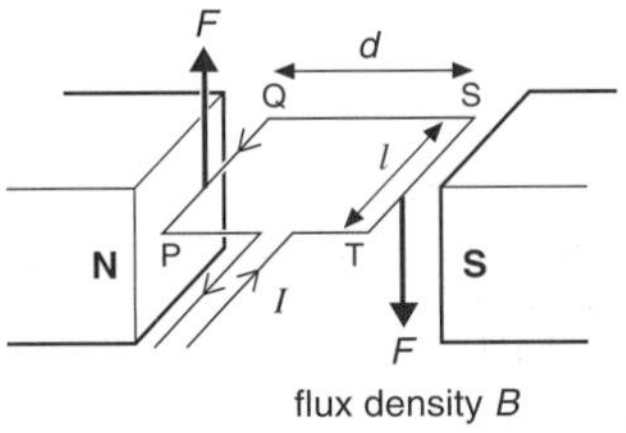

A rectangular coil in a magnetic field.

- The fact that a force is produced on a straight wire in a magnetic field can be used in a coil to produce rotation, which is used in electric motors and meters. If Fleming's left-hand rule is applied to the coil shown above right, the force on PQ is upwards and the force on ST is downwards. This produces a couple which causes the coil to rotate.

The force on either PQ or ST is $F = BIl$ and the torque produced by the couple is $\boldsymbol{T} = Fd$. Putting these together gives

$\boldsymbol{T} = BIld$ or $\boldsymbol{T} = BIA$, where A = area of coil

For N turns of coil: $\boldsymbol{T} = \boldsymbol{BIAN}$.

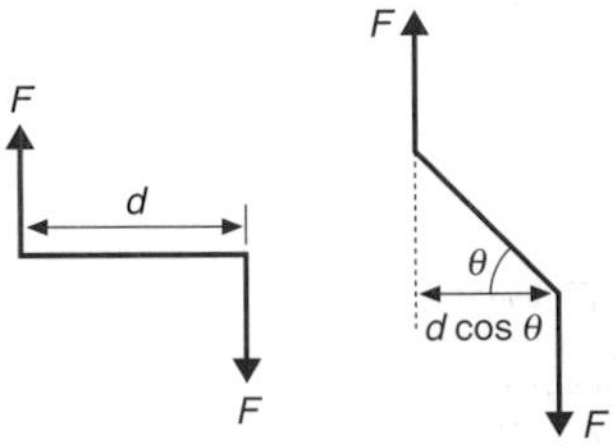

There is a reduction in torque as the coil rotates.

torque on a coil:
$\boldsymbol{T} = \boldsymbol{BIAN}$

Look at the figure above right. As the coil rotates the torque is reduced to $\boldsymbol{T} = BIAN \cos\theta$ because the perpendicular distance between the two forces is reduced. This can be overcome in two ways (see right): by having curved magnetic poles and a cylindrical iron core which produce a radial field so the force is always at right angles to the field and so continuously giving $\boldsymbol{T} = BIAN$ or by having several different coils in different orientations. In direct current circuits, to ensure continuous rotation, the direction of current in the coil must be reversed twice every rotation; this is achieved with a **split-ring commutator** (see below). For alternating current (see unit 23) a **slip-ring commutator** must be used (see below).

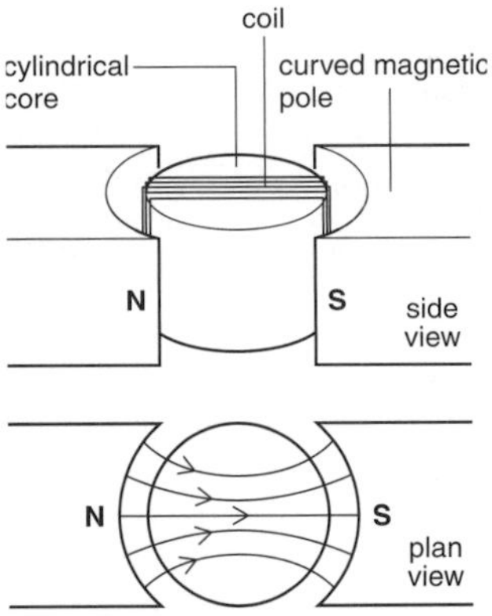

Using a radial field.

Multiple coils.

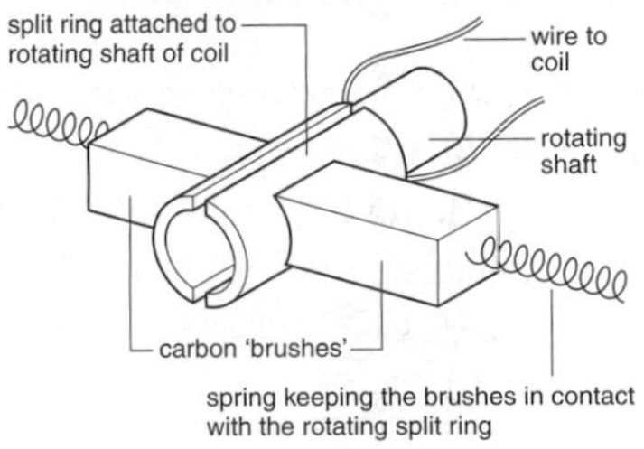

A split-ring commutator.

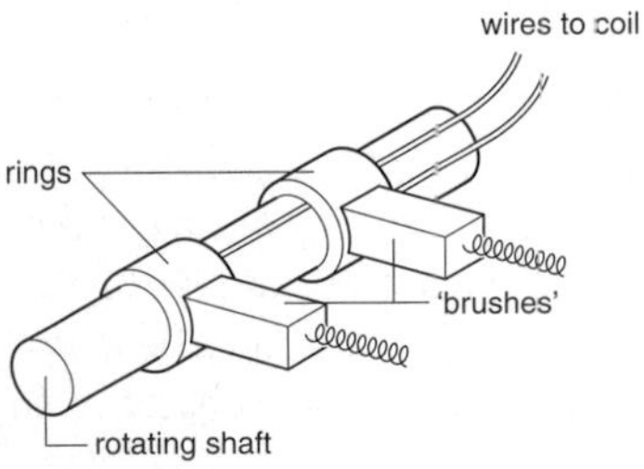

A slip-ring commutator.

Unit 22

TESTS

RECALL TEST

1 Define the tesla.

(2)

2 What is Fleming's left-hand rule used for?

(2)

3 Define the ampere.

(2)

4 Upon what factors does the force acting on a current-carrying conductor lying in a magnetic field depend?

(2)

5 State the direction of the force between two parallel current-carrying conductors.

(2)

6 Where would you use a split-ring commutator and a slip-ring commutator?

(2)

7 How is it possible to ensure that a motor coil rotates continuously?

(2)

8 What happens to a charged particle when it enters a magnetic field acting perpendicular to its direction of motion?

(2)

9 What factors determine the rate of rotation of a rectangular coil in a magnetic field?

(2)

10 What is a velocity selector?

(2)

(Total 20 marks)

CONCEPT TEST

Take $\mu_0 = 4\pi \times 10^{-7}\,\text{N/A}^2$, electron charge $= -1.6 \times 10^{-19}\,\text{C}$, and $g = 9.8\,\text{m/s}^2$

1 A 2.0 m length of wire is lying perpendicular to a magnetic field of flux density $1.2 \times 10^{-4}\,\text{T}$ and carrying a current of 3.0 A. What is the force acting on it? If the wire is moved so that it is at 30° to the direction of the field, what is the new force?

(4)

2 An electron is travelling at $1.0 \times 10^{2}\,\text{m/s}$ when it enters a magnetic field of flux density $2.0 \times 10^{-2}\,\text{T}$. What is the force acting on the electron? In relation to the original direction of motion, in what direction does this force act?

(4)

3 Two parallel wires have currents 4 A and 7 A respectively in the same direction. If they are 0.50 m apart what is the magnetic flux density at the midpoint between them? Do they move towards each other or away?

(6)

4 The power cable for an electrical heater carries a current of 6.0 A. If the live and neutral wires are separated by 3.0 mm, what is the force per unit length between them? Is the force attractive or repulsive? Sketch the resulting magnetic field pattern.

(6)

5 A 6.0 A current is flowing in the rectangular coil shown right, which has 200 turns and lies in a magnetic field of flux density 4.0×10^{-2} T. What is the torque produced when the plane of the coil is at 0° to the field, and at 30° to the field? Explain why the torque changes.

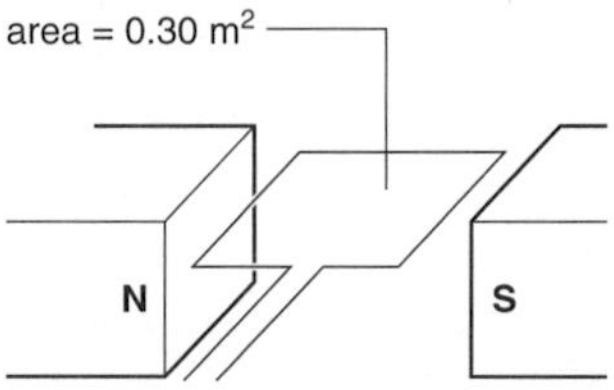

(6)

6 A steel tube of mass 10 g is connected to a cell in a circuit and is free to roll up a pair of rails as shown right. What current is required to keep the tube stationary if the rails are at an angle of 30° to the horizontal?

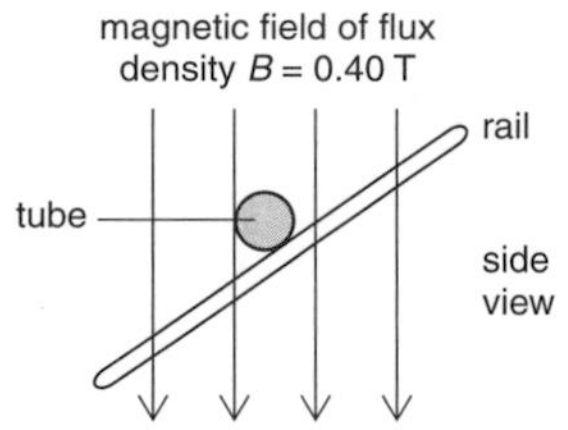

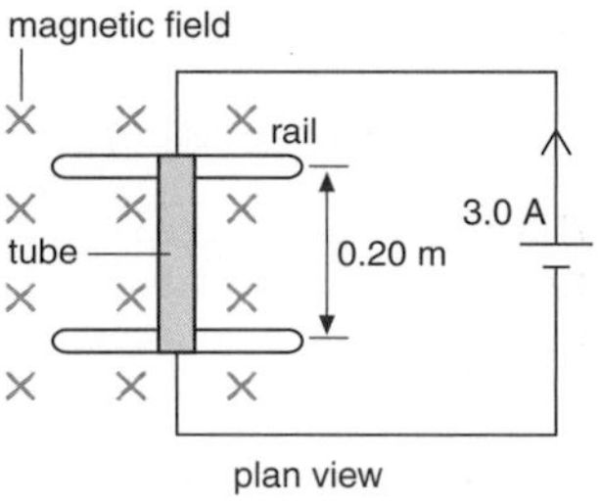

(8)

7 An alpha particle travelling at velocity 20 m/s enters a magnetic field of 4.0×10^{-3} T, acting perpendicular to the direction of motion. The particle moves into a circular path. Explain why it does this, and determine the force acting on the alpha particle and the time taken for one revolution. (Mass of nucleon = 1.7×10^{-27} kg.)

(8)

8 In a mass spectrometer, a velocity selector is used to make sure that all of the charged particles entering it are travelling at the same speed. Explain what a velocity selector is, and determine the velocity of charged particles that pass through a velocity selector consisting of two parallel plates 4.0 cm apart, with a p.d. of 10 000 V between the plates and a perpendicular magnetic field of flux density 3.0×10^{-3} T.

(8)

(Total 50 marks)

Unit 23 ELECTROMAGNETIC INDUCTION

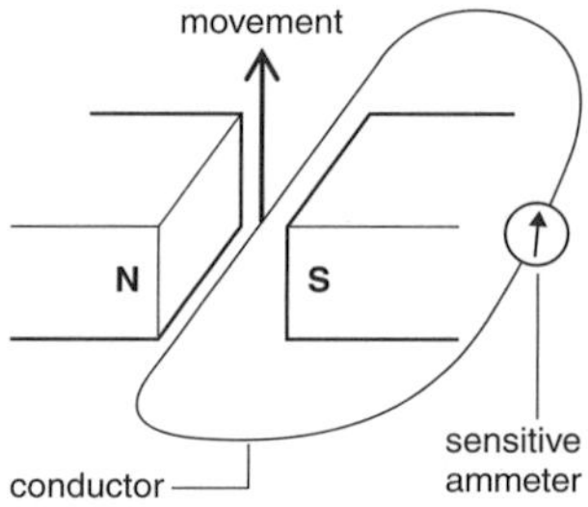

Wire moving through a magnetic field.

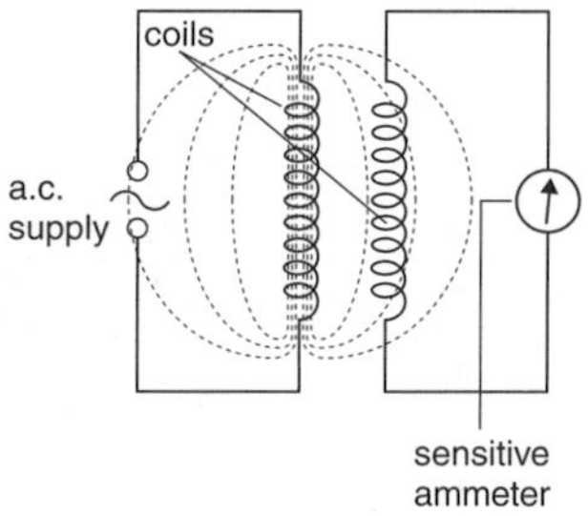

A stationary conductor in a changing magnetic field.

- If a wire is moved through a magnetic field, a current is induced in the wire (see left). (The wire must cut through the field lines, not move in the same direction as them.) The current is greater if the field strength is increased, or the motion faster, or the length of the conductor greater, or the direction of motion is perpendicular to the field direction. A current can also be induced if the conductor is stationary and the strength or direction of the field is changing. This would happen if the conductor lay near another one with an a.c. current flowing through it (see left).

  This can be summarized by saying that any conductor that experiences a change in the magnetic flux passing through it will have an e.m.f. induced in it. Current only flows if there is a complete circuit, so we generally refer to induced e.m.f.s rather than induced currents. This is embodied in **Faraday's law of electromagnetic induction**: the size of the induced e.m.f. is directly proportional to the rate of change of **flux linkage**.

  Electromagnetic induction will take place in any conductor, but is usually associated with a coil of wire, like a solenoid, called an **inductor**. **Flux linkage** is the product of the magnetic flux (ϕ) passing through such a coil and the number of turns of the coil, N. **Flux linkage =** $\boldsymbol{N\phi}$. (Flux linkage is sometimes represented by Φ, giving $\boldsymbol{\Phi = N\phi}$.)

flux linkage: $\boldsymbol{\Phi = N\phi}$

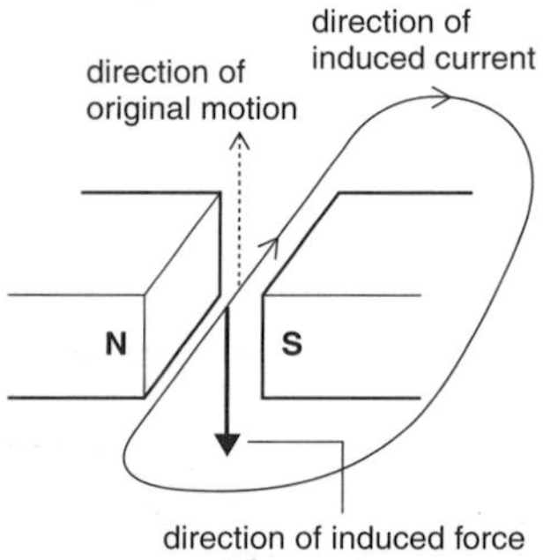

The motion of the wire produces a current which produces a force opposing the original motion.

- The direction of the induced current is determined by applying the principle of conservation of energy. For an e.m.f. to be produced and enable a current to flow, energy must have come from somewhere. Applying Fleming's left-hand rule (see unit 22) to the top figure on this page, we see that the direction of current flow would produce a force which is opposite to the direction of the motion of the conductor (see left). When the conductor is moved, work is done against this force and so energy is put into the system. This is summarized in **Lenz's law**.

  Lenz's law states that the direction of an induced e.m.f. is such that its effects oppose the change that produced it. This change could be an increase or decrease in flux density, a change in the position of the conductor or the field, or a combination of both. The direction of the induced e.m.f. is given by the **right-hand dynamo rule** (see right.)

- Both of these laws can be brought together in **Neumann's equation**, which is used to calculate induced e.m.f.s:

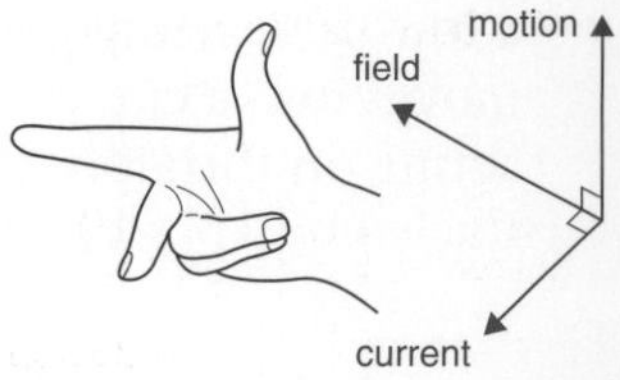

The right-hand dynamo rule.

$$\boldsymbol{E = -\frac{\Delta(N\phi)}{\Delta t}} \text{ or } \boldsymbol{E = -\frac{\Delta(NBA)}{\Delta t}}$$

  where ϕ is flux, N is the number of turns of a coil, A is area, and B is field strength. These equations are not as bad as they look: they just divide the total change in magnetic flux linkage by the time taken.

electromagnetic induction:
$\boldsymbol{E = -\frac{\Delta(NBA)}{\Delta t}}$

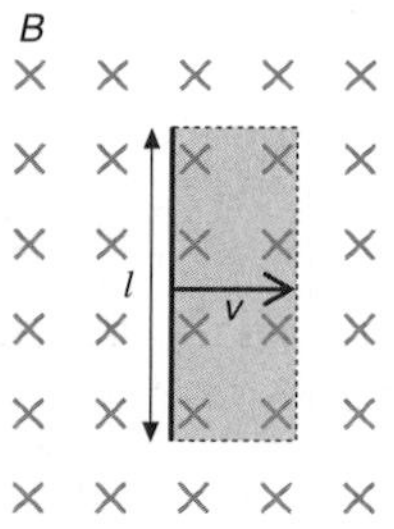

area swept out per second $A = lv$

A straight conductor moving through a magnetic field perpendicular to its direction of motion.

- A straight conductor moving through a magnetic field acting perpendicular to its direction of motion (see left) will have an e.m.f. E induced between its ends. As the conductor moves it sweeps out an area of $A = lv$ each second. Flux $= BA$ so it cuts through flux of $\phi = Blv$ each second:

$$E = -\frac{\Delta(\phi)}{\Delta t} = -\frac{Blv}{1} \text{ so } \boldsymbol{E = -Blv}$$

induced e.m.f. in straight conductor:
$\boldsymbol{E = -Blv}$

- If a conductor has a changing current flowing through it, as in an a.c. circuit or a d.c. one which has just been switched on, a changing magnetic field is produced around the conductor (see right). Any conductor lying in a changing magnetic field will have an e.m.f. induced in it, so the conductor induces an e.m.f. in itself. This is called **self-induction**. The size of the **induced e.m.f.**, or **back e.m.f.**, E_B, is directly proportional to the rate of change of current. This gives the equation

$$\boldsymbol{E_B = -L\frac{\Delta I}{\Delta t}}$$

where L is a constant, called the **self-inductance**, of the conductor and depends upon its size and shape. The unit is the henry (H). The negative sign is a consequence of Lenz's law.

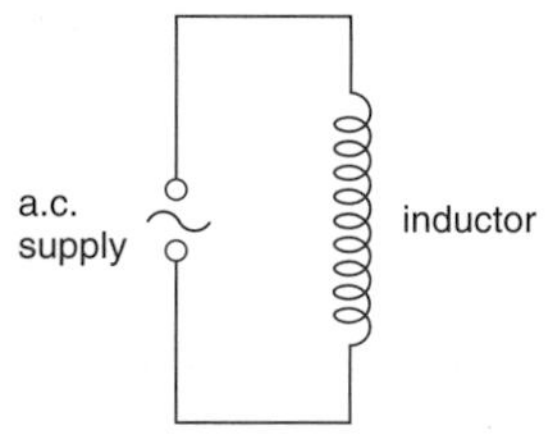

A circuit that will produce self-induction.

self-induction:

$$E_B = -L\frac{\Delta I}{\Delta t}$$

- A **transformer** uses induction to convert one a.c. voltage into another, either increasing or decreasing it. It consists of two coils wound around a soft iron core (see figure). An a.c. current passes through the primary coil P, which produces a changing magnetic field. This field is concentrated by the iron core and passes through the secondary coil S, inducing an e.m.f. in it. The e.m.f.s and numbers of turns of each coil are linked by the equation $\boldsymbol{E_P/E_S = N_P/N_S}$ where N_S and N_P are the numbers of turns on each coil and E_P and E_S are the respective e.m.f.s. If the transformer is 100% efficient then power in = power out, so $\boldsymbol{I_PE_P = I_SE_S}$ so $\boldsymbol{I_P/I_S = N_P/N_S}$.
- In reality the transformer is not 100% efficient, and it is necessary to identify the sources of energy loss and look at how they might be reduced.
  1. Not all of the magnetic flux produced by the primary coil passes through the secondary coil. This is called **flux loss** and is reduced by the soft iron core which concentrates the magnetic flux.
  2. Small, circular **eddy currents** are induced in the core; this dissipates power because of the resistance of the core. This is reduced by using a laminated core (sheets of conductor separated by sheets of insulator).
  3. The coil wires have a resistance which dissipates power (**copper losses**). This is reduced by using thick wires with low resistivity.
  4. Every time the magnetic field changes direction, the magnetic domains in the material have to change orientation, which dissipates some energy. This is called **hysteresis loss**.

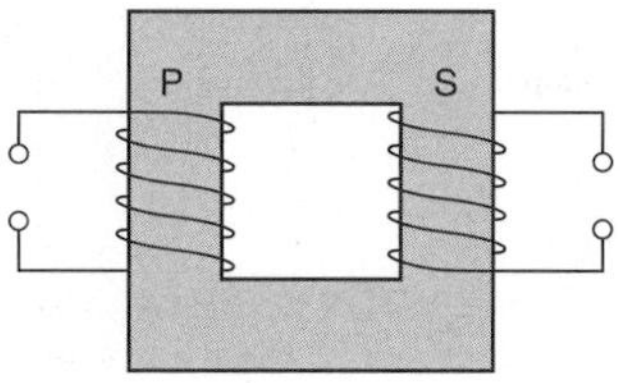

A transformer.

transformers: $\dfrac{E_P}{E_S} = \dfrac{N_P}{N_S}$

$$I_PE_P = I_SE_S$$

- When power is transmitted over long distances, transformers are used to increase and decrease voltages. The power lost due to the resistance of a wire is given by the equation $P = I^2R$, so if the current is reduced the power loss is reduced. When the voltage is increased by a transformer, the current is reduced (as $P = IV$), so we use transformers to increase the voltage to a high value when transmitting power over long distances, and then to reduce it to a low value nearer the consumer.

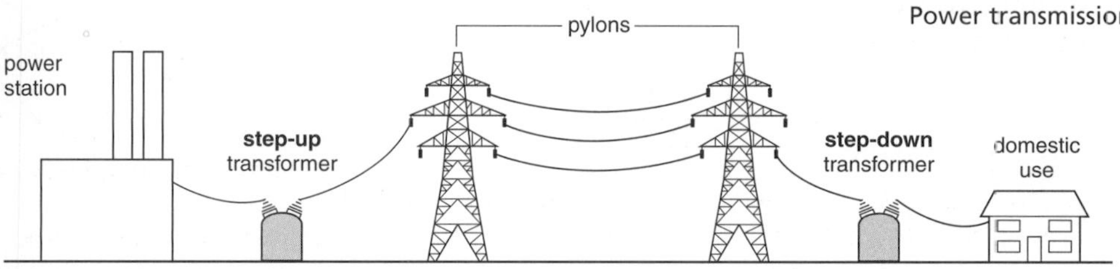

Power transmission.

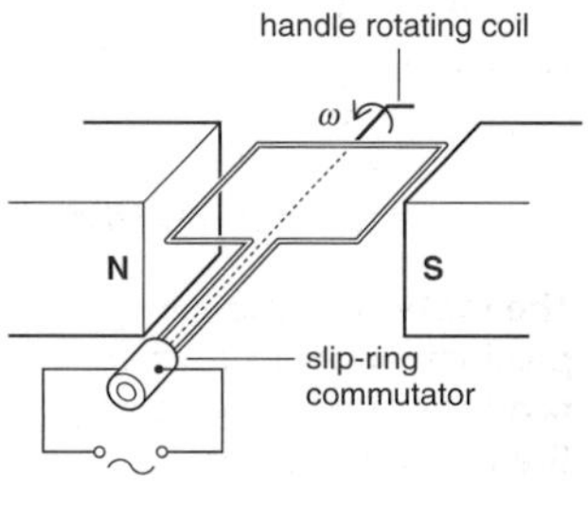

A simple alternator.

e.m.f in alternator:
$E = E_0 \sin(2\pi ft)$

- **Dynamos** and **alternators** use the principle of electromagnetic induction to convert mechanical energy into electrical energy. They are essentially identical to the motor (see unit 22), except that the coil is rotated by some force. The e.m.f. produced in an alternator is given by the equation $\boldsymbol{E = E_0 \sin(2\pi ft)}$, where the maximum e.m.f. $\boldsymbol{E_0 = BAN\omega}$. f is frequency, t is time, B is flux density, A is area of coil, N is number of turns of coil, and ω is angular velocity. E will vary in the form of a sine curve with an amplitude of E_0 (see right). This produces an alternating voltage or a.c. supply. The average value of an a.c. supply is zero, so we use **root mean square (r.m.s.)** values to represent it. An r.m.s. value is defined as the d.c. value that would produce the same power dissipation as the a.c. supply. $\boldsymbol{E_{r.m.s.} = E_0/\sqrt{2}}$ or $\boldsymbol{V_{r.m.s.} = V_0/\sqrt{2}}$ or $\boldsymbol{I_{r.m.s.} = I_0/\sqrt{2}}$. E_0, V_0, and I_0 are peak values.

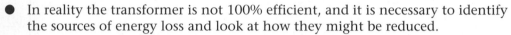

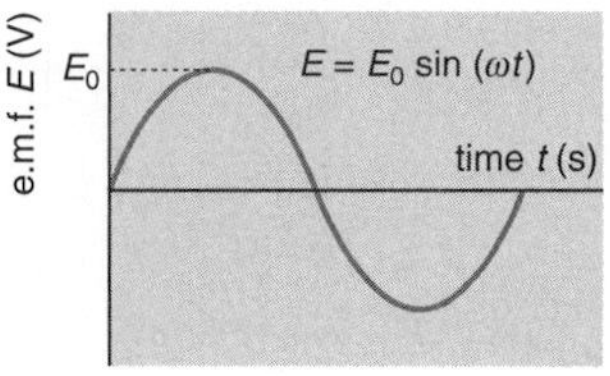

Unit 23

TESTS

RECALL TEST

1 What is 'electromagnetic induction'?

(2)

2 What is Faraday's law?

(2)

3 What is Lenz's law?

(2)

4 What is meant by 'flux linkage'?

(2)

5 What is the right-hand dynamo rule used for?

(2)

6 What is a transformer?

(2)

7 What is an eddy current?

(2)

8 Why do we transmit power over long distances at high voltages?

(2)

9 What is an alternator?

(2)

10 What is meant by 'self-induction'?

(2)

(Total 20 marks)

CONCEPT TEST

Take $\mu_0 = 4\pi \times 10^{-7}$ N/A^2 and electron charge = -1.6×10^{-19} C

1 What is the e.m.f. induced in a coil of 50 turns and cross-sectional area 0.080 m^2 lying in a field that changes from 2.0×10^{-3} T to 8.0×10^{-3} T in 0.50 s? If its resistance is 20 Ω what current will flow in the coil?

(4)

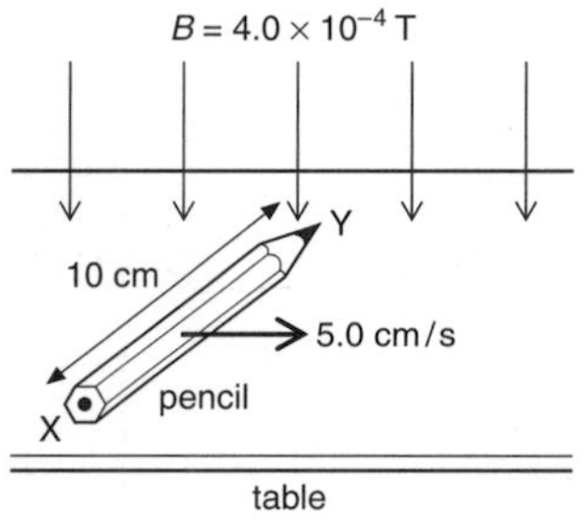

2 A pencil, 10 cm long, is rolling across a table as shown left. The central 'lead' is conducting. If there is a vertical magnetic field acting downwards what is the e.m.f. induced across the ends of the pencil? Which end is positive, X or Y?

(4)

3 A 2.0 V supply is connected to a 10 mH inductor which has a changing current passing through it. If the current changes from 0 to 3.0 A in 0.020 s, what is the back e.m.f. induced in the inductor and what is the p.d. across it?

(4)

4 A transformer steps down a mains supply voltage of 240 V to 12 V in an appliance. There are 2000 turns on the primary coil; how many are there on the secondary coil? If the input cable has a 0.050 A fuse in it before the transformer, which is 88% efficient, what current flows into the appliance?

(6)

5 An overhead transmission cable 100 km long has a resistance of 50 Ω per km. If the input power is 2.0 MW and the transmission voltage is 400 000 V, what will be the output power? Why do we transmit power at high voltage?

(6)

6 In the circuit shown right, the switch is closed and the current builds as shown in the graph. What is the initial back e.m.f. produced by the inductor? What is the gradient of the graph at this point if the inductor has an inductance of 2.5 mH? What is the final steady current? What is the p.d. across the inductor when the rate of change of current is 3000 A/s?

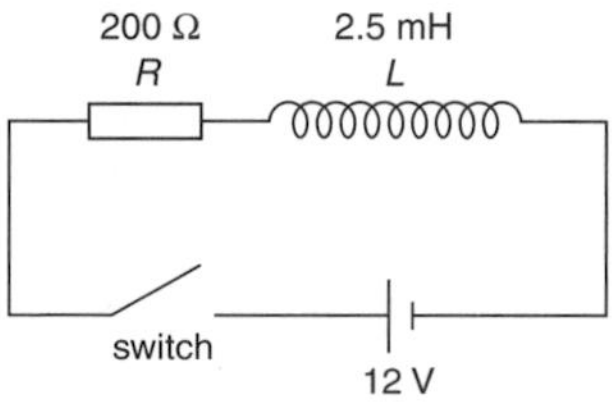

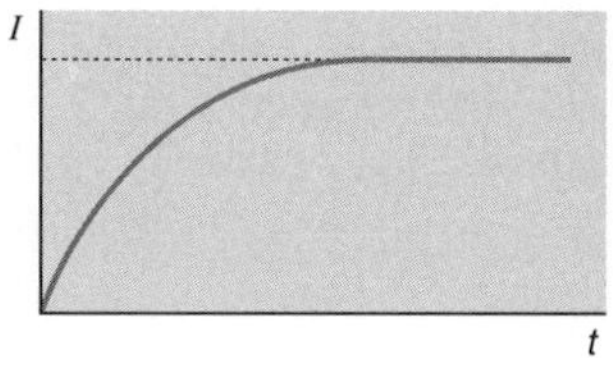

(6)

7 When an electric motor is switched on it accelerates up to a steady speed. Explain in terms of electromagnetic induction why it reaches a steady speed. If the motor becomes jammed in some way and is unable to move, it becomes hot and can catch fire. Explain why this happens.

(6)

8 The figure right shows a circuit containing two circular loops of conductor, A and B. State the direction of the current in loop A in each of the following situations.

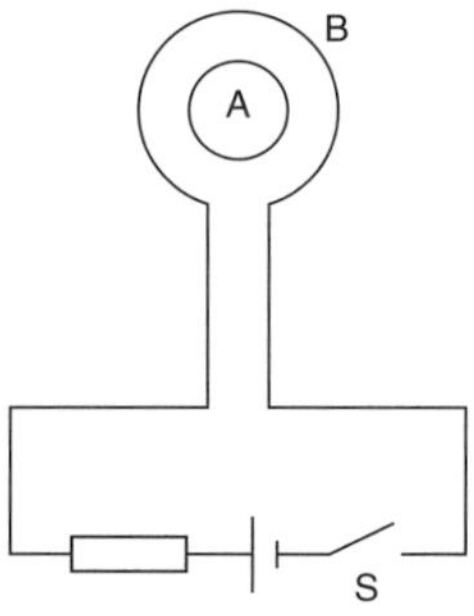

a The switch S is closed.

(2)

b The switch S is opened.

(2)

c Loop A is raised vertically above loop B with S closed.

(2)

9 An aircraft with a wingspan of 10 m is flying horizontally at a velocity of 200 m/s (see right). The Earth's magnetic field is 2.0×10^{-4} T at an angle of 60° to the horizontal at this point in the direction shown. What is the e.m.f. induced between the wing tips of the aircraft? Could this e.m.f. be used to power the cabin lights? The aircraft then does a 'loop-the-loop'. Sketch a graph on the axes right showing the variation of e.m.f. with the angle the aircraft makes with the horizontal.

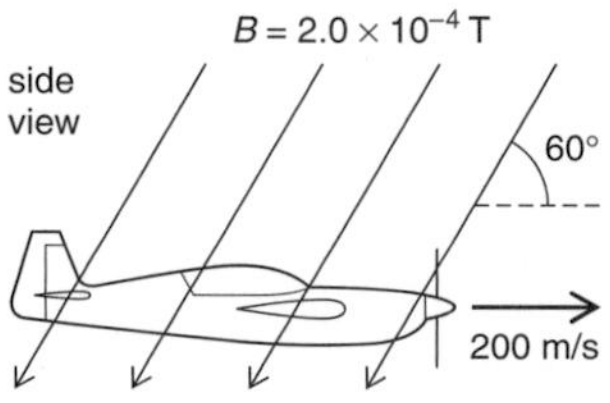

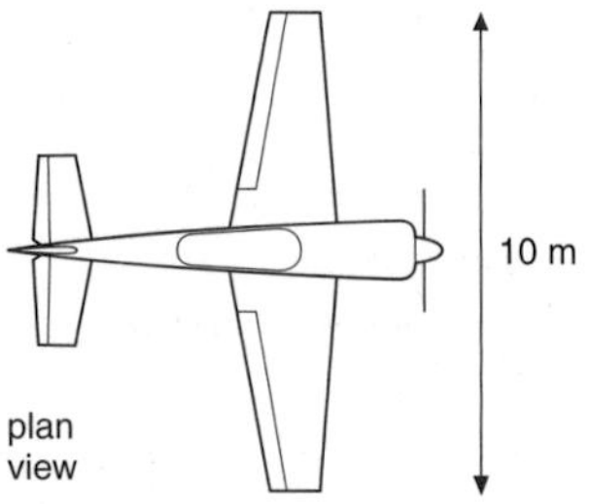

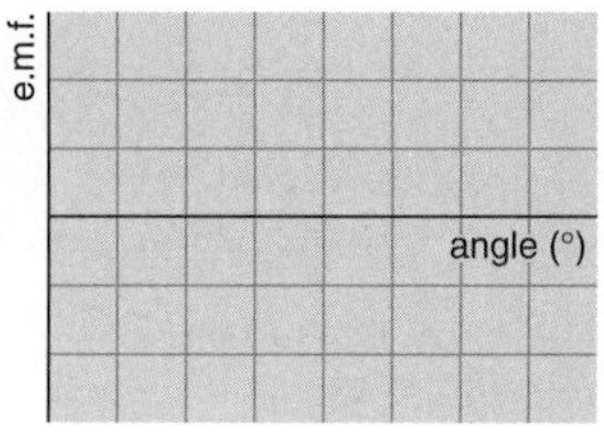

(8)

(Total 50 marks)

Unit 24

SIMPLE HARMONIC MOTION (SHM)

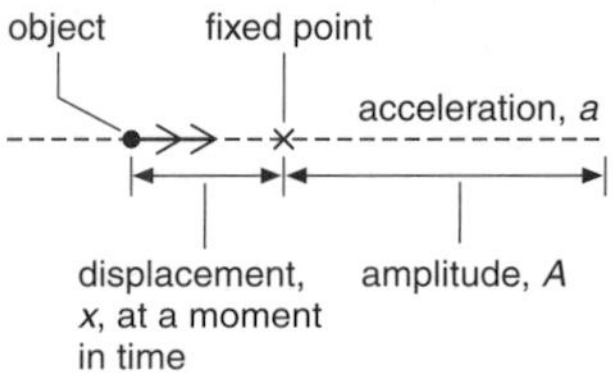

An object performing simple harmonic motion (SHM).

SHM acceleration:
$a = -\omega^2 x$

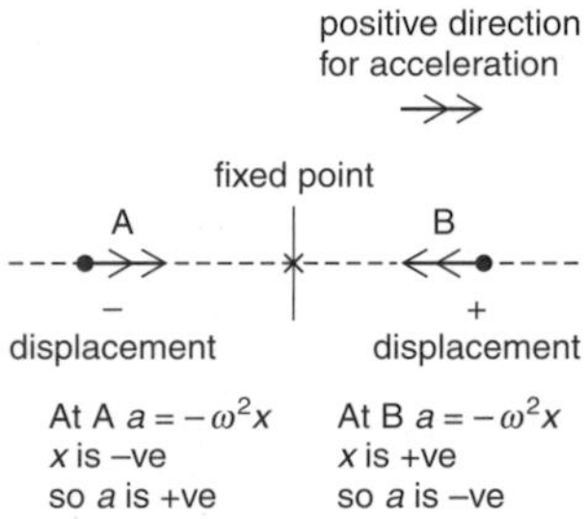

How the minus sign works in the SHM equation.

- When something vibrates from side to side or up and down with a repeated cycle of motion it is said to be **oscillating**. If the time taken for each oscillation is constant it is called the **time period**, T, and the oscillator can then be used for the measurement of time. A good example of this is a pendulum clock. For an oscillating object to do this, it has to move in a particular way. Its acceleration must always act towards the centre of the oscillation, and be directly proportional to its displacement from the centre of the oscillation. If an oscillator does this then it is said to be performing **simple harmonic motion** (**SHM**) (see left).

  Simple harmonic motion (**SHM**) takes place if an object is oscillating in such a way that its acceleration is directly proportional to its displacement from a fixed point and is always directed towards that fixed point.

  This can be represented by the equation

  $$a = -\omega^2 x$$

  where a is acceleration, x is displacement, and ω is a constant called the **angular frequency**. ω is similar to angular velocity (see unit 8a) and is used here because of the cyclic nature of an oscillation, which means that its back-and-forth motion can be modelled by an object going around a circle. The negative sign means that the acceleration will always be towards the centre (see left). From unit 8a: $\omega = 2\pi/T$.

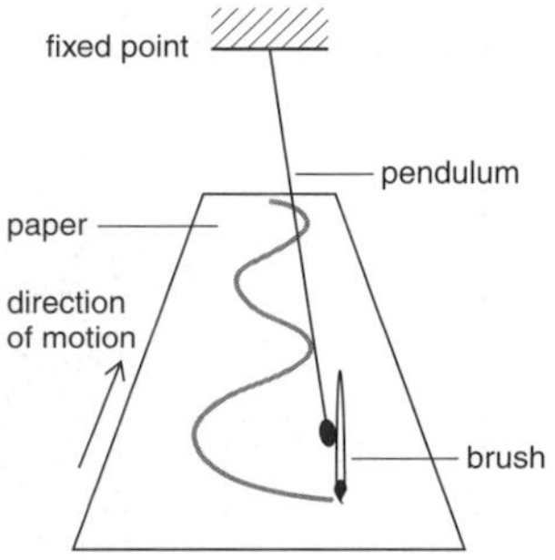

A simple pendulum producing a graph of displacement against time.

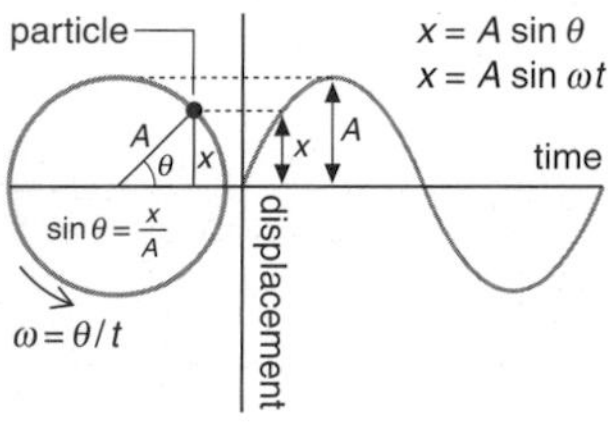

- To investigate the motion of an SHM oscillator in more detail, take a simple pendulum, with a paintbrush attached. Draw a roll of paper along underneath the pendulum at a constant speed, so that the brush just touches the paper throughout its swing. The pendulum is set swinging and the paper is set moving, producing the trace shown left. This is a sine curve if we take the motion as starting at the centre, and a cosine curve if we take the motion as starting at one extreme of the oscillation. We will look at the sine curve in more detail. It is possible to map this sine curve onto a circle (which is why the oscillations are called 'cyclic') and obtain an equation to represent it (see left). The gradient of the displacement–time graph gives the variation of velocity with time. The gradient of this curve produces the acceleration–time graph. The corresponding equations describe the graphs, and can be determined by calculus.

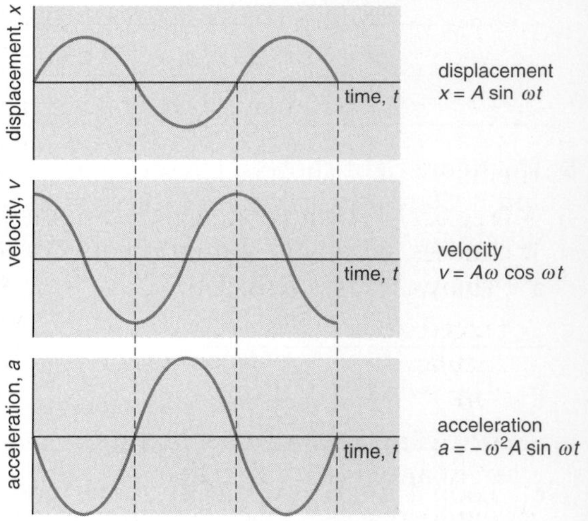

Graphs of motion against time.

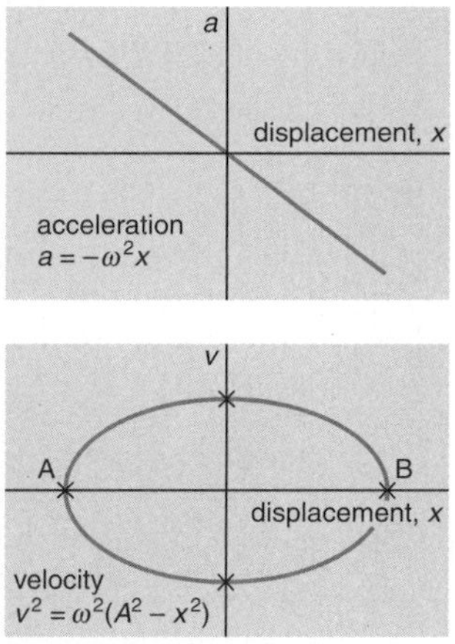

Graphs of motion against displacement.

- It is also useful to be able to show how the oscillator's motion varies with displacement from the centre. From the original SHM equation $a = -\omega^2 x$, and the equation of a straight-line graph $y = mx + c$, $c = 0$, and the gradient is negative. Therefore a graph of acceleration versus displacement is a straight line passing through the origin with a negative gradient (see left).

  The velocity graph is obtained by looking at the motion during one complete cycle. At A the velocity is zero. The oscillator accelerates as it moves towards the centre, and has a maximum velocity towards the right at the centre (take motion towards the right as positive). It then decelerates until it has zero velocity at B. It then accelerates to the left as it moves back again. This gives maximum negative velocity at the centre, and then zero velocity again at A (see left). This produces an ellipse or circle. The equation for such a curve is $\boldsymbol{v^2 = \omega^2 (A^2 - x^2)}$ where A is the amplitude, x is displacement, ω is angular frequency, and v is velocity. This equation is not used very often in A-level physics any more, but the equation for maximum velocity is. Maximum velocity will occur when x is zero, so we get $\boldsymbol{v = \omega A}$.

maximum SHM velocity:
$\boldsymbol{v = \omega A}$

- Like most other areas in physics, SHM must also be considered in terms of energy. If a system is not dissipating any energy because of frictional effects, its total energy is constant, so there is a continuous interchange between kinetic energy (KE) and potential energy (PE) (see right). Clearly, as the oscillator is moving fastest in the middle, KE is greatest there. So from the equations above we get

$$\boldsymbol{KE = \tfrac{1}{2}mv^2 = \tfrac{1}{2}m\omega^2(A^2 - x^2)}$$

and maximum KE is given by

$$\boldsymbol{KE_{max} = \tfrac{1}{2}m\omega^2A^2}, \text{ where } x = 0.$$

The total energy (TE) equals the KE at the centre so

$$\boldsymbol{TE = \tfrac{1}{2}m\omega^2A^2} \text{ all the time.}$$

$TE = KE + PE$ so from simple algebra

$$\boldsymbol{PE = \tfrac{1}{2}m\omega^2x^2}$$

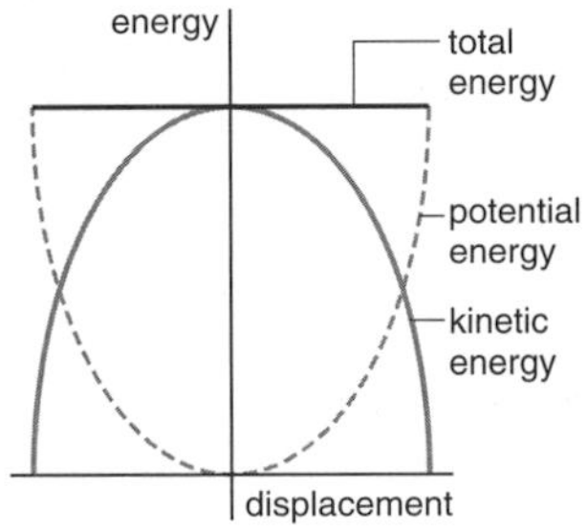

Variation of forms of energy with displacement.

- There are lots of different objects that perform SHM, but there are two main examples: the **simple pendulum**, and the **spring oscillator** system. Both of these examples have been, and still are, used to measure time. The time period, T, of oscillation of each system is given by the equations

$$\boldsymbol{T = 2\pi\sqrt{\frac{l}{g}}} \text{ (pendulum) and } \boldsymbol{T = 2\pi\sqrt{\frac{m}{k}}} \text{ (spring oscillator)}$$

where l is length of pendulum, g is the acceleration due to gravity, m is mass of object, and k is the spring constant (see unit 27a).

For any SHM system to operate it must have some mechanism of storing potential energy, such as elasticity, and a mass to enable it to have kinetic energy (this is because any oscillation involves a continuous interchange between potential and kinetic energy, as we saw before).

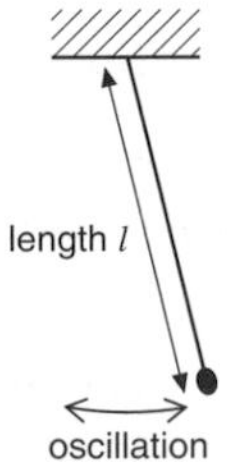

A simple pendulum.

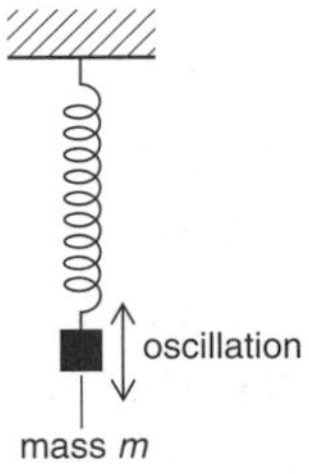

A spring oscillator.

- There are many different types of oscillation, some of which are SHM and some of which are not. There are several general terms that apply to all of them:

A **free oscillation** is one where an object is free to oscillate at any frequency it chooses, usually its natural frequency. It is not constrained or forced to move by any external agent. Example: simple pendulum.

A **forced oscillation** occurs when an object is forced to oscillate at a particular frequency by an external oscillator giving it energy. Example: car piston.

Natural frequency, f_0, is the frequency at which an object oscillates with greatest amplitude (see right).

Resonance occurs when the driving frequency of an external vibrator corresponds to the natural frequency of an object. Example: opera singer shattering a glass.

Damping is the reduction in amplitude of an oscillation due to frictional forces which dissipate energy as heat. The amplitude decreases exponentially (see right). Heavy damping reduces the natural frequency slightly (see above right). Example: shock absorbers in car suspension systems.

Critical damping occurs when a displaced oscillator returns to its equilibrium position, but does not move beyond (i.e. just fails to oscillate) (see right).

Overdamping occurs when the damping is so great that a displaced oscillator does not return to its equilibrium position.

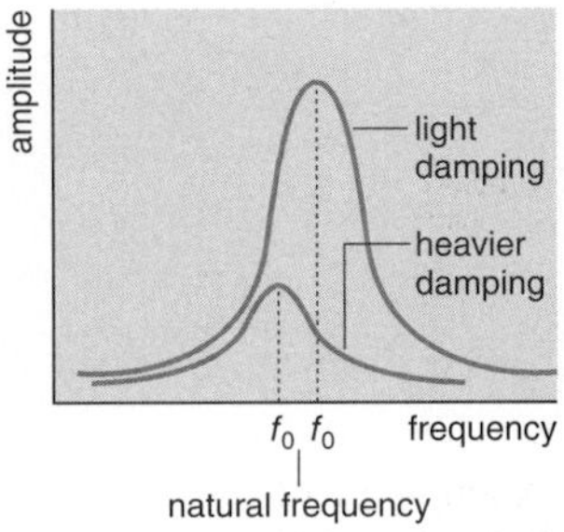

Variation of amplitude with frequency. The different curves are for different amounts of damping.

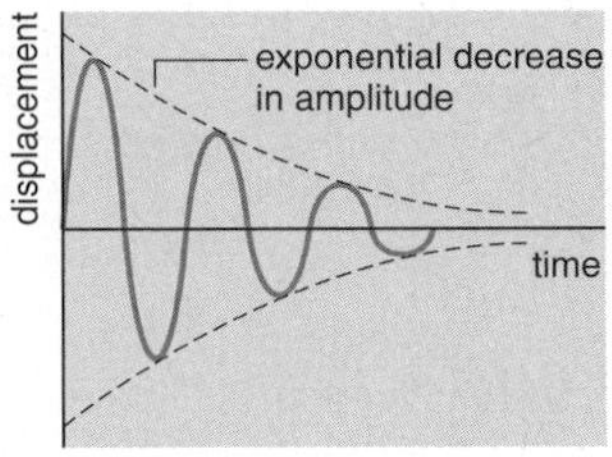

The reduction in amplitude of an oscillation due to damping.

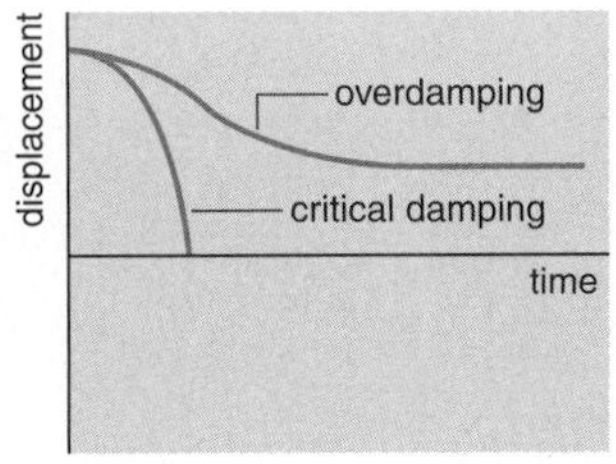

Critical damping and overdamping.

TESTS

RECALL TEST

1 Define 'SHM'.

(2)

2 Where is the acceleration greatest in an SHM oscillation?

(2)

3 Where is the velocity greatest in an SHM oscillation?

(2)

4 What is meant by the 'natural frequency' of an oscillator?

(2)

5 When does resonance occur?

(2)

6 Where is the potential energy greatest in an SHM oscillation?

(2)

7 Where is the kinetic energy greatest in an SHM oscillation?

(2)

8 What two properties are required for a system to perform SHM?

(2)

9 What is the most useful application of SHM oscillators?

(2)

10 What is meant by 'critical damping'?

(2)

(Total 20 marks)

CONCEPT TEST

Take $g = 9.8\,\text{m/s}^2$

1 An object is oscillating from side to side, performing SHM. The time period is 2.0 s, and the amplitude is 3.0 cm. What are the maximum velocity and acceleration, and the acceleration 1.0 cm from the centre?

(6)

2 A pendulum of length 2.0 m is swinging between two points 40 cm apart. What is the time period of oscillation and the maximum KE of the pendulum if the mass of the pendulum bob is 200 g?

(6)

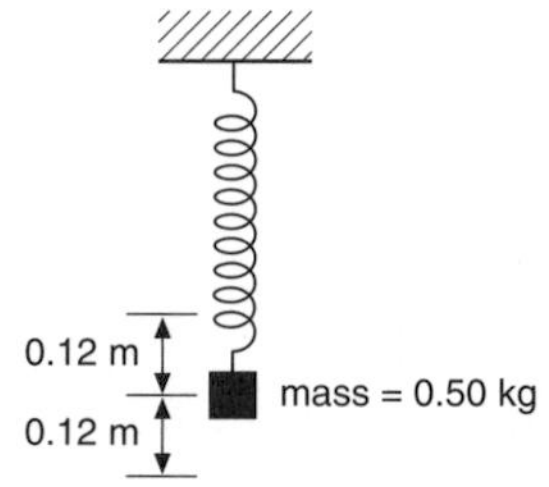

3 An oscillating spring system as shown left is performing SHM. If the spring constant, k, of the system is 12 N/m and its mass is 0.50 kg, what is the frequency of the oscillation and the maximum acceleration?

(4)

4 A laboratory dynamics trolley of mass 2.0 kg is attached to two springs with a combined spring constant of 25 N/m, as shown right. The trolley is displaced 5.0 cm to the left. Calculate the maximum velocity of the trolley and the PE stored when the trolley is 2.0 cm from the centre.

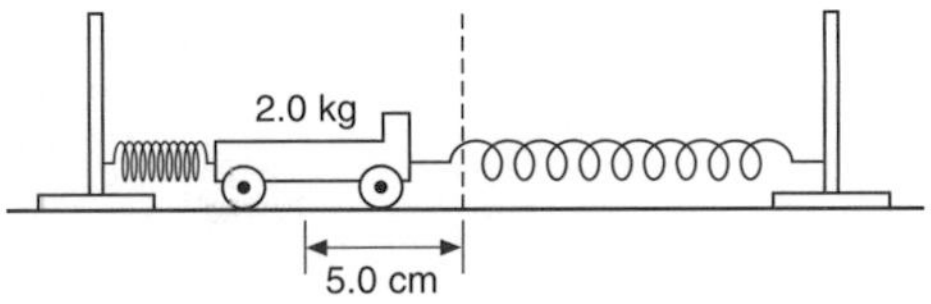

(6)

5 The propeller shaft of a boat causes the engine of mass 2.0 kg to vibrate with SHM. When the vibration equals the natural frequency of vibration of the engine, resonance occurs and the engine shakes itself apart. Assuming the engine behaves like a spring oscillator system with a spring constant of 4.0×10^5 N/m and the propeller rotates at the frequency of vibration of the engine, what is the maximum number of revolutions per second the propeller can turn at before the engine breaks up? What is the KE of the vibrating engine at resonance if the maximum amplitude of oscillation is 5.0 cm?

(6)

6 The simple pendulum shown right is performing SHM. What is the acceleration at B? What is the velocity at B? How long does it take to move from point A to point C?

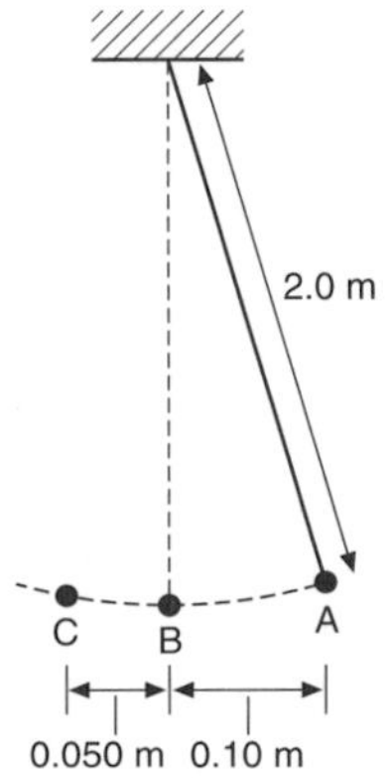

(6)

7 A mass of 0.15 kg is placed on the end of a spring which extends 0.60 m (x). The mass is then pulled downwards 0.60 cm and released. Using the equation $F = kx$, determine the spring constant of the spring, k. Also calculate the time period of the resulting oscillations and their maximum velocity. What is the upward tension in the spring when it is 0.20 cm below the equilibrium position? (For the spring constant see unit 27a.)

(8)

8 At an airport, a moving belt carries the luggage away from the check-in and goes over several equally spaced rollers which cause the luggage to oscillate up and down with an amplitude of 0.040 m. What is the maximum frequency of oscillation at which the luggage remains in contact with the belt? If the rollers are 20 cm apart, what is the maximum speed the belt can move at before this happens?

(8)

(Total 50 marks)

Unit 25

THERMAL PHYSICS

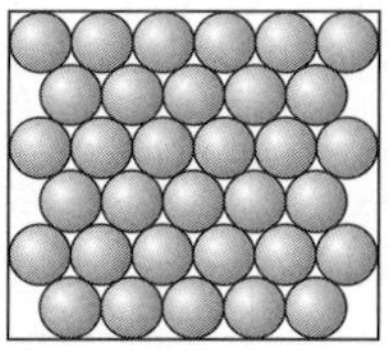

A simple model of a solid.

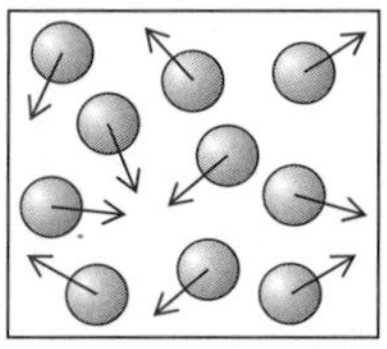

A simple model of a liquid.

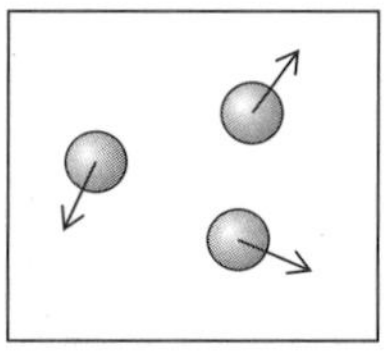

A simple model of a gas.

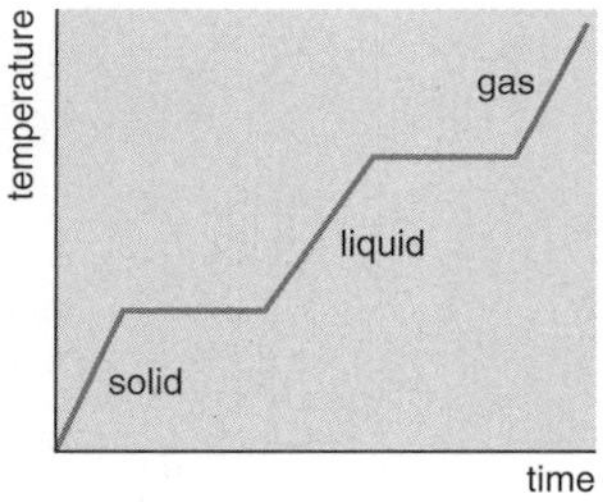

Variation of temperature with time for a material heated at a constant rate.

specific heat capacity:
$Q = mc\Delta\theta$

$P = \dfrac{mc\Delta\theta}{\Delta t}$

latent heat:
$Q = ml$
$P = ml/\Delta t$

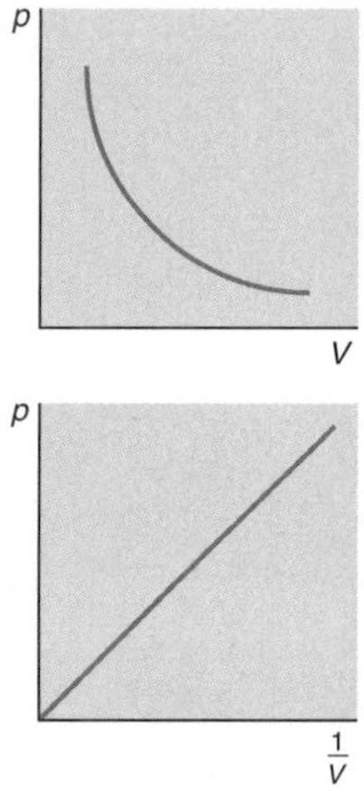

Boyle's law.

- A mug of hot coffee could be described in thermal terms as having a high **temperature** and containing **heat** energy.

  Temperature is a measure of how hot or cold something is. It is represented by a number on one of several possible scales. It is a fundamental quantity, like length or time (see unit 27b), so the choice of unit in which it is measured is entirely arbitrary.

  Heat is the energy that will flow from one object to another when they are at different temperatures. It is easiest to just think of heat as energy.

- If a solid object is heated its particles (e.g. atoms or molecules) will move around more and vibrate more rapidly about their normal fixed positions. If the heating continues the particles will eventually gain enough energy to partly break free of each other and increase their mutual separation, and the solid becomes a liquid. The particles still exert forces on each other, but no longer have fixed positions. If the heating continues the particles will then gain enough energy to totally break free of each other and the liquid becomes a gas. In a gas the particles only exert significant forces on each other during collisions. The particles move around randomly. The energy form associated with motion is kinetic energy (KE), so when an object's temperature increases, so does its particles' average KE.

- If an object is heated at a constant rate its temperature will change in the manner shown below left. When it is being heated normally the heat energy going into the object, Q, increases its temperature at a uniform rate. This is represented by the equation

  $$Q = mc\Delta\theta$$

  where m is mass, $\Delta\theta$ is temperature change, and c is the **specific heat capacity** of the material.

  The **specific heat capacity** of a material is the heat energy required to raise the temperature of 1 kg of the material by 1 °C.

  Also $P = Q/\Delta t$ so $P = mc\Delta\theta/\Delta t$ where P is power and Δt is change in time.

- When the material changes state the temperature remains constant. Energy is still going into it, but is being used to increase the separation of molecules, not the temperature. This increases the PE of the atoms with respect to each other and is represented by the equation

  $$Q = ml$$

  where l is the **specific latent heat**, of which there are two types: **vaporization** (liquid to gas), and **fusion** (solid to liquid). Also $P = Q/\Delta t$ so $P = ml/\Delta t$; both c and l are different for different materials.

  Specific latent heat is the energy required to change the state of 1 kg of material.

- The behaviour of a gas is governed by pressure, temperature, and volume. There are a series of laws linking them. The first is **Boyle's law**.

  Boyle's law states that the pressure, p, of a fixed mass of gas is inversely proportional to the volume, V, of the gas if temperature is constant.

  $$p \propto \frac{1}{V} \text{ or } pV = \text{constant or } p_1V_1 = p_2V_2$$

  The next is **Charles's law,** which states that the volume, V, of a fixed mass of gas is directly proportional to the temperature, T, if the pressure is constant.

  $$V \propto T \text{ or } \frac{V}{T} = \text{constant or } V_1/T_1 = V_2/T_2$$

  The last is the **pressure law**, which states that the pressure, p, of a fixed mass of gas is directly proportional to the temperature if the volume, V, is constant.

  $$p \propto T \text{ or } \frac{p}{T} = \text{constant or } \frac{p_1}{T_1} = \frac{p_2}{T_2}$$

These laws can be combined to produce the **ideal gas equation**, which is

$$\frac{pV}{T} = \text{constant or } \frac{p_1V_1}{T_1} = \frac{p_2V_2}{T_2} \text{ or } pV = nRT$$

where n is the number of moles (see below), and R is the **molar gas constant**. ($R = 8.3\,\text{J/(K mol)}$.) It is called the 'ideal' gas equation because it is an approximation which only works for gases at low pressures and at temperatures well above those at which they liquefy.

One **mole** is the amount of substance that contains as many atoms as there are in 12 g of carbon-12 (6.0×10^{23}). When dealing with gases the number of atoms present is more important than the mass in kg.

The graphs for Charles's law and the pressure law both hit the temperature axis at –273.16 °C or 0 K on the **Kelvin scale – absolute zero.** The Kelvin scale unit size is the same as the Celsius scale, but the zero point is different. 0 °C = 273 K.

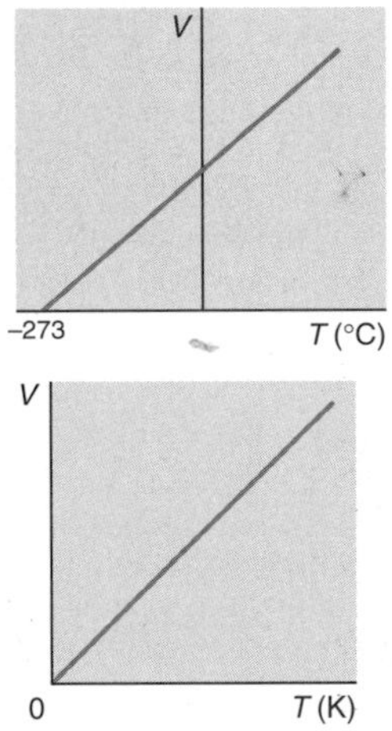

Charles's law.

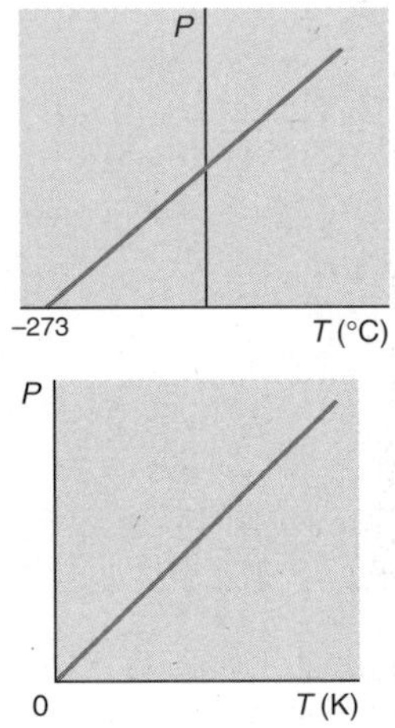

The pressure law.

ideal gas equation:

$$\frac{p_1V_1}{T_1} = \frac{p_2V_2}{T_2}$$

$$pV = nRT$$

- Gases consist of lots of small particles moving around randomly with a relatively large amount of KE. This model is called the **kinetic theory of gases**, which is based upon two pieces of evidence: **diffusion** and **Brownian motion**.

Diffusion is the way that two gases or liquids mix together; for example, in the way that the smoke from a lit match is soon smelt some distance away.

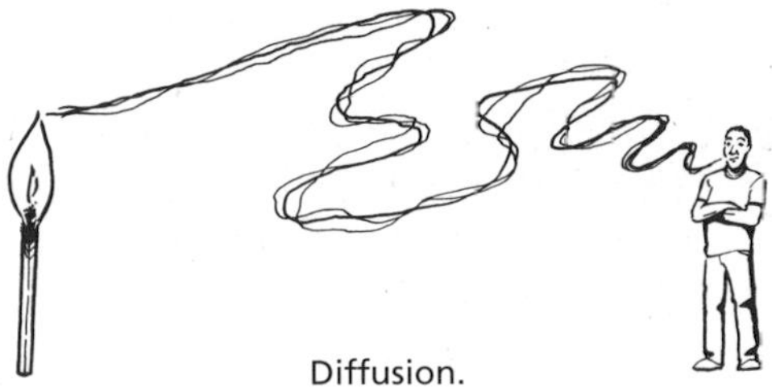

Diffusion.

Brownian motion is the erratic movement of gas or liquid particles as seen under a microscope. When this was first observed, in smoke particles, they were thought to be alive, but what is actually happening is that air molecules are hitting the particles and giving some of their KE to them.

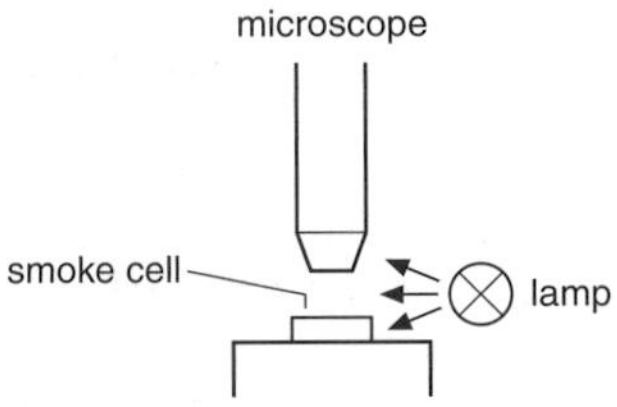

Apparatus for observing Brownian motion.

Brownian motion: the path taken by a smoke particle.

The kinetic theory of gases is based upon several key assumptions:

1. A gas consists of particles (called molecules) which move about randomly.
2. Newtonian mechanics applies to molecular collisions, and they are perfectly elastic.
3. Intermolecular forces are only significant during a collision and the duration of a collision is negligible compared with the time between collisions.
4. The volume of the gas molecules is negligible compared with the volume occupied by the gas, and gravitational effects can be ignored.

- From these assumptions it is possible to derive equations for the pressure of a gas of density ρ, occupying volume V, containing N particles of mass m moving with velocity c. $\langle c^2 \rangle$ means average (mean) squared speed.

$$p = \tfrac{1}{3}\rho\langle c^2\rangle \text{ or } pV = \tfrac{1}{3}Nm\langle c^2\rangle$$

or root mean square (r.m.s.) speed $\sqrt{\langle c^2\rangle} = \sqrt{\dfrac{3p}{\rho}}$

Combining $pV = nRT$ and $pV = \frac{1}{3}Nm\langle c^2\rangle$ for one mole of gas gives

$$\tfrac{1}{2}m\langle c^2\rangle = \tfrac{3}{2}kT$$

where k is the **Boltzmann constant**. ($k = R/N_A$) where N_A is the Avogadro constant. So the temperature of a gas is proportional to the KE of its particles (**$KE \propto T$**).

Kinetic theory equations:

$$p = \tfrac{1}{3}\rho\langle c^2\rangle$$

$$pV = \tfrac{1}{3}Nm\langle c^2\rangle$$

$$\sqrt{\langle c^2\rangle} = \sqrt{\frac{3p}{\rho}}$$

$$\tfrac{1}{2}m\langle c^2\rangle = \tfrac{3}{2}kT$$

Unit 25

TESTS

RECALL TEST

1 What is meant by 'temperature'?

(2)

2 What is meant by 'heat energy'?

(2)

3 What is 'specific heat capacity'?

(2)

4 What is meant by 'specific latent heat of vaporization'?

(2)

5 What is Boyle's law?

(2)

6 What is Charles's law?

(2)

7 What is the pressure law?

(2)

8 When is a gas no longer regarded as ideal?

(2)

9 What evidence is there for the kinetic theory of gases?

(2)

10 What is a mole?

(2)

(Total 20 marks)

CONCEPT TEST

Take $R = 8.3$ J/(K mol)

1 4.0kg of copper at 25 °C are heated and increase in temperature to 80 °C; how much heat energy is required to do this? (Specific capacity of copper = 380 J/(kg K).) Assume no heat is lost to the surroundings.

(2)

2 A 12 V electric heater draws 2.0 A for 12 min to raise the temperature of a 3.0 kg block of metal. If the initial temperature is 20 °C, what would the final temperature be? Assume no heat is lost to the surroundings. (Specific heat capacity of the metal = 500 J/(kg K).)

(4)

3 A car of mass 1400 kg is travelling at 30 m/s. When it applies its brakes, each of mass 26 kg, what will be the increase in the temperature of the brakes? What assumption have you made? (Specific heat capacity of brake material = 600 J/(kg K).)

(4)

4 What is the temperature difference between the top and the bottom of the waterfall shown right? Would it be possible to detect this difference? (Take the specific heat capacity of water as 4200 J/(kg K).)

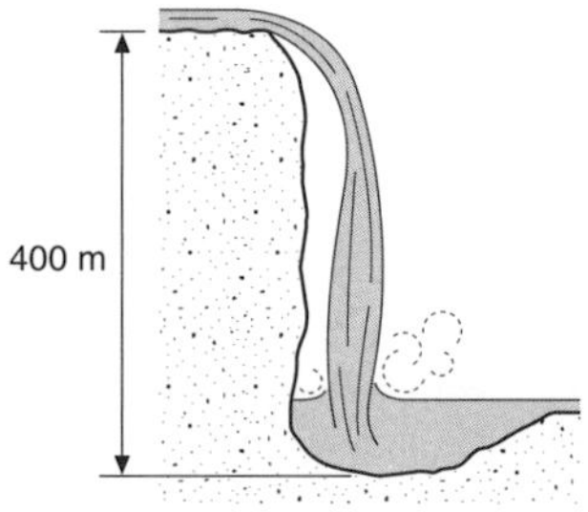

(4)

5 200 g of ice at 0 °C are dropped into a copper container of mass 100 g containing 300 g of water at 21 °C. What is the final temperature of the mixture? (Specific heat capacity of copper = 400 J/(kg K), specific heat capacity of water = 4200 J/(kg K), latent heat of fusion of ice = 3.3×10^3 J/kg.)

(6)

6 4.0 m^3 of helium is at a temperature of 25 °C and a pressure of 4.5 kPa. It is compressed to 1.0 m^3 and heated simultaneously to 56 °C. What is its new pressure?

(4)

7 10 m^3 of a gas is at a pressure of 2.4×10^4 Pa and temperature of 27 °C. How many moles of gas are present, how many atoms are present, and what mass of gas is present? (Molar mass of gas = 0.029 kg/mol.)

(6)

8 200 g of gas occupies 3.0 m^3 at a pressure of 2.0×10^3 Pa. What is the root mean square speed of the gas's molecules?

(4)

9 A runner has just stopped after a race. Perspiration is used by her body (mass 60 kg) to cool her down. If her temperature is dropping by 0.75 K per second, what is the rate of evaporation of water? (Take the specific heat capacity of the human body to be the same as water, 4200 J/(kg K), and the latent heat of vaporization to be 2.4×10^6 J/kg.)

(8)

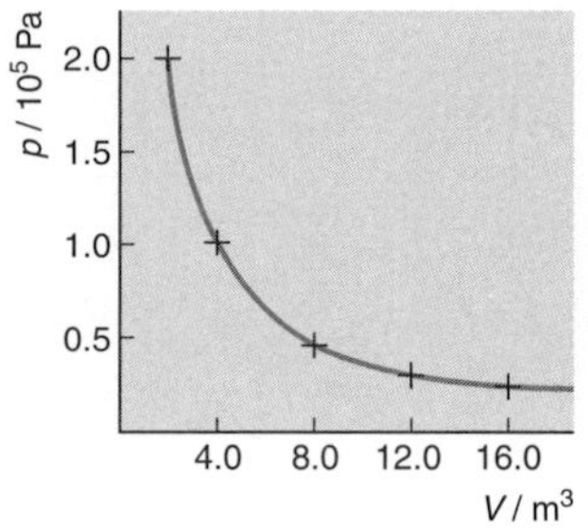

10 The graph above right shows a gas changing. Use the information in the graph to see if this change is taking place at a constant temperature (isothermal). The graph below right shows a cyclic process for 4.0 mol of gas. Determine the maximum temperature of the gas during the process.

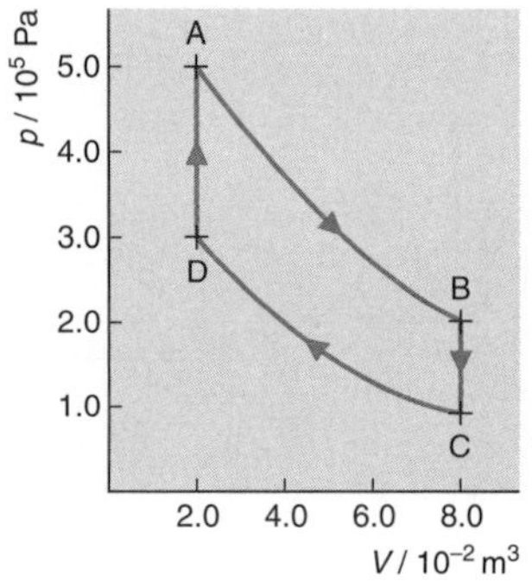

(8)

(Total 50 marks)

Unit 26 THERMODYNAMICS

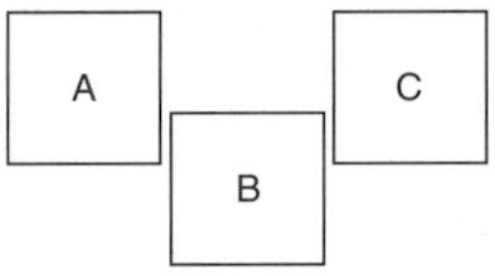

An illustration of the zeroth law of thermodynamics. A is in contact with B and therefore in thermal equilibrium with it. B is in contact with C and therefore in thermal equilibrium with it. So A must be in thermal equilibrium with C.

- The word 'thermodynamics' is made up of two parts: thermo (heat), and dynamics (movement). This is exactly what it deals with: the movement of heat energy. There are three laws of thermodynamics.

- The **zeroth law of thermodynamics** is the basis of **thermometry** (the theory and use of thermometers). It is called the zeroth law because the first and second laws of thermodynamics were formulated before it was realized that another more fundamental law needed to be stated before the others.

  The **zeroth law of thermodynamics** states that if object A is in thermal equilibrium with object B, and object B is in thermal equilibrium with object C, then object A is in thermal equilibrium with object C. Any objects in thermal equilibrium have no overall (**net**) transfer of heat energy between them and have the same temperature.

- There are many different types of thermometer, each with different characteristics. To make a thermometer you need two things: a **thermometric property** and '**fixed points**'.

  A **thermometric property** is something which changes consistently with temperature, e.g. volume, resistance, pressure, or e.m.f. **Fixed points** are single temperatures at which a particular physical event will always take place, e.g. the ice and steam points of water, or absolute zero and the triple point of water (the temperature at which water can exist with all of its three states in equilibrium).

  The values of the thermometric properties at the fixed points can be used to calculate the temperature.

temperature equations:

$\theta = 100 \times \frac{X_\theta - X_0}{X_{100} - X_0}$ (°C)

$\theta = 273.16 \times \frac{p_\theta}{p_{tr}}$ (K)

$$\theta = 100 \times \frac{X_\theta - X_0}{X_{100} - X_0} \text{ (°C) and } \theta = 273.16 \times \frac{p_\theta}{p_{tr}} \text{ (K)}$$

where X_0, X_{100}, and X_θ are the respective values of the thermometric property at 0 °C, 100 °C, and the temperature being determined. The Kelvin scale assumes that pressure is proportional to temperature: p_θ is the pressure of a fixed volume of gas at the temperature being measured, and p_{tr} is the pressure at the triple point of water (set at 273.16 K).

first law of thermodynamics:

$\Delta U = \Delta Q + \Delta W$

- The **first law of thermodynamics**, an application of energy conservation, states that the change in internal energy, ΔU, of a system is equal to the sum of the changes of heat energy (ΔQ or Q) and work done on or by the system (ΔW or W).

  The internal energy is the sum of the particles' kinetic energy (KE) and potential energy (PE). ΔU is +ve if the temperature increases (KE increases) or if there is a change of state from solid to liquid or liquid to gas (PE increases). ΔQ is +ve if the heat energy enters the system, and –ve if it leaves. ΔW is +ve if the system is compressed, and –ve if it expands. The first law is best explained through a series of different applications:

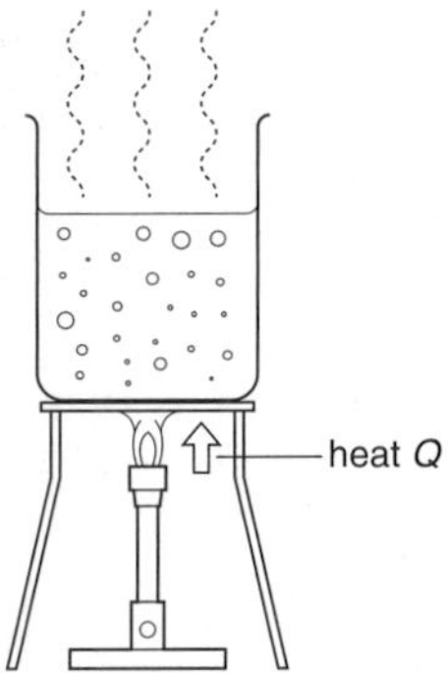

Boiling water.

- **1** Boiling water   Take the first law and look at each term in turn.

$\Delta U = \Delta Q + \Delta W$	ΔQ is +ve as heat is going into the system.
+ + 0	ΔW is –ve as water expands when it changes state.
	ΔU is +ve: the system is changing state so PE increases.

- **2** Heating a gas with a constant volume (**isovolumetric** heating)

$\Delta U = \Delta Q + \Delta W$	ΔQ is +ve as heat is going into the system.
+ + 0	ΔW is 0 because the volume is fixed.
	ΔU is +ve as the temperature is increasing.

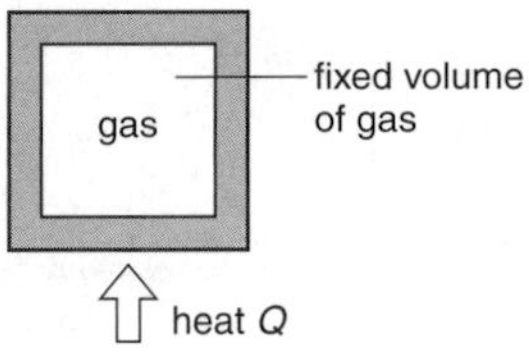

Isovolumetric heating.

  Therefore $\boldsymbol{\Delta U = \Delta Q}$. In the gas $\boldsymbol{\Delta Q = nC_V\Delta T}$, where n is the number of mol, C_V is the **molar heat capacity at constant volume**, and ΔT is temperature change.

  The **molar heat capacity at constant volume** is the heat energy required to raise the temperature of 1 mol of a substance by 1 K at constant volume.

- **3** Heating a gas at constant pressure (**isobaric** heating)

  $\Delta U = \Delta Q + \Delta W$ ΔQ is +ve as heat is going into the system.
  $+ \quad + \quad -$ ΔW is –ve as the volume is increasing.
  ΔU is +ve as the temperature is increasing.

  For this gas $\boldsymbol{\Delta Q = nC_p\Delta T}$, where n is the number of moles, C_p is the **molar heat capacity at constant pressure**, and ΔT is the temperature change. It is also possible to determine an equation for the work done, W. Work done = force × distance ($W = Fs$) and pressure = force/area; putting these together gives

  $W = pAs$ so $\boldsymbol{\Delta W = p\Delta V}$

  where ΔV is change in volume.

Combining all of the above together gives $nC_V\Delta T = nC_p\Delta T - p\Delta V$. Also $pV = nRT$, so if p is constant and temperature changes it becomes $p\Delta V = nR\Delta T$ which gives

$nC_V\Delta T = nC_p\Delta T - nR\Delta T$, and so $\boldsymbol{C_p - C_V = R}$

The molar heat capacity at constant pressure, C_p, is the heat energy required to raise the temperature of one mole of a substance by 1 K at constant pressure. C_p is greater than C_V, because when a gas is heated at constant pressure it expands and so work must be done on the surroundings.

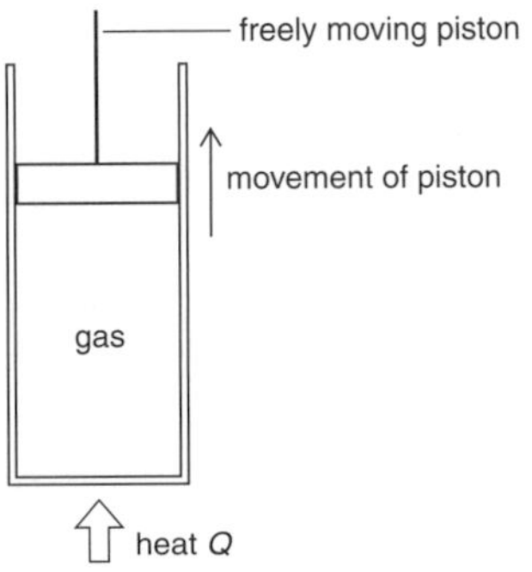

Isobaric heating. The piston moves up so the pressure inside stays the same as the pressure outside.

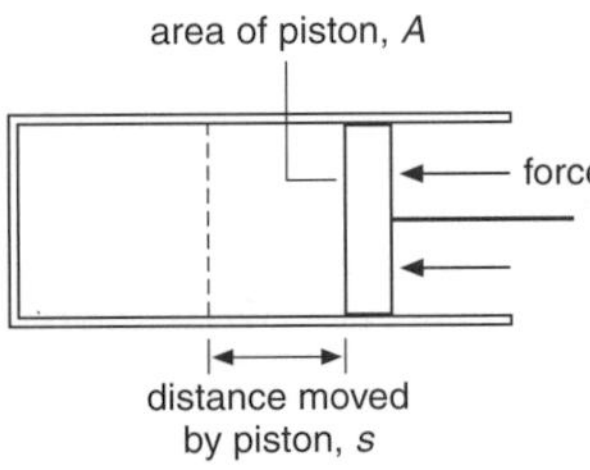

Work done by a piston on a gas.

- **4** Compression of a gas at constant temperature (**isothermal** compression)

  $\Delta U = \Delta Q + \Delta W$ ΔQ is –ve as heat is leaving the system.
  $0 \quad - \quad +$ ΔU is 0 as the temperature is constant.
  ΔW is +ve as the volume is decreasing.

  So $\boldsymbol{\Delta W = \Delta Q}$ for isothermal changes. Isothermal changes must take place slowly to allow the heat energy to escape. Boyle's also law applies (see unit 25).

work done in a gas:

$\boldsymbol{\Delta W = p\Delta V}$

Molar heat capacities:

$\boldsymbol{C_p - C_V = R}$

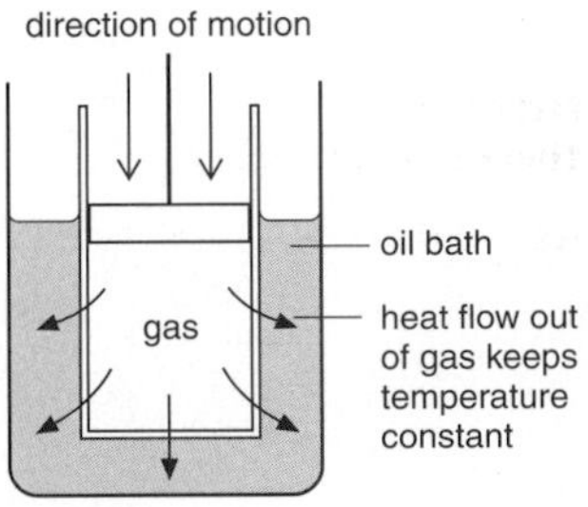

Isothermal compression.

- **5** **Adiabatic** compression of a gas (no heat energy enters or leaves the system)

  $\Delta U = \Delta Q + \Delta W$ ΔQ is 0 as no heat enters or leaves the system.
  $+ \quad 0 \quad +$ ΔW is +ve as the volume is decreasing.
  ΔU is +ve as the temperature is increasing.

  So $\boldsymbol{\Delta U = \Delta W}$ for an adiabatic change. Adiabatic changes must be done quickly so there is no time for the heat energy to escape. As the piston is pushed downwards, the gas molecules bounce off the piston with greater KE than they had before, so compressing the gas puts energy into the system.

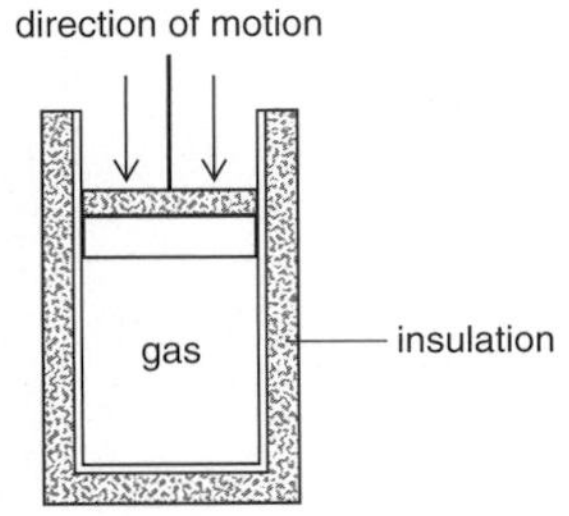

Adiabatic compression.

- Such changes are shown on pressure–volume graphs (p–V graphs). The graph bottom right shows a cycle of changes. A–B = adiabatic compression. B–C = isovolumetric decrease in pressure. C–D = isothermal expansion. D–A = isobaric expansion. Area enclosed by graph = net work done.

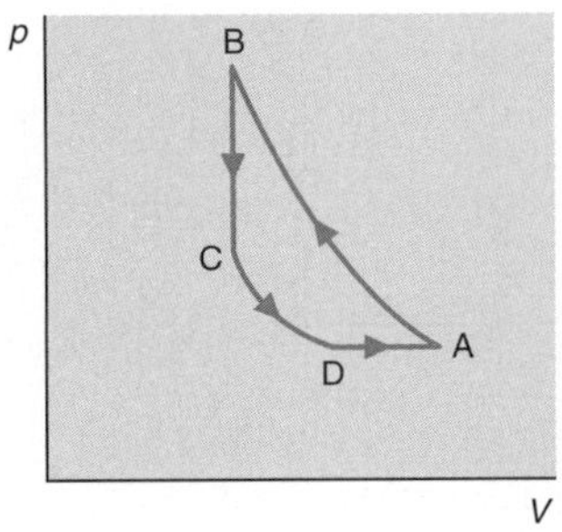

- The **second law of thermodynamics** states that a **heat engine** can never be 100% efficient. A **heat engine** is a device that converts heat energy into mechanical energy when heat moves from a hot place to a cold one. The efficiency of energy transfer is given by $\boldsymbol{100 \times (T_1 - T_2)/T_1}$ where T_1 is the absolute (thermodynamic) temperature (temperature in K) of the material entering the heat engine and T_2 is the thermodynamic temperature of the material leaving the heat engine (see right).

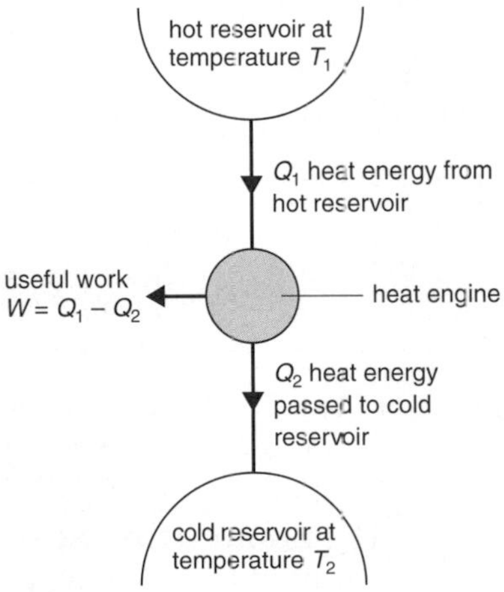

A heat engine.

TESTS

RECALL TEST

1 What is the zeroth law of thermodynamics?

(2)

2 What is the first law of thermodynamics?

(2)

3 What is the second law of thermodynamics?

(2)

4 What is meant by 'internal energy'?

(2)

5 What is meant by an 'isothermal change'?

(2)

6 What is meant by an 'adiabatic change'?

(2)

7 What is meant by an 'isobaric change'?

(2)

8 What is meant by 'molar heat capacity at constant volume'?

(2)

9 What is a heat engine?

(2)

10 Why is C_p greater than C_V?

(2)

(Total 20 marks)

CONCEPT TEST

Take $R = 8.3\,\text{J/(mol K)}$; for all substances take $C_V = 20.8\,\text{J/(mol K)}$ and $C_p = 29.1\,\text{J/(mol K)}$

1 A platinum resistance thermometer has a resistance of 20 Ω at 0 °C and 65 Ω at 100 °C. What is the temperature when it has a resistance of 42 Ω?

(4)

2 A gas thermometer has a pressure of 1.0×10^5 Pa at 0 °C and 2.4×10^6 Pa at 100 °C. What is the temperature when it has a pressure of 4.2×10^6 Pa?

(4)

3 2 mol of a gas is heated at a constant volume and its temperature increases from 239 K to 400 K. What is the increase in internal energy? Why is there no work done?

(4)

4 **a** What two processes could result in an increase in the internal energy of a gas? Explain how the first law of thermodynamics applies to **b** heating a fixed mass of gas at constant pressure and so increasing its temperature, and **c** heating the same mass of gas at constant volume.

(8)

5 4.5 mol of a gas is heated at a constant pressure of 1.6×10^4 Pa and the gas expands by $1.5\,m^3$. If the temperature increases by 330 K what is the increase in internal energy?

(6)

6 In the *p*–*V* diagram shown right, describe what is happening in each of the changes.

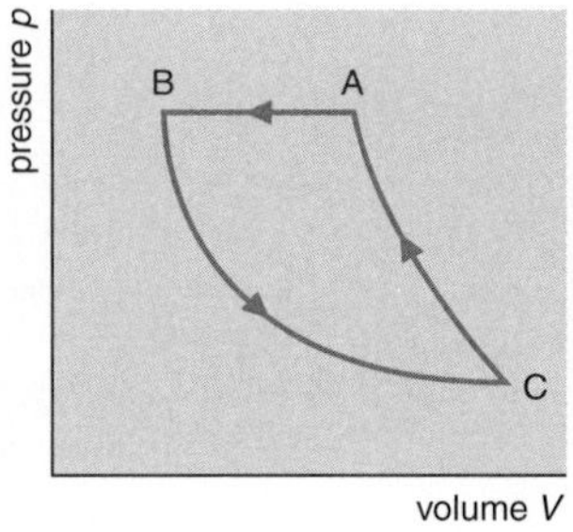

(6)

7 In an adiabatic compression of 3.5 mol of a gas at a constant pressure of 2.4×10^3 Pa, the temperature of the gas increases from 200 K to 567 K. What is the corresponding change in volume?

(8)

8 A fridge is placed in a small sealed room. Describe how the first law of thermodynamics applies to the fridge as a system and then to the room as a system, given that the electrical energy going into the pump can be considered as work being done on the fridge and the temperature of the fridge is constant.

(6)

9 A heat engine converts heat energy into mechanical energy. What is its efficiency if the temperature of the material going into it is 200 °C and the temperature of the material coming out is 50 °C?

(4)

(Total 50 marks)

Unit 27a MATERIALS SCIENCE

This area deals with the structure of different types of material and their properties when placed under forces.

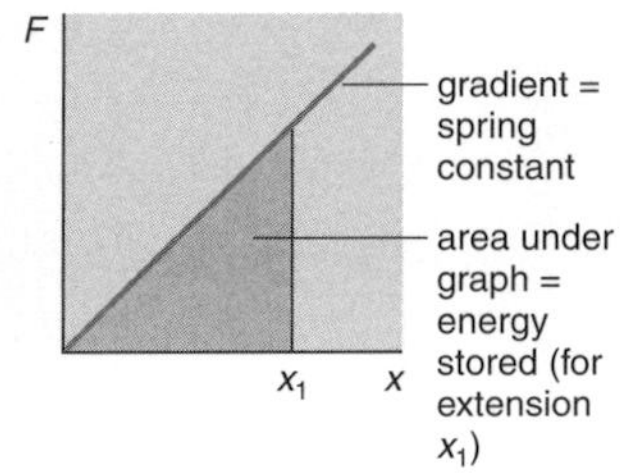

Hooke's law.

Hooke's law:
$F = kx$
$F = ke$
$F = k\Delta l$

- When a piece of material has a force acting on it, it stretches or is compressed. (In this unit we shall consider only **tensile** (stretching) forces.) If it stretches uniformly it is said to obey **Hooke's law** (see left).

  Hooke's law is obeyed when the extension, x, e, or Δl, produced by a force, F, is directly proportional to the applied force; $\mathbf{F = kx}$ where k is a constant called the **spring constant**.

  From the definition of work done ($W = Fs$) in unit 5 and the graphs in unit 2 the area under the graph equals the energy stored (see above left). This gives

  $$E = \tfrac{1}{2}Fx \text{ or by substituting in } F = kx,\ E = \tfrac{1}{2}kx^2$$

$$\text{stress} = \frac{\text{force}}{\text{area}}$$

$$\sigma = \frac{F}{A}$$

$$\text{strain} = \frac{\text{extension}}{\text{original length}}$$

$$\varepsilon = \frac{x}{l_0}$$

- When a force, F, is applied to a piece of material, the amount that it stretches, x, depends upon its original length, l_0, cross-sectional area, A, and the type of material. To be able to compare differently shaped objects we use stress, σ, strain, ε, and the Young modulus of elasticity, E.

  Stress is equal to force per unit area (units Pa or N/m^2).

  Strain is equal to the fractional increase in length. It has no units.

  These two quantities can be combined together to get a measure of the elasticity of the material. This is called the **Young modulus of elasticity**.

  The Young modulus is equal to the ratio of stress to strain (units Pa).

  The area under a stress–strain graph is equal to the energy stored per unit volume (see left).

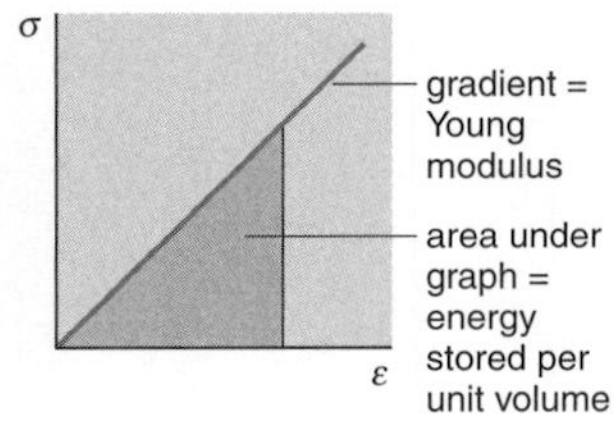

The Young modulus.

Young modulus:

$$E = \frac{\text{stress}}{\text{strain}} = \frac{\sigma}{\varepsilon}$$

  A material which will show a lot of plastic deformation is called **ductile**, and one which will show very little plastic deformation is called **brittle**. When a ductile material is placed under tension, a stress–strain graph for the material shows several stages which need to be explained in detail:

  A is the **limit of proportionality**, the point after which stress is no longer directly proportional to strain.

  B is the **elastic limit**: up to this point the object undergoes **elastic deformation** and will return to its original shape when the force is removed. Beyond this point **plastic deformation** takes place, i.e. the material will not return to its original shape when the force is removed, but be left with a **permanent deformation**.

  C is the **yield point**, beyond which the material shows a large increase in strain for a small increase in stress.

  D is the **breaking point** and gives the **ultimate tensile strength** (**UTS**) of the sample: the maximum stress that can be applied before it breaks.

  O to B is the **elastic region** and B to D is the **plastic region**.

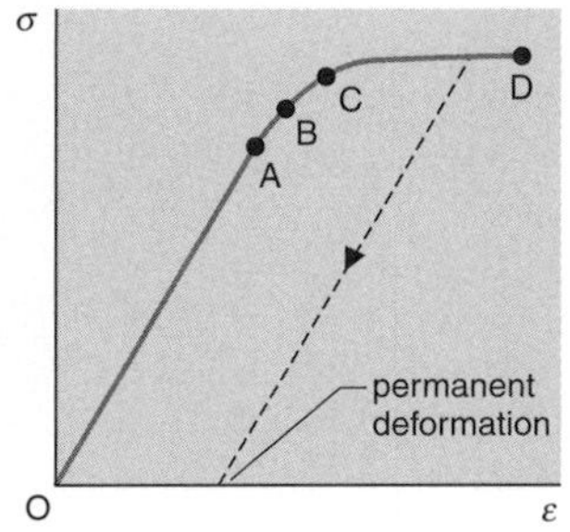

Stress–strain graph for an elastic material.

- It is often necessary to draw stress–strain graphs for a type of material under tension or compression. Some common examples are shown below.

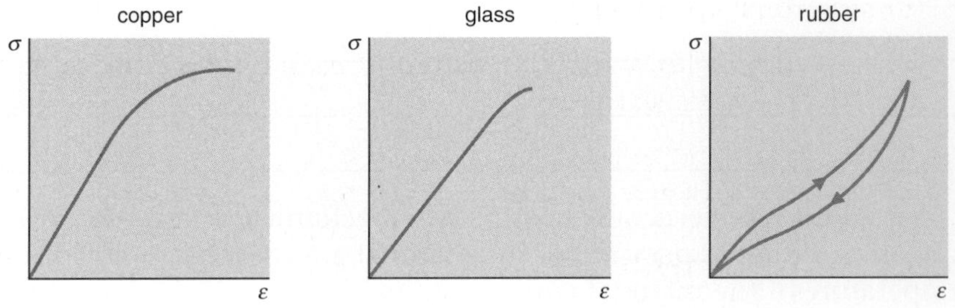

Common stress–strain graphs.

BASE UNITS

Unit 27b

There is a basic universal set of units called the SI ('Système International') system. This is agreed between all countries and used throughout science, including in this book.

- Some of these units are the fundamental units from which all other units can be derived, and are called the **base units**. Each of these base units has an equivalent **dimension**, but it is usually base units that are used.

Quantity	Base unit	Dimension
time	second (s)	[T]
mass	kilogram (kg)	[M]
length	metre (m)	[L]
current	ampere (A)	[I]
temperature	kelvin (K)	[θ]
amount of substance	mole (mol)	[N]

The base units.

There are several things that you could be asked to do with base units.

- **Expressing quantities in their base units.** Take an equation that defines the required quantity in terms of fundamental base units, and combine the units together. It is also possible to use equations that contain quantities with known base units to determine the base units of another quantity.

Example *Determine the base units of each of these quantities: velocity, acceleration, force, charge, energy.*

$\text{velocity} = \dfrac{\text{displacement}}{\text{time}} \rightarrow \mathbf{m/s}$ $\text{acceleration} = \dfrac{\text{change in velocity}}{\text{time}} \rightarrow \mathbf{m/s^2}$

force = mass × acceleration → $\mathbf{kg\,m/s^2}$ charge = current × time → **A s**

work done = force × distance = $(\text{kg m/s}^2) \times (\text{m}) \rightarrow \mathbf{kg\,m^2/s^2}$

- **Deriving the units of constants.** Several equations contain constants. Some of these constants, such as g, have dimensions, and some, such as π, do not.

Example *Determine the dimensions of the universal gravitational constant, G, in the equation $F = Gm_1m_2/r^2$.*

Rearrange the equation: $G = Fr^2/m_1m_2$.

List the base units: $F \rightarrow (\text{kg m/s}^2)$, $r \rightarrow (\text{m})$, $m_1 \rightarrow (\text{kg})$, $m_2 \rightarrow (\text{kg})$.

Substitute: $G \rightarrow (\text{kg m/s}^2)(\text{m})^2/(\text{kg})(\text{kg}) \rightarrow (\text{kg})^{-1}(\text{m})^3(\text{s}^{-2})$, giving $\mathbf{kg^{-1}\,m^3\,s^{-2}}$.

- **Checking the homogeneity of equations.** For an equation to be correct the base units of each term in the equation must be the same.

Example *Show that the equation $v = u + at$ is homogeneous.*

List the base units of each quantity, then substitute them into the equation.

v and u = (m/s), a = (m/s^2), t = (s)

(m/s) → (m/s) + (m/s^2) (s)

(m/s) → (m/s) + (m/s) the dimensions of each term are the same, **so the equation is homogeneous**.

- The most important application of base units is in the derivation of equations for a particular phenomenon when all the variables are known. This is no longer required at A level, but you may meet it if you go on to study physics at university.

TESTS

RECALL TEST

1 What is Hooke's law?

(2)

2 What is 'stress'?

(2)

3 What is 'strain'?

(2)

4 What is the 'Young modulus'?

(2)

5 What is meant by 'elastic deformation'?

(2)

(Total 10 marks)

CONCEPT TEST

1 A 200 g mass is added to a spring and it extends 0.04 m. What is the spring constant and how much energy is stored in the spring?

(2)

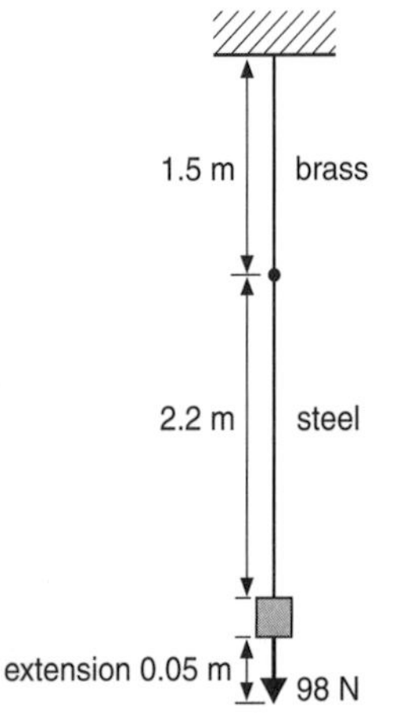

2 A piece of steel wire of length 2.0 m and cross-sectional area $3.0 \times 10^{-6}\,m^2$ has a 40 kg weight attached to it. How far does it extend? (Young modulus of steel = 2.0×10^{11} Pa.)

(4)

3 A 1.5 m piece of brass wire is attached to a 2.2 m length of steel wire in a vertical direction (see left). A 10 kg mass is attached to one end and they extend by 0.05 m. By how much does each wire extend? (Young moduli: brass = 9.2×10^{10} Pa, steel = 2.0×10^{11} Pa.)

(4)

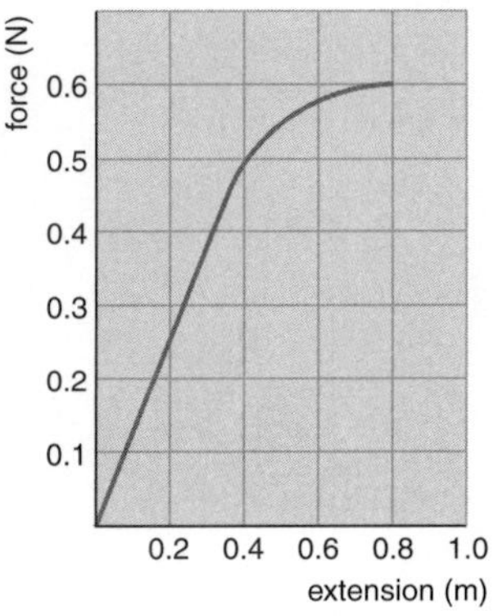

4 The figure left shows the stress–strain graph for a material under tension. What is the energy stored for a piece of this material with an extension of 0.80 m and of thickness 2.0 mm?

(4)

5 Draw stress–strain graphs for: a brittle material such as steel, a ductile material such as copper, and a polymer such as rubber.

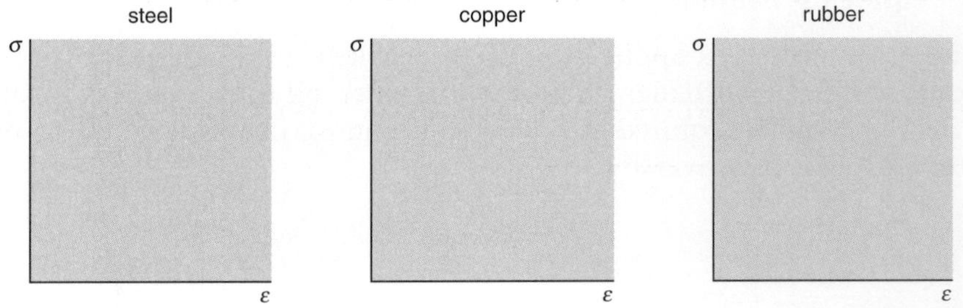

(6)

(Total 20 marks)

TESTS

RECALL TEST

1 What is a base unit?

(2)

2 What is a dimension?

(2)

3 What is meant by 'homogeneity' in an equation?

(2)

4 Why must some constants have base units?

(2)

(Total 8 marks)

CONCEPT TEST

1 Determine the base units of **a** power, **b** potential difference, **c** resistance, **d** electric field strength, **e** magnetic flux density, and **f** capacitance.

(6)

2 Determine the base units of the following constants: **a** radioactive decay constant, **b** Planck constant, **c** specific heat capacity, **d** ideal gas constant, **e** permittivity of free space, **f** permeability of a vacuum.

(6)

3 Show that the following equations are homogeneous: **a** $v^2 = u^2 + 2as$, **b** $T = 2\pi(l/g)^{0.5}$, **c** $v = f\lambda$, **d** $F = (mv - mu)/t$, **e** $I = nAve$, and **f** $W = \frac{1}{2}CV^2$.

(6)

(Total 18 marks)

Unit 28

PRACTICAL WORK

- Practical work is an integral part of scientific method. Scientific advances come from formulating a theoretical idea (hypothesis), testing it experimentally, and then, if necessary, adjusting the hypothesis in the light of the results. If the experimental results support the theory, both are published within the scientific community, and other experimenters try to replicate the results. When everyone agrees that the theory is correct it is accepted as fact until further work supersedes it.

  Practical work is done at A level for two main reasons:

  1 To help explain and illustrate different aspects of theory in the syllabus. Such practicals may be concerned with the determination of a physical constant, the verification of a particular law, or the determination of the relationship between two quantities.

  2 To equip students with the knowledge and expertise required for working in scientific environments, by making it part of their A-level assessment. This is done using practical exams or coursework.

  Coursework and practical exams are designed to build upon the approach used in GCSE coursework. Assessment is targeted at four general areas: a) planning, b) implementing, c) analysing evidence and drawing

- **1 Variables** A **variable**, as its name suggests, is something that changes. It is the name given to the various physical quantities that can change in an experiment, such as length and temperature. In a particular experiment variables may be classified under two types: **dependent** and **independent**. The value of a **dependent variable** depends upon the value of an **independent variable**, which is fixed by the experimenter. For example, you could drop a ball from a particular height (independent variable) and measure the height it bounces to (dependent variable). Both values are measured and recorded, but as the independent value is fixed, it is the dependent variable that produces a set of different values when readings are repeated.

  In GCSE the expression 'a **fair test**' is used a lot, which means that when investigating the effect of changing one variable all other variables are kept constant. In this way only the effects of changing the one variable will be seen.

  In some situations it is not possible to keep constant all the variables affecting an experiment, so we use the idea of a '**control**'. A **control** is a second experiment, identical to the first one, run under identical conditions. A variable is changed in one experiment and its effect is measured; the other experiment is left unchanged. The two experiments are then compared and any changes in the control are set off against those in the actual experiment. This isolates effects due to variations in the surroundings, such as temperature, from the actual changes produced by the experiment. A good example of this would be an experiment to determine the latent heat of ice using an electric heater. The two experiments are set up as shown left. The water melted in the control is melted by heat from the surroundings, and the amount is subtracted from that melted in the actual experiment to give the amount of ice melted by the heater alone.

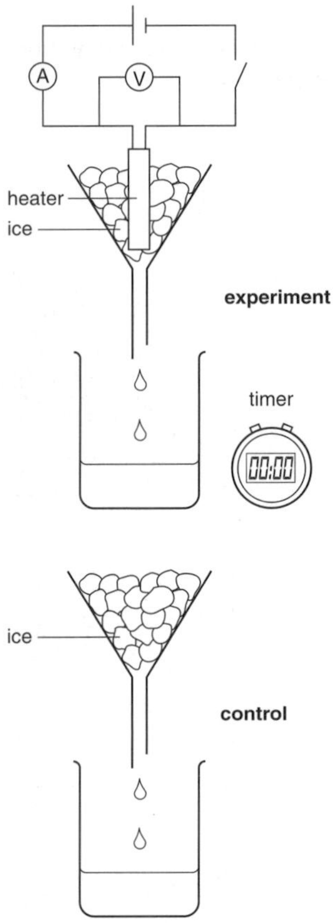

Experiment using a control.

- **2 Errors** In all experiments there are errors in the actual readings that have been taken. They fall into two categories: **random** and **systematic**. **Systematic** errors, as the name implies, are errors in the system or apparatus, such as an ammeter which is not zeroed properly adding a fixed amount to all its readings. With some systematic errors all of the readings are out by the same amount. These errors cannot be reduced by repeated readings, and if detected can only be corrected by adjusting the instrument or correcting all the readings by the appropriate amount. Constant systematic errors do not affect values derived from the gradient of a straight-line graph, but do affect values derived from the y intercept.

  Random errors are variations from a true reading and can be smaller or larger than the actual value. These are often due to the input of the human doing the measurements and can be reduced by repeat readings and taking averages.

In the case of timing or measuring experiments, the longer the time over which the measurement takes place, or the larger the measurement, the smaller the effect of the error. When a measurement is stated, its accuracy is often given too, for example 2.8 ± 0.2. Such quoted errors can come from an estimate of the inaccuracy in taking a measurement, or are dictated by the smallest size of unit on the measuring device. For example, a ruler calibrated in millimetres could produce a reading of 24 ± 1mm.

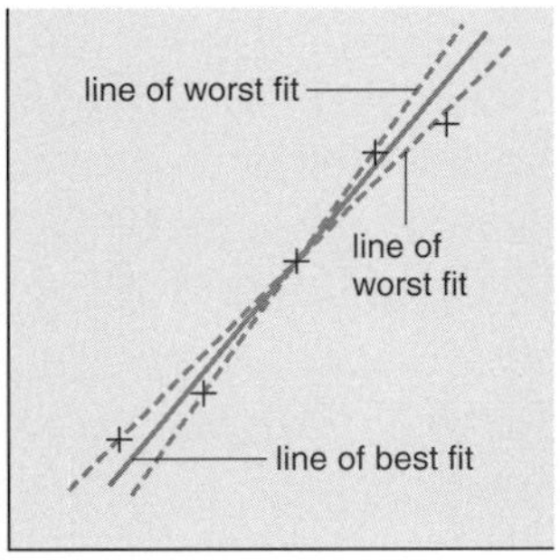

Graph with lines of best and worst fit.

- **3 Significant figures** In unit 2 it was stated that the number of significant figures in a set of values limits the number of significant figures in any number calculated from these values. The same applies to the derived results in an experiment. Once data is produced all readings are recorded in a table with the appropriate number of significant figures, together with any averages and other derived results. The table below shows a sample set of data with derived results.

Length (m)	Time for 20 oscillations (s) 1	2	3	Average	Time period for one swing T (s)	T^2 (s^2)
0.14	15.6	15.9	15.6	15.7	0.785	0.616
0.35	23.6	23.4	23.4	23.5	1.18	1.39
0.49	27.4	27.6	27.9	27.6	1.38	1.90
0.56	29.3	30.9	29.2	29.8	1.49	2.22
0.63	31.0	31.1	30.9	31.0	1.55	2.40
0.70	32.5	32.6	32.9	32.7	1.64	2.69
0.84	37.4	36.8	36.7	37.0	1.85	3.42

Data for a simple pendulum experiment.

- **4 Graphs** In most experiments graphs of the data are plotted and straight lines or curves of best fit are drawn. The reasons for plotting a graph are, firstly, to further reduce the effect of errors by averaging them out graphically, secondly, to determine a mathematical relationship between variables, or thirdly, to use the intercept or gradient to determine the values of constants. A **line of best fit** is a line drawn in such a way that the sum of all the perpendicular distances from the line to the points on one side of the line is the same as on the other. In practice this is done by eye and trial and error, but a computer program using a 'least squares fit' package will do it for you. If you do A level maths, you can use 'regression' and 'correlation' to help you. It is also possible to plot **worst-fit lines** in order to get an error value for the gradient (see above right). Lines of worst fit are normally drawn in pairs, one either side of the line of best fit. They are the two lines that could be drawn to satisfy the points yet deviate most from the line of best fit.

  Generally speaking, a graph should fill a minimum of half the sheet of graph paper and simple scales should be used. The axes do not have to start at zero, but this will affect intercept values taken from the axes. Many different-shaped graphs can be produced by experiment. The most useful is that of a straight line, because the intercept and gradient can be used to provide information. When **logs** are taken of some equations which are not linear, a linear equation is formed, which can yield values for constants or powers. For example, $y = x^z$ gives $\log y = z \log x$, so a graph of $\log y$ against $\log x$ gives a straight line with gradient equal to z. $y = wx^z$ gives $\log y = z \log x + \log w$ and $y = ze^{-kx}$ gives $\ln y = \ln z - kx$ (see right).

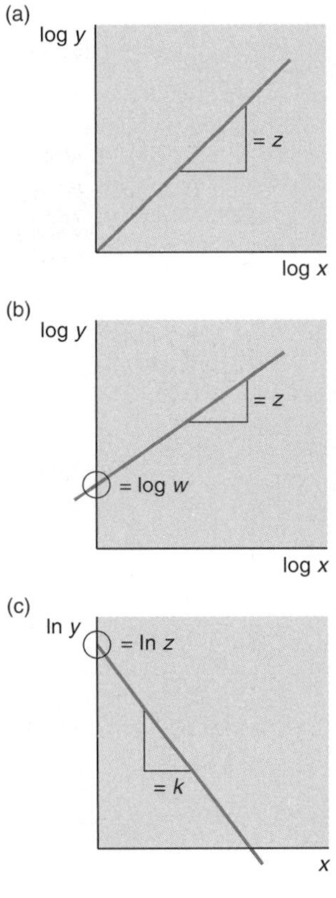

(a) A graph of $\log y = z \log x$.
(b) A graph of $\log y = z \log x + \log w$.
(c) A graph of $\ln y = \ln z - kx$.

- **5 Datalogging** Most of the data gathered at GCSE is done by the student, observing and recording data by hand. It is a requirement of A level that students should have some knowledge of **datalogging**. **Datalogging** is the use of electronic sensors to measure and record data. The data can then be downloaded into a computer, and tables and graphs produced accordingly. The advantage of this is obvious: it is not necessary to monitor the experiment continuously. It also reduces the degree of random error due to the observer.

Unit 28

TEST

RECALL TEST

1 What is meant by 'scientific method'?

(2)

2 What are the three ways in which practical work can be used to explain physics?

(2)

3 What is a dependent variable?

(2)

4 What is an independent variable?

(2)

5 What is meant by a 'fair test'?

(2)

6 What is meant by a 'control'?

(2)

7 What is a systematic error?

(2)

8 What is a random error?

(2)

9 In experimental readings, why should the number of significant figures be noted?

(2)

10 What is meant by a 'line of best fit'?

(2)

11 What can be derived from a straight-line graph?

(2)

12 Why are logs taken of some data?

(2)

13 What is 'datalogging'?

(2)

14 How does a constant systematic error affect a set of data?

(2)

15 How can the effects of random error be reduced?

(2)

(Total 30 marks)

ANSWERS

UNIT 1

RECALL TEST

1 The velocity that an object has in relation to another object.
2 A vector has both size and direction; a scalar only has size.
3 Mass remains constant whatever the direction of an object's motion.
4 The length of the path between two points in a straight line.
5 The length of the path between two points where the direction of travel changes.
6 The direction of motion matters for velocity, but not speed.
7 Vertical and horizontal motions are independent of each other. There is always acceleration due to gravity in the vertical direction, but if friction is ignored there is no acceleration horizontally.
8 Rate of change of velocity.
9 Take the change in the quantity, and divide it by the time taken for the change.
10 As gravity acts downwards, there can be no acceleration acting horizontally.

CONCEPT TEST

1 $v = \Delta s/\Delta t = 1000/300 = \mathbf{3.3\ m/s}$.
$a = \Delta v/\Delta t = (30 - 3.3)/8.0 = \mathbf{3.3\ m/s^2}$.

2 $s = 200$ m, $u = 0$, $a = 9.8$ m/s², $v = ?$, $t = ?$
$s = ut + \frac{1}{2}at^2$, $200 = 0 + (\frac{1}{2} \times 9.8 \times t^2)$, $\mathbf{t = 6.4\ s}$.
$v = u + at = 0 + (9.8 \times 6.4) = \mathbf{63\ m/s}$.

3 $s = ?$, $u = 200$ m/s, $a = ?$, $v = 280$ m/s, $t = 10$ s
$a = \Delta v/\Delta t = (280 - 200)/10 = \mathbf{8.0\ m/s^2}$.
$s = ut + \frac{1}{2}at^2$, $s = (200 \times 10) + (\frac{1}{2} \times 8 \times 10^2) = \mathbf{2400\ m}$.
Or $v^2 = u^2 + 2as$, $280^2 = 200^2 + (2 \times 8 \times s)$, $\mathbf{s = 2400\ m}$.

4 $s = ?$, $u = 0$ m/s, $a = 9.8$ m/s², $v = ?$, $t = 4$ s
$s = ut + \frac{1}{2}at^2 = 0 + (\frac{1}{2} \times 9.8 \times 4^2) = \mathbf{80\ m}$.

5 $s = ?$, $u = 34.0$ km/h, $a = ?$, $v = 60.0$ km/h, $t = 5.00$ s
$u = 34$ km/h $= 34\,000/(60 \times 60) = 9.44$ m/s.
$v = 60$ km/h $= 60\,000/(60 \times 60) = 16.7$ m/s.
$v = u + at$, $a = (v - u)/t = (16.7 - 9.44)/5 = \mathbf{1.45\ m/s^2}$.
$s = ut + \frac{1}{2}at^2 = (9.44 \times 5) + (\frac{1}{2} \times 1.45 \times 5^2) = \mathbf{65.3\ m}$.
Or $v^2 = u^2 + 2as$
$16.7^2 = 9.44^2 + (2 \times 1.45 \times s)$, $\mathbf{s = 65.4\ m}$.

6 $v = u + at$, $a = (v - u)/t = (-15 - 10)/0.2 = \mathbf{-125\ m/s^2}$.

7 $s = 1$ m, $u = 200$ m/s, $a = 9.8$ m/s², $v = 0$ m/s, $t = ?$
$v^2 = u^2 + 2as$
$0^2 = 200^2 + (2 \times 9.8 \times s)$, $s = 2040 = \mathbf{2.0\ km}$.
$v = u + at$, $0 = 200 - 9.8t$, $t = 20.4$.
Time to return, $T = t \times 2 = \mathbf{41\ s}$.

8 Vertically: $s = ?$, $u = 0$ m/s, $a = 9.8$ m/s², $v = ?$, $t = ?$
$s = ut + \frac{1}{2}at^2$, $1 = 0 + (\frac{1}{2} \times 9.8 \times t^2)$, $t = 0.45$ s.
Horizontally: $s = ?$, $u = 4$ m/s, $a = 0$, $v = u$, $t = 0.45$ s
$s = ut + \frac{1}{2}at^2 = (4 \times 0.45) + 0 = \mathbf{1.8\ m}$.

9 Vertically: $s = 1.0$ m, $u = ?$, $a = -9.8$ m/s², $v = 0$ m/s, $t = ?$
$v^2 = u^2 + 2as$, $0^2 = u^2 - (2 \times 9.8 \times 1)$, $\mathbf{u = 4.4\ m/s}$.
$v = u + at$, $0 = 4.4 - 9.8t$, $t = 0.45$ s.
Horizontally: $s = 5.0$ m, $u = ?$, $a = 0$, $v = u$, $t = 0.45$ s.
$s = ut + \frac{1}{2}at^2$, $5.0 = (u \times 0.45) + 0$, $\mathbf{u = 11\ m/s}$.

10 $s = ?$, $u = 0$ m/s, $a = 9.8$ m/s², $t = 4.0$ s
$v = 0.9 \times 3 \times 10^8 = 2.7 \times 10^8$ m/s.
$v = u + at$, $a = (v - u)/t$
$a = (2.7 \times 10^8 - 0)/4.0 = 6.75 \times 10^7$ m/s².
$s = ut + \frac{1}{2}at^2$
$s = 0 + (\frac{1}{2} \times 6.75 \times 10^7 \times 4^2) = \mathbf{5.4 \times 10^8\ m}$.

11 Horizontally: $s = 1.2$ m, $u = 0$ m/s, $a = 20\,000$ m/s², $t = ?$, $v = ?$
$v^2 = u^2 + 2as$
$v = 0 + (2 \times 20\,000 \times 1.2) = 219$ m/s $= 220$ m/s.
$s = ut$, $200 = 219t$, $t = 0.91$ s.
Vertically: $s = ?$, $u = 0$ m/s, $a = 9.8$ m/s², $v = ?$, $t = 0.91$ s.
$s = ut + \frac{1}{2}at^2 = 0 + (\frac{1}{2} \times 9.8 \times 0.91^2) = \mathbf{4.1\ m}$.

UNIT 2

RECALL TEST

1 Acceleration.
2 Velocity.
3 Change in velocity.
4 Displacement or distance travelled.
5 Constant velocity.
6 Constant acceleration.
7 Acceleration.
8 Constant deceleration.
9 Constant acceleration.
10 Plot a graph of height against 1/fillings. The gradient is the acceleration due to gravity.

CONCEPT TEST

1
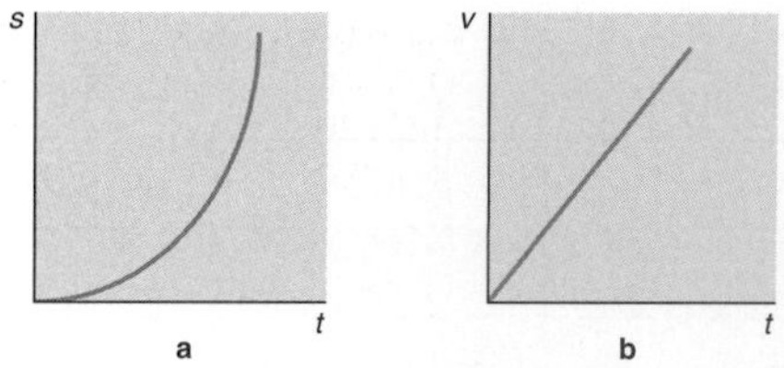

2 A–B: accelerating uniformly at $a = \Delta v/\Delta t = 20/10 =$ $\mathbf{2.0\ m/s^2}$; moving forwards.
B–C: constant velocity of 20 m/s; moving forwards.
C–D: decelerating uniformly at $a = \Delta v/\Delta t = -20/3 =$ $\mathbf{-6.7\ m/s^2}$; moving forwards.
D–E: accelerating uniformly at $a = \Delta v/\Delta t = 10/2 = \mathbf{5.0\ m/s^2}$; moving backwards.
E–F: decelerating uniformly at $a = \Delta v/\Delta t = -10/3 =$ $\mathbf{-3.3\ m/s^2}$; moving backwards.

3
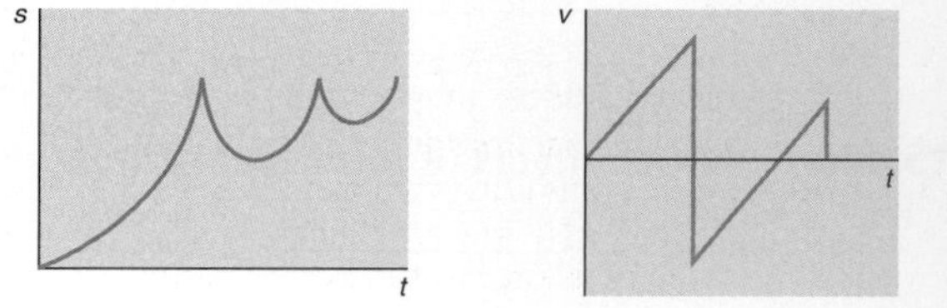

4
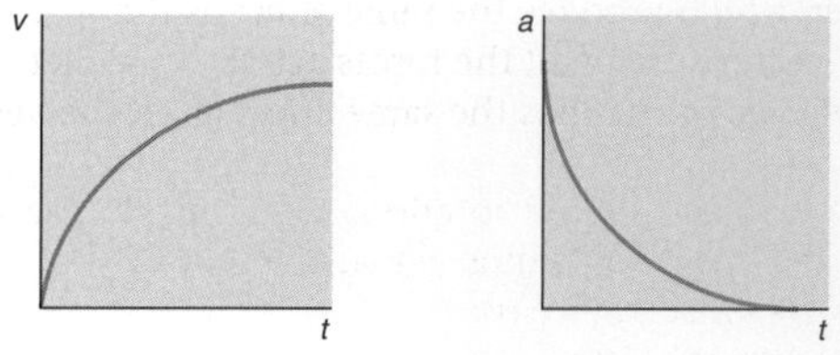

5 Area above axis, $A = (\frac{1}{2} \times 4 \times 8) + (\frac{1}{2} \times 8 \times 8) = 48$ m.
Area below axis, $B = (\frac{1}{2} \times 6 \times 6) + (\frac{1}{2} \times 6 \times 3) = 27$ m.
Total distance, $d = A + B = \mathbf{75\ m}$.
Total displacement, $s = A - B = \mathbf{21\ m}$.

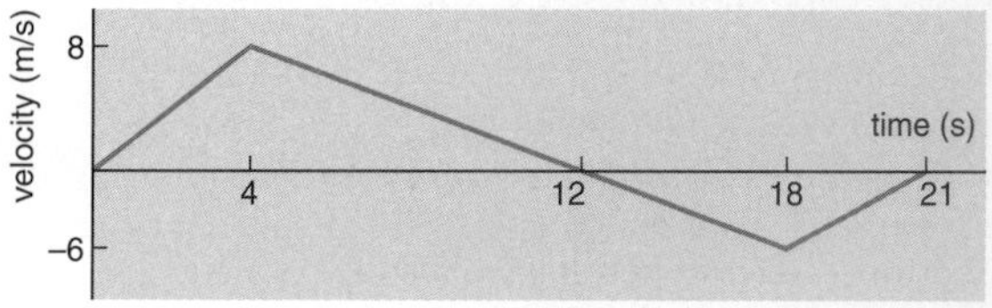

6
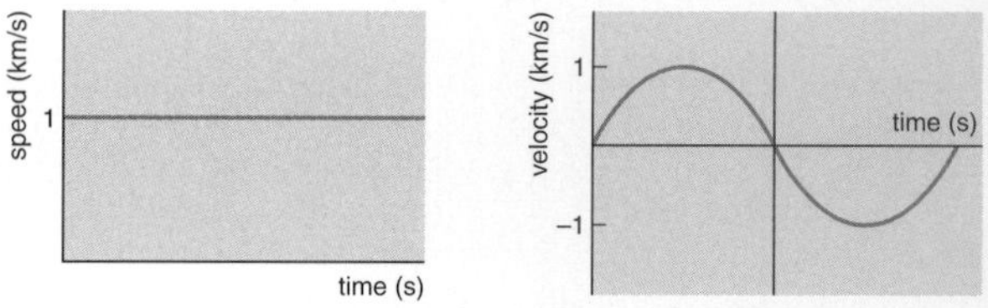

7 Average speed = total distance/total time
$= 10\,000/300 =$ **33 m/s**.
Maximum velocity at steepest point,
$v = \Delta s/\Delta t = (9000 - 6000)/25 =$ **120 m/s**.
Accept values around $v = 120$ m/s.

8

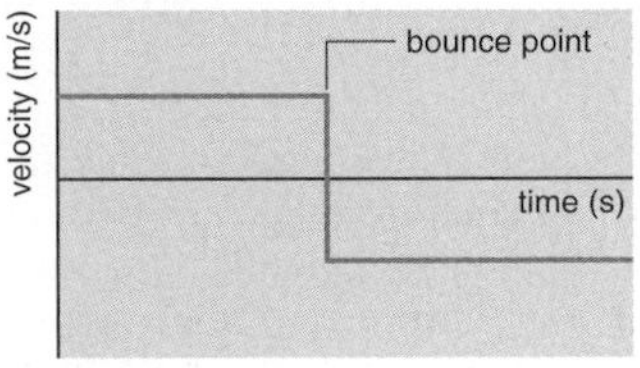

9 Maximum acceleration at steepest point,
$a = \Delta v/\Delta t = (2 - -1)/0.3 =$ **10 m/s²**.
Distance travelled equals the area under the graph. The best way to do this is to count the number of squares (N). Take one square as being $0.5 \times 0.1 = 0.05$. Area above axis equals 36 squares, area below equals 20 squares, so the net N is 16.
Distance, $d = N \times 0.05 = 16 \times 0.05 =$ **0.8 m**.

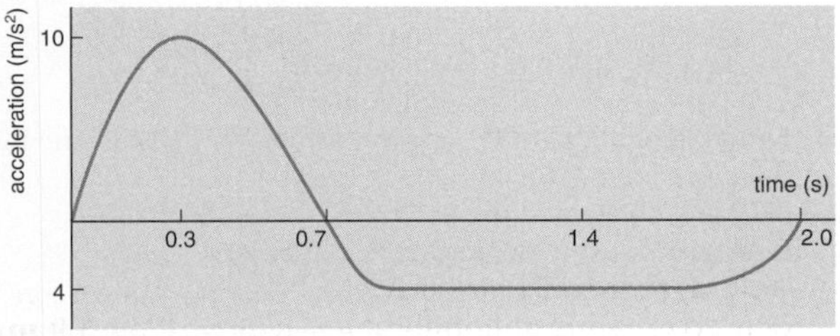

UNIT 3

RECALL TEST

1 The maximum velocity that can be reached by an object moving through a liquid or gas.
2 The object is stationary or moving with a constant velocity.
3 A measure of the amount of material present in a object.
4 It is the force times the perpendicular distance from the line of action of the force to the axis of rotation.
5 The components are two vectors, perpendicular to each other, which produce the same effect as the original vector.
6 The vector sum of all the forces acting on something, or a single vector that has the same effect as all the vectors present.
7 For an object to be in rotational equilibrium the sum of the clockwise moments must equal the sum of the anticlockwise moments.
8 Force per unit area.
9 The turning effect produced by two identical anti-parallel forces (its torque is equal to one of the forces times the perpendicular distance between them).
10 It will cause the object to accelerate; by how much depends upon its mass and the size of the force. ($F = ma$)

CONCEPT TEST

1 $F = ma = 1500 \times 2.0 =$ **3.0 kN**.
2 $v = u + at$, $a = (v - u)/t = (10 - 6)/4 = 1$ m/s2.
$F = ma = 6 \times 1 =$ **6 N**.
3 Tension – weight = resultant force ($T - W = R$)
$20\,000 - (1500 \times 9.8) = 1500a$, $\boldsymbol{a} =$ **3.5 m/s²**.
4 Find resultant force R using vector sum: $R^2 = 12^2 + 20^2$, $R = 23$ N.
$F = ma$, $23 = 4a$, $a = 6$ m/s².
5 Take moments about pivot:
Σ (anticlockwise moments) = Σ (clockwise moments)
$(0.20 \times 9.8 \times 0.50) + (0.40 \times 4.0) = 0.30 \times 9.8 \times A$, $A = 0.878$,
A = 880 g.
6 Resultant force vertically = 10 – 10 = 0.
Horizontally, 36 – 50 = –14. (Taking +ve as to the right.)
$F = ma$, $-14 = 2a$, $\boldsymbol{a} =$ **–7 m/s²**.
7 Tension – weight = resultant force ($T - W = R$)
$T - (500 \times 9.8) = 500 \times 3.0$, $\boldsymbol{T} =$ **6400 N**.
8 Moment = force × perpendicular distance.
$M = Fs = (0.25 \cos 35°) \times 12 =$ **2.5 N m**.
9 Upward force = downward force
$14 = (9.8 \times 0.30) + W$, $W = 11$ N.
Take moments about A:
Σ (clockwise moments) = Σ (anticlockwise moments)
$(11 \times s) + (0.30 \times 9.8 \times 1.1) = (10 \times 0.50) + (4.0 \times 2.9)$
s = 1.2 m.
10 Take moments about elbow:
Σ (anticlockwise moments) = Σ (clockwise moments)
$20 \times 0.40 = 0.34 \times W$, $\boldsymbol{W} =$ **24 N**.
11 Moment = force × perpendicular distance
$M = Fs = (12 \sin 80°) \times 0.35 =$ **4.1 N m**.
12 Resolve the 25 N force into two components vertically and horizontally. Find the resultants of the vertical and horizontal forces, and then the overall resultant using Pythagoras.
Vertically: $V = 26 - (24 \sin 30°) = 14$ N.
Horizontally: $H = 10 - (25 \cos 30°) = -11$ N.
$R^2 = 14^2 + (-11)^2$, $\boldsymbol{R} =$ **18 N**.
$\tan \theta = 14/11$, $\boldsymbol{\theta} =$ **52°** to the left, above the horizontal.
13 Take moments about the pivot.
Σ (anticlockwise moments) = Σ (clockwise moments)
$50 \times 9.8 \times 0.25 \cos 25° = F \times 0.85 \times \cos 25°$, $\boldsymbol{F} =$ **140 N**.
14 Draw a scale diagram, and use the parallelogram of forces. Alternatively, resolve the 6 N force into two components, one in the direction of the 10 N force. Then use Pythagoras to get the resultant.
$((6 \cos 55°) + 10)^2 + (6 \sin 55°)^2 = R^2$, $\boldsymbol{R} =$ **14 N**.
$\tan \theta = (6 \cos 55°)/(10 + 6 \sin 55°)$, $\boldsymbol{\theta} =$ **13°** above the horizontal.
15 $P = F/A = 60 \times 9.8/1.0 =$ **590 Pa**.

UNIT 4

RECALL TEST

1 The force of gravity acting on a mass.
2 The force that opposes motion, or probable motion, produced between surfaces or fluids and surfaces.
3 The force acting in solids which are being stretched that opposes the stretching force.
4 The force that occurs between objects in contact, and acts at 90° to a surface, from the point of contact.
5 The force acting on objects in fluids; it is equal to the weight of the displaced fluid.
6 A diagram that shows the forces acting on an object.
7 The point at which, if all the mass could be concentrated, the object would behave in the same way.
8 We need to classify forces into different types in order to model situations and predict the behaviour of objects.
9 Weight is gravitational; all the others are electromagnetic.
10 Gravitational, electromagnetic, strong nuclear force, and weak nuclear force.

CONCEPT TEST

1

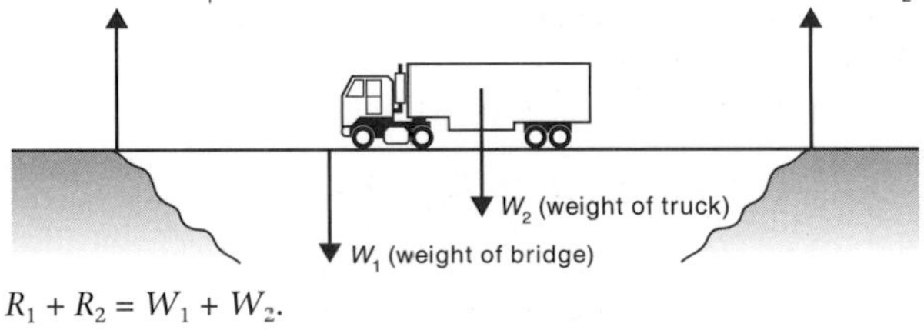

$R_1 + R_2 = W_1 + W_2$.

2

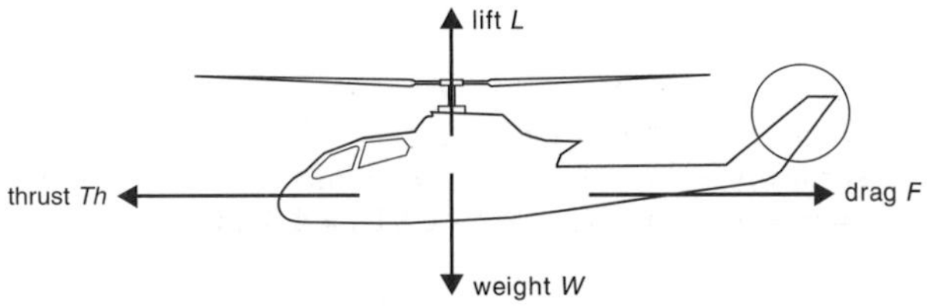

$Th = F$, $L = W$.

3 Resolve forces parallel to and perpendicular with the plane: $W \sin \theta = F$, $W \cos \theta = R$.

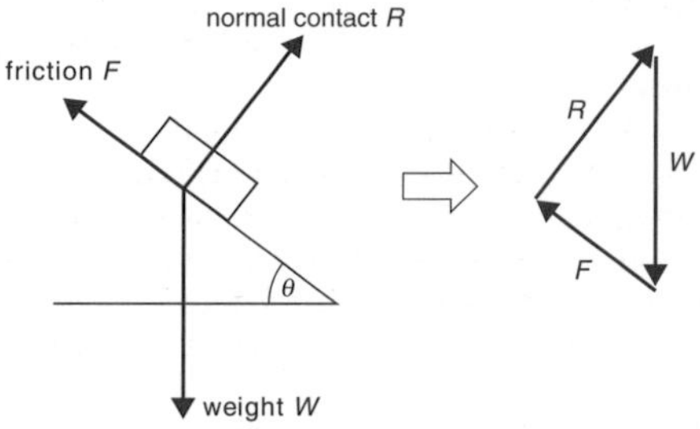

4 Resolve the tension force vertically and horizontally: $T \cos 40° = W$, $T \sin 40° = R$.
$T = 60 \times 9.8/\cos 40° =$ **770 N**.
$770 \sin 40° = R$, $R =$ **490 N.**

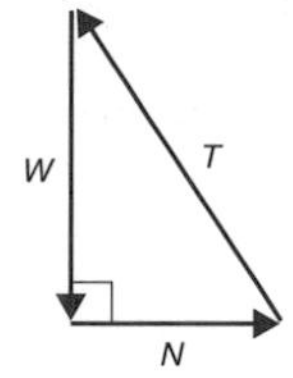

5 F must equal the resultant of the 600 N and 800 N forces. To do this question, resolve the 600 N force vertically and horizontally. Then use Pythagoras.
$800 + 600 \cos 50° = 1186$.
$600 \sin 50° = 459.6$.
$F^2 = 459.6^2 + 1186^2$ so $F =$ **1270 N.**

6 Velocity is constant so there is no resultant force. Resolve the weight parallel to and perpendicular with the plane.
$W \cos 20° = R$, $5000 \times 9.8 \cos 20° = R$,
$N = 46\,000$ N.
$Th = F + W \sin 20°$
$20\,000 = F + 5000 \times 9.8 \times \sin 20°$,
$\boldsymbol{F} =$ **3200 N.**

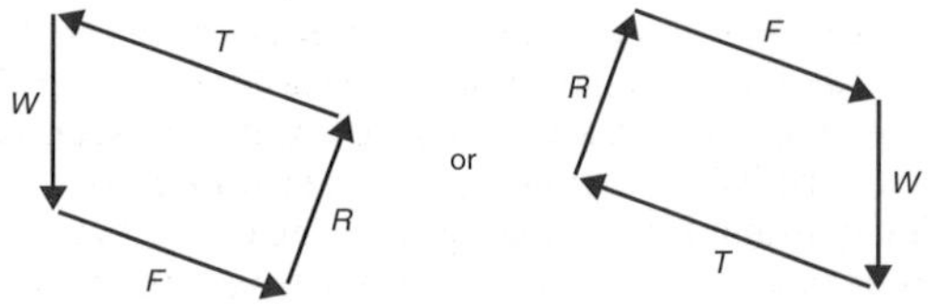

7 Resolve forces in the direction of motion:
$G \cos 30° + A \cos 40° - F = ma$
$(20\,000 \cos 30°) + (15\,500 \cos 40°) - F = 75\,000 \times 0.3$,
$\boldsymbol{F} =$ **6700 N.**

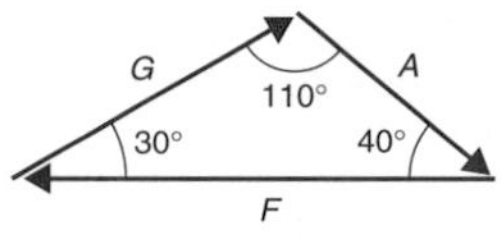

8 Resolve T and D vertically and horizontally to get two equations.
$D \cos 40° = T \cos 30°$.
$D \sin 40° = T \sin 30° + W$.
Then solve:
$D = 1.13T$, $D \times 0.643 = 0.500T + (70 \times 9.8)$,
$0.643 \times 1.13T = 0.500T + 686$, $\boldsymbol{T} =$ **3000 N.**
$\boldsymbol{D} =$ **3400 N.**

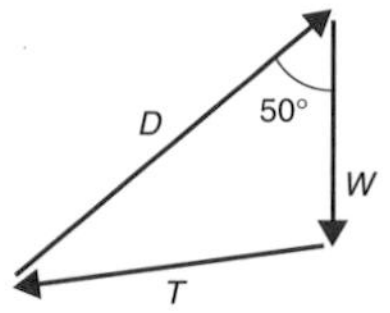

$T \cos 30° + 1000 = F$,
$F = 3600$ N.

UNIT 5

RECALL TEST

1 Energy is the capacity to perform work, or the amount of work that has been done.
2 Energy cannot be destroyed or created, only transformed from one form to another.
3 Gravitational, chemical, and elastic.
4 Work done equals the product of force and displacement in the direction of the force.
5 Power is the rate of doing work.
6 The ratio of energy output to input. Efficiency = energy output/energy input.
7 The power realized at a moment in time.
8 It requires more power, because the energy is used in less time.
9 The loss in KE is the work done against friction in the brakes and is changed into heat.
10 Work done.

CONCEPT TEST

1 a $PE = mg\Delta h = 4.0 \times 9.8 \times 6.0 = 235$. $\boldsymbol{PE} =$ **240 J**.
b $KE = \frac{1}{2}mv^2$, PE lost $= KE$ gained.
$\frac{1}{2} \times 4.0 \times v^2 = 235$, $\boldsymbol{v} =$ **11 m/s.**

2 Gravitational PE → KE → gravitational PE, with some lost to heat and sound because of friction.
PE lost $= KE$ gained, $mgh = \frac{1}{2}mv^2$,
$v = (2gh)^{0.5} = (2 \times 9.8 \times 0.50)^{0.5} =$ **3.1 m/s.**

3 They each use the same amount of energy, so in energy terms there is no difference between them. Clare uses the energy in less time, which requires more power. Initially she will feel more tired at the pub.

4 Heat gained $= PE$ lost,
heat $= mgh_1 - mgh_2$ or heat $= mg\Delta h$
heat energy $= 0.20 \times 9.8 \times 4.0 =$ **7.8 J**.

5 Total force $F = 6 \times 20 = 120$ N.
$W = F \times s = 120 \times 15 = 1800$ J.
$W = \Delta KE = \frac{1}{2}mv^2$
$1800 = \frac{1}{2} \times 280 \times v^2$, $\boldsymbol{v} =$ **3.6 m/s.**

6 $F = mg$, $W = F \times s$ (or $PE = mgh$)
$W = 2000 \times 9.8 \times 10 = 196\,000$ J, $W = 196$ kJ.
Efficiency = output/input
$0.6 = 196\,000/\text{input}$, input = 330 kJ.

7 a Gravitational PE → KE → gravitational PE → KE and some heat and sound due to friction.
bi From the equation PE lost $= KE$ gained, $mgh = \frac{1}{2}mv^2$; mass cancels leaving $gh = \frac{1}{2}v^2$, so it is irrelevant to the velocity at any instant. Therefore the mass does not affect the time taken.
bii If the track is initially steeper the toy will gain velocity more quickly and so the travel time will be reduced.
c PE lost $= KE$ gained, $mgh = \frac{1}{2}mv^2$, $v = (2gh)^{0.5}$.
At C $v = (2 \times 9.8 \times 0.30)^{0.5} =$ **2.4 m/s.**
At D $v = (2 \times 9.8 \times 1.2)^{0.5} =$ **4.8 m/s.**

8 Work done – PE gained = energy dissipated by friction.
PE gained $= mg\Delta h = 200 \times 9.8 \times 6.0 = 11\,760$ J.
$W = F \times s = 1000 \times 15 = 15\,000$ J.
W (friction) $= 15\,000 - 11\,760 = 3240$ J $=$ **3200 J.**

9 $P = Fv = 15 \times 20 =$ **300 W.**

10 Area under graph $= W$
Area $= (\frac{1}{2} \times 20 \times 5.0) + (5 \times 20)$, $W =$ **150 J.**

11 Work done by locomotive,
$W = F \times s = 2.00 \times 10^8 \times 300 = 6.00 \times 10^{10}$ J.
$\Delta KE = \frac{1}{2}m(v^2 - u^2)$
$\Delta KE = \frac{1}{2} \times 5.00 \times 10^6 \times (25.0^2 - 4.00^2) = 1.52 \times 10^9$ J.
$W - \Delta KE$ = work done against friction
$6.00 \times 10^{10} - 1.52 \times 10^9 = 5.85 \times 10^{10}$ J.
$W = F \times s$, so
$5.85 \times 10^{10} = F \times 300$, $\boldsymbol{F} =$ **1.95 × 10⁸ N.**

UNIT 6

RECALL TEST

1 When it is moving at constant velocity or at rest.
2 It will accelerate.
3 The frictional force of the case acting on the ground.
4 Change in momentum or impulse (see unit 7).
5 A single force that represents the combined effect of all the forces present in a given situation.
6 A resultant force is directly proportional to the rate of change of momentum it produces.
7 An object will remain in a state of uniform motion or at rest unless acted upon by a resultant force.
8 For every force that acts there is an equal and opposite reaction force.
9 It may accelerate, decelerate, and/or change direction. (If the force acts at 90° to the direction of motion its speed will remain constant, but the object will change its direction of motion. If the force acts in the same direction as the original motion the object will accelerate. If it acts in the opposite direction the object will decelerate. If the direction is at any other angle the object will accelerate/decelerate and change its direction of motion.)
10 1 N is the force that will accelerate a mass of 1 kg by 1 m/s^2.

CONCEPT TEST

1 Initially the ball is stationary, so from Newton's first law all the forces must cancel each other out. When the ball is released there is a resultant force equal to the weight, so it accelerates. As its velocity increases, the frictional force increases until it equals the weight; there is then no resultant force, so the velocity is constant.

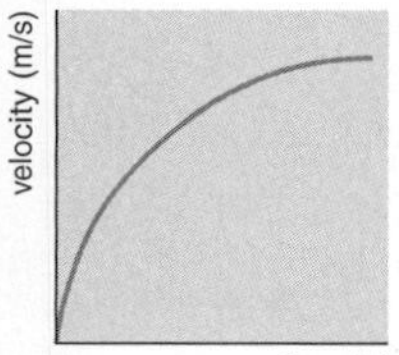

2 a $D - F = ma$, $30 - 26 = 8.0a$, $\boldsymbol{a} = \mathbf{0.50\ m/s^2}$.
b $W = mg = 8.0 \times 9.8 = 78.4$ N.
$78.4 \times 0.25 = 19.6$.
$D - F = ma$, $30 - 19.6 = 8.0a$, $\boldsymbol{a} = \mathbf{1.3\ m/s^2}$.
3 Horizontally: $F_H = 60 \cos 30° - 60 \sin 45° = 9.54$ N.
Vertically: $F_V = 60 \cos 45° - 5.0 - 60 \sin 30° = 7.43$ N.
$F_H^2 + F_V^2 = F_R^2$, $9.54^2 + 7.43^2 = F_R^2$, $F_R = 12$ N.
$F = ma$, $a = 12/0.50 = \mathbf{24\ m/s^2}$.
4 $T - W = ma$, $4000 - mg = 4.0m$, $4000 = m(9.8 + 4.0)$, $\boldsymbol{m} = \mathbf{290\ kg}$.
5 If it is moving with constant velocity at a constant depth, the drag force equals the driving force produced by the fish, and its weight equals the upthrust. So Newton's first law applies here as there is no resulting acceleration or associated force.

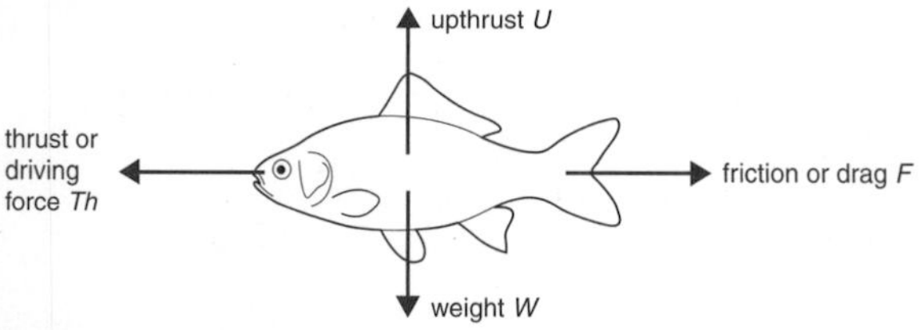

6 Change in momentum $\Delta p = (mv - mu)$
$(0.20 \times 15) - (-0.20 \times 8.0) = 4.6$ kg m/s, $\Delta p = 4.6$ kg m/s.
$F = (mv - mu)/t = 4.6/0.10 = \mathbf{46\ N}$.

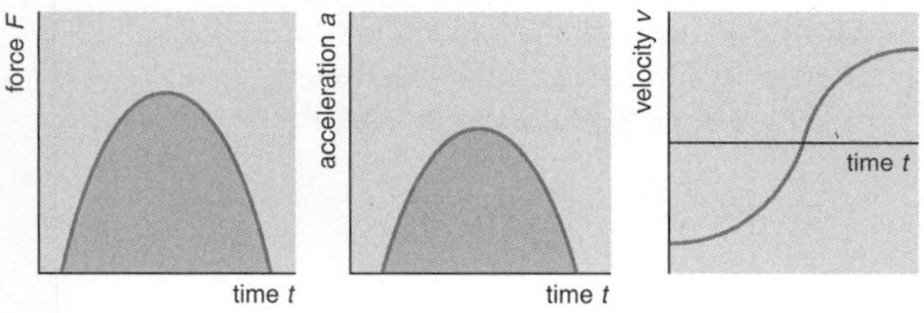

7 Take moving towards the left as negative:
Use $F = (mv - mu)/t$, $Ft = mv - mu$, so
$(-8.0 \times 10^{10} \times 10) = (2.0 \times 10^6 \times v) - (2.0 \times 10^6 \times 3.0 \times 10^4)$
$((-8.6 \times 10^{11}) + (6.0 \times 10^{10}))/2.0 \times 10^6 = \mathbf{-4.0 \times 10^5\ m/s}$.
(To the left.)
8 Use $F = (mv - mu)N/t$:
$F = ((5 \times 10^{-5} \times 16) - 0) \times 500/1 = \mathbf{0.4\ N}$.
9 W is the gravitational pull of the Earth on the brick; its reaction force is the gravitational pull of the brick on the Earth. R is the normal contact force of the ground on the brick; its reaction force is the normal contact force of the brick on the ground.
10 A hovering helicopter is obviously stationary, so from Newton's first law the upward forces must equal the downward forces. The rotor blades are pushing the air downwards, so from Newton's second law the downward force must equal the rate of change of momentum of the air. The rotors blades push down on the air, and from Newton's third law the air pushes back upwards with an equal and opposite force. The weight equals the upward force of the air on the blades.

UNIT 7

RECALL TEST

1 The product of mass and velocity.
2 In a system of colliding bodies the total momentum before a collision is equal to the total momentum after the collision, in a given direction.
3 In an elastic collision KE is always conserved, in an inelastic collision it is not.
4 Change in momentum.
5 N s or kg m/s.
6 The momentum change Δp is fixed, so from $Ft = \Delta p$, if t is prolonged, F is reduced. The parachutist does this by bending and rolling.
7 Inelastic, because there is no KE before the explosion, but there is after the explosion.
8 Momentum is a vector, KE is a scalar.
9 Yes. (In a collision some of the kinetic energy is always passed on as heat, so completely elastic collisions do not really happen. On a atomic level, heat is vibrational KE, so elastic collisions can take place. (See unit 25.))
10 The crumple zones prolong the time over which the force is applied, and so reduce the force acting, and any injury.

CONCEPT TEST

1 Total momentum before = total momentum after
$0 = (400 \times 0.030) - (3.0 \times v)$, $\boldsymbol{v} = \mathbf{4.0\ m/s}$.
2 Total momentum before = total momentum after
$0.400 \times 25 = 1.4 \times v$, $\boldsymbol{v} = \mathbf{7.1\ m/s}$.
3 Total momentum before = total momentum after
$0.02 \times 400 = 2.02 \times v$, $v = 4$ m/s.
KE lost = PE gained
$\frac{1}{2}mv^2 = mgh$, $h = v^2/2g = 4^2/2 \times 9.8 = \mathbf{0.8\ m}$.
4 Impulse $I = \Delta p$, $I = mv - mu = m(v - u)$
$-750 = 200(v - 2.00)$,
$\boldsymbol{v} = \mathbf{-1.75\ m/s}$. Moves off backwards.
5 Total momentum before = total momentum after
$(2500 \times 4) + 0 = 5000 \times v$, $\boldsymbol{v} = \mathbf{2\ m/s}$.
6 Taking the original direction of motion as positive:
$I = m(v - u)$, $-2.00 \times 10^6 = 2.00 \times 10^4(v - 500)$, $\boldsymbol{v} = \mathbf{400\ m/s}$.
7 Taking the original direction of motion as positive:
$I = m(v - u) = 0.20(-9.0 - 12) = \mathbf{-4.2\ N\ s}$.
$F = m(v - u)/t = -4.2/0.15 = \mathbf{-28\ N}$.

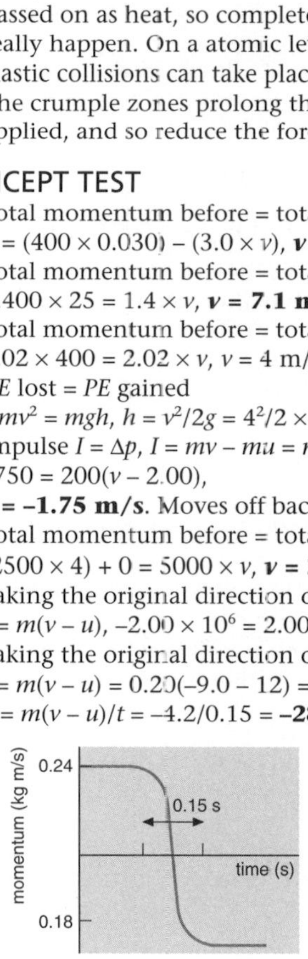

8 Heating gas: heated atoms move faster, and hit the side of the container more often, with greater momentum, so force and pressure increases. Decreasing volume: as temperature is constant the average velocity remains constant, but as volume is decreased the atoms hit the sides more frequently, thereby increasing the average force and pressure.
9 $I = m(v - u)$, $-1.65 \times 10^9 = 4000(v - 3.12 \times 10^5)$, $\boldsymbol{v = -1.01 \times 10^5}$ **m/s**. So it moves backwards.
10 Total momentum before = total momentum after $(40 \times 0.5) - N(1 \times 2.5) = 0$, $\boldsymbol{N = 8}$ lumps of earth.
11 To spread out the time of the impact, and so reduce the force acting on the parachutist. Weight of man, $W = 80 \times 9.8 = 784$ N. Maximum force that can be produced by legs before breaking, $L = 2 \times 600 = 1200$ N. So force, F, left over for deceleration acting in the opposite direction to motion is $F = L - W = 1200 - 784 = 416$ N.
$t = (mv - mu)/F = (0 - (80 \times 12))/(-416) = $ **2.3 s**.

UNIT 8a

RECALL TEST

1 It is the force acting on an object that enables it to move in a circular path.
2 It is the apparent force that you feel pushing you outwards as you go around a circle. (It is due to the fact that you are just trying to carry on moving in a straight line and obey Newton's first law.)
3 Towards the centre of the circle.
4 The rate of change of angular displacement.
5 They are not a force in their own right, they are the resultant of the forces present.

CONCEPT TEST

1 Centripetal force is provided by the friction of the tyres with the road.
$F = mv^2/r$, $3000 = 800v^2/25$, $\boldsymbol{v = 9.7}$ **m/s**.
2 Centripetal force is equal to the resultant force:
$W - R = mv^2/r$; for a minimum velocity $R = 0$:
$mg = mv^2/r$, $v^2 = rg$
$v^2 = 0.50 \times 9.8$, $v = 2.2$ m/s.
PE lost = KE gained, $mgh_1 = \frac{1}{2}mv^2$, $h_1 = v^2/2g$
$h_1 = 2.2^2/2 \times 9.8 = 0.25$, $h_T = h_1 + 2r = $ **1.25 m**.
3 $\omega = 2\pi/T = 2\pi/4.0 = 1.57$ rad/s.
$F = mr\omega^2 = 40 \times 6.0 \times 1.57^2 = 592$ N.
$W = mg = 40 \times 9.8 = 392$ N.
$T^2 = F^2 + W^2 = 592^2 + 392^2$, $\boldsymbol{T = 710}$ **N**.
4 The centripetal force is provided by the horizontal component of the normal contact force.
$\tan\theta = v^2/rg = 14^2/30 \times 9.8$, $\boldsymbol{\theta = 34°}$.
5 Assume the centre of mass is halfway along the blade.
Convert rev/s to rad/s:
5.0 rev/s $= 5.0 \times 2\pi = 10\pi$ rad/s.
$F = mr\omega^2 = 30 \times 2.5 \cos 20° \times (10\pi)^2 = $ **70 kN**.

UNIT 8b

RECALL TEST

1 There must be no resultant force, and no resultant moment.
2 A vector diagram of the forces acting on an object, which forms a polygon.
3 If an object is in rotational equilibrium, the sum of the clockwise moments equals the sum of the anticlockwise moments.
4 A point through which an unknown force acts.
5 Equate forces in two directions at right angles to each other (usually vertically and horizontally), and take moments about one or more points.

CONCEPT TEST

1 Take moments about O:
Σ (clockwise moments) = Σ (anticlockwise moments)
$82 \times 9.8 \times 1.0 = F \times 1.6$, $F = 500$.
Force in each arm = **250 N**.
2 Take moments about O:
Σ (clockwise moments) = Σ (anticlockwise moments)
$10 \times 9.8 \times 0.050 = T \times 0.040$, $\boldsymbol{T = 120}$ **N**.
3 Take moments about A:
Σ (clockwise moments) = Σ (anticlockwise moments)
$30 \times 9.8 \times 1.0 = 2.5 \times T \sin 40°$, $T = 183$ N, $\boldsymbol{T = 180}$ **N**.
To find force H at hinge A, equate forces vertically:
$T \sin 40° + H_V = mg$
$183 \sin 40° + H_V = 30 \times 9.8$, $H_V = 176 = 180$ N.
Equate forces horizontally:
$T \cos 40° = H_H$, $H_H = 183 \cos 40° = 140$ N.
$H^2 = H_V^2 + H_H^2 = 140^2 + 176^2$, $H = 224.9$, $\boldsymbol{H = 220}$ **N**.
4 The force X will equal the weight of the dancer, and must act upwards to balance the weight.
$X = 9.8 \times 75 = $ **740 N**.

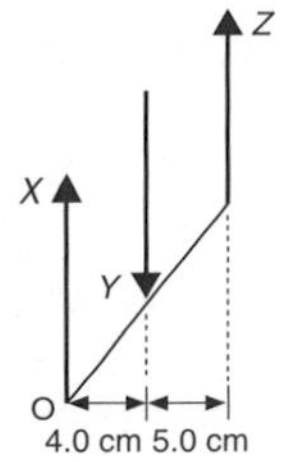

The other forces have to take the directions shown (see figure above). If they go in the opposite directions, and you take moments about Y, it cannot balance.
Take moments about Y:
Σ (anticlockwise moments) = Σ (clockwise moments)
$740 \times 0.040 = Z \times 0.050$, $\boldsymbol{Z = 590}$ **N**.
Equate forces vertically:
$X + Z = Y$, $Y = 590 + 740 = $ **1300 N**. Z is in tension.
Y is under compression.
5 Put the forces and distances in the diagram as shown here.

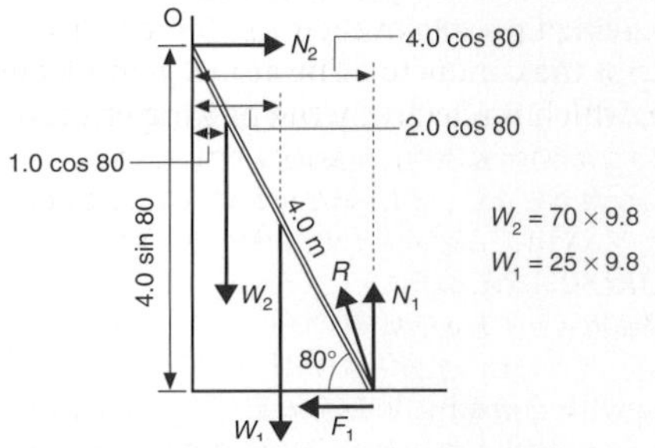

Take moments about O:
Σ (clockwise moments) = Σ (anticlockwise moments)
$(70 \times 9.8 \times \cos 80°) + (25 \times 9.8 \times 2.0 \cos 80°) + (F_1 \times 4.0 \sin 80°) = (N_1 \times 4.0 \cos 80°)$, $204 + 3.9F_1 = 0.69N_1$.
Equate forces vertically: $W_1 + W_2 = N_1$
$(70 \times 9.8) + (25 \times 9.8) = N_1$, $N_1 = 931$ N.
Equate forces horizontally: $F_1 = N_2$.
So $204 + 3.9F_1 = 0.69 \times 931$, $F_1 = 112$ N.
R is the resultant of F_1 and N_1:
$R^2 = F_1^2 + N_1^2 = 112^2 + 931^2$, $R = 938$ N, $R = 940$ N.
$\tan\theta = N_1/F$, $\boldsymbol{\theta = 83°}$.
6 Take moments about O:
Σ (anticlockwise moments) = Σ (clockwise moments)
$F_y = R_x$ and $\tan\theta = x/y$.
Combine together to get $F = R \tan\theta$.
Also we know that $F = mv^2/r$ and $W = R$, giving us $mv^2/r = mg \tan\theta$; rearranging produces
$\tan\theta = v^2/rg = 20^2/15 \times 9.8$, $\boldsymbol{\theta = 70°}$.

UNIT 9

RECALL TEST

1 It is a property that electrons have which can produce a force. (On a large scale it may be considered as a measure of the presence or absence of electrons.)
2 Current is the rate of flow of charge.
3 Potential difference is the energy used in moving unit charge between two points.
4 Resistance is the opposition to the flow of charge, and is equal to the ratio of potential difference to current.
5 In a conductor p.d. is directly proportional to the current if temperature is constant.
6 A semiconductor is a material which only partially conducts and whose resistance decreases as temperature increases.
7 A superconductor is a material which has zero resistance at a particular temperature, usually very low.
8 It is defined as the resistance of a unit cube of that material.
9 Temperature, length, cross-sectional area, and type of material.
10 The energy given to unit charge as it passes through a cell or other e.m.f. source.

CONCEPT TEST

1 $I = Q/t$, $Q = It = 2.0 \times 3.0 = \mathbf{6.0\ C}$.
$n = Q/e = 6.0/(1.6 \times 10^{-19}) = \mathbf{3.8 \times 10^{19}}$.
2 $V = E/Q$, $E = VQ = 4.0 \times 3.0 = \mathbf{12\ J}$.
3 $I = Q/t = 40/5 = \mathbf{8\ A}$.
$V = E/Q = 200/40 = \mathbf{5.0\ V}$.
4 $I = V/R = 6.0/20 = \mathbf{0.30\ A}$.
$P = I^2R = 0.30^2 \times 20 = \mathbf{1.8\ W}$.
5 $R = V/I = 24/6.0 = \mathbf{4.0\ \Omega}$.
$P = IV = 6.0 \times 24 = \mathbf{140\ W}$.
$P = 3 \times 144 = 432$ W.
$P = IV$, $432 = I \times 24$, $\mathbf{I = 18\ A}$.
6 Total resistance $R = 20 \times 10 = 200\ \Omega$.
$I = V/R = 240/200 = \mathbf{1.2\ A}$.
$P = IV = 1.2 \times 240 = \mathbf{290\ W}$.
New $R = 19 \times 10 = 190\ \Omega$, $P = V^2/R = \mathbf{300\ W}$.
7 When electrons move through a conductor they collide with atoms, giving up some of their KE; these collisions are the resistance of the conductor. The atoms gain KE and vibrate more, which is effectively the heating effect of the current. This increases the chance of a collision, so resistance increases. Electrons throughout the whole conductor start to move when the switch is closed, so a bulb lights immediately.
8 Resistance of 1 km = 12 000 Ω.
$P = I^2R$, $500 = I^2 \times 12\,000$, $I = 0.204$ A.
Output $P = IV$, $P = 0.204 \times 200\,000 = \mathbf{41\ kW}$.
9 Power input = power output/0.4, $P = 60/0.4 = 150$ W.
$R = V^2/P = 240^2/150 = 384\ \Omega$, $\mathbf{R = 380\ \Omega}$.
$R = \rho l/A$, $l = RA/\rho = 380 \times \pi \times (0.25 \times 10^{-3})^2/(2.0 \times 10^{-3})$
$\mathbf{l = 0.04\ m}$.
10 A typical torch current is 0.15 A (usually written on bulb), and a typical wire radius is 1.0 mm.
$I = nAve$, $v = I/nAe$
$v = 0.15/(5.0 \times 10^{25} \times \pi \times (1.0 \times 10^{-3})^2 \times 1.6 \times 10^{-19})$
$\mathbf{v = 6 \times 10^{-3}\ m/s}$.
Values will vary depending upon the values of I and A, but should be around 10^{-3}.

UNIT 10

RECALL TEST

1 A complete conducting path in a loop containing cells and electrical components.
2 Charge is always conserved. The total is always the same.
3 The current is the same at all points, and the sum of the p.d.s equals the e.m.f. of the cell.
4 The p.d. is the same across parallel components, and current splits at a junction, the majority taking the path of least resistance.
5 The resistance of a cell due to its internal chemical structure.
6 The p.d. dropped across the internal resistance of the cell.
7 It is a device used for providing a p.d. different from the supply.
8 The resistance.
9 The sum of the currents at a junction is zero. This is based upon the principle of conservation of charge.
10 The sum of the e.m.f.s equals the sum of the p.d.s in any closed loop. This is based upon the conservation of energy.

CONCEPT TEST

1 The energy supplied to unit charge by a cell equals the sum of all the energies used by that unit charge in moving through each component in the circuit.
$V = IR = 6 \times 5 = \mathbf{30\ V}$.
$E = VIt = 6 \times 30 \times 4 = 720$, $\mathbf{E = 700\ J}$.
2 $(1/R_1) + (1/R_2) = 1/R_T$
$(1/30.0) + (1/20.0) = 1/R_T$, $R_T = 12.0\ \Omega$.
Total $R = 12.0 + 50.0 = 62.0\ \Omega$.
$V = IR$, $I = V/R = 12.0/62.0 = 0.1935$. $\mathbf{A_1 = 0.194\ A}$.
$V = IR = 12.0 \times 0.1935 = 2.32$ V.
$I = V/R = 2.32/20.0 = 0.116$. $\mathbf{A_2 = 0.116\ A}$.
$I = V/R = 2.32/30.0 = 0.077$. $\mathbf{A_3 = 0.077\ A}$.
3 $P = I^2R = 10^2 \times 500 = \mathbf{50\ kW}$.
$(1/R_1) + (1/R_2) = 1/R_T$
$(1/500) + (1/500) = 1/R_T$, $R_T = 250\ \Omega$.
(Two identical resistors in parallel always have a combined resistance of half the value of one of them.)
So $P = I^2R = 10^2 \times 250 = \mathbf{25\ kW}$.
4 For a 1 metre length:
$(1/R_1) + (1/R_2) = 1/R_T$
$(7 \times 1/5.0) + (20 \times 1/10) = 1/R_T$, $R_T = 5/17\ \Omega$.
For 1 km: $R = (5/17) \times 1000 = 294$, $\mathbf{R = 290\ \Omega}$.
5 $(1/R_1) + (1/R_2) = 1/R_T$, $(1/10\,000) + (1/200) = 1/R_T$, $R_T = 196\ \Omega$.
$I = V/R = 10/196 = \mathbf{0.051\ A}$.
$P = V^2/R = 10^2/196 = \mathbf{0.51\ W}$.
$(1/500) + (1/200) = 1/R$, $R = 143\ \Omega$.
$I = V/R = 10/143 = \mathbf{0.070\ A}$.
$P = V^2/R = 102/143 = \mathbf{0.70\ W}$.
6 The internal resistance of the cell uses up some of the e.m.f.:
$P = IV$, $I = P/V = 10.0/8.60 = 1.16$ A.
$V = E - Ir$, $r = (E - V)I = (12.0 - 8.6)/1.16 = 2.93\ \Omega$.
$R = V/I = 8.60/1.16 = 7.41\ \Omega$.
Tot $R = 7.41 + 2.93 = \mathbf{10.3\ \Omega}$.
7 The keys short out the terminals of the battery. The keys have low resistance, so the current is high, and from $P = I^2R$ the power dissipated by the cell's internal resistance will be high. Hence the high temperature and burns.
$I = V/R = 6.5/20 = 0.33$A.
$V = E - Ir$, $E = V + Ir$
$E = 6.5 + (0.33 \times 6.0) = \mathbf{8.5\ V}$.
8 p.d across the 20 Ω resistor: 20 – 3.0 = 17 V.
$I = V/R = 17/20 = 0.85$ A.
$R = V/I = 3.0/0.85 = 3.5\ \Omega$.
$(1/R_1) + (1/R_m) = 1/R_T$, $(1/50) + (1/R_m) = 1/3.5$, $\mathbf{R_m = 3.8\ \Omega}$.
9 Using Kirchhoff's second law: take loop PTXQ:
Σ e.m.f.s = Σ p.d.s, $8 = I \times 20$, $I = 0.4$ A. $\mathbf{A_1 = 0.4\ A}$.
Take loop QXYS:
Σ e.m.f.s = Σ p.d.s, $24 = I \times 30$, $I = 0.8$ A. $\mathbf{A_2 = 0.8\ A}$.
Using Kirchhoff's first law: $A_1 + A_3 = A_2$
$0.4 + A_3 = 0.8$, $\mathbf{A_3 = 0.4\ A}$.
$P = I^2R = 0.4^2 \times 20 = \mathbf{3\ W}$.
$P = 0.8^2 \times 30 = \mathbf{20\ W}$.
10 No current flows through the meter, so just do a Kirchhoff's second law loop around the outside of the circuit:
Σ e.m.f.s = Σ p.d.s, $12 - 8.0 = I(1.0 + 2.0)$, $4.0 = 3.0I$,
$I = 1.3$ A.
$V = E - Ir = 12 - (1.3 \times 1.0) = 10.7$ V, $\mathbf{V = 11\ V}$.
Or $V = E + Ir = 8.0 + (1.3 \times 2.0) = 10.7$ V, $\mathbf{V = 11\ V}$.

UNIT 11

RECALL TEST

1 Electromagnetic and mechanical.
2 In a transverse wave the direction of oscillation is perpendicular to the direction of wave motion; in a longitudinal wave it is the same as the direction of wave motion.
3 Energy.
4 The number of waves passing a point each second, or the number of oscillations per second.
5 The distance between two adjacent positions on a progressive wave that are in phase with each other.
6 The maximum displacement.
7 It is a measure of how far along one wave cycle a point is.
8 If two or move waves of the same type occupy the same place at the same time their displacements add.
9 They must be in phase with each other and have the same wavelength and frequency.
10 They must be π out of phase with each other and have the same wavelength, frequency, and amplitude.

CONCEPT TEST

1 Time period $T = 2/6 = \mathbf{1/3\ s}$.
$f = 1/T = 1/(1/3) = \mathbf{3\ Hz}$.
2
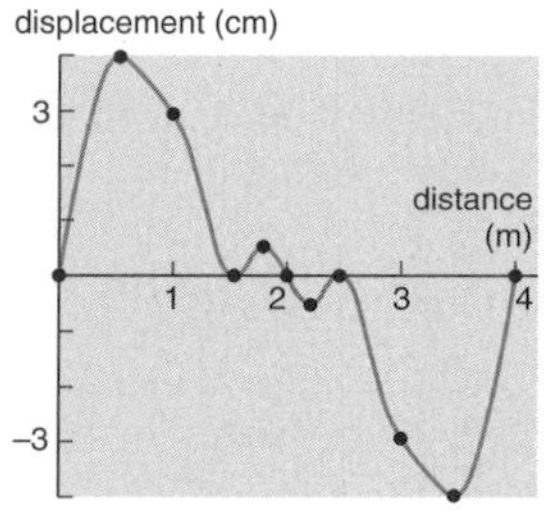

3 $v = f\lambda$, $\lambda = v/f = 300/140 = \mathbf{2.14\ m}$.
$x/\lambda = \phi/2\pi$, $\phi = 2\pi x/\lambda = 2\pi \times 0.600/2.14 = \mathbf{1.76\ rad}$.
4 $v = f\lambda$, $\lambda = v/f = 30/50 = \mathbf{0.60\ m}$.
Path difference (PD) = 38 – 30 = **8 m**.
PD = 8/0.6 = $13\frac{1}{3}$ wavelengths. $(1/3) \times 0.60 = 0.20$.
$x/\lambda = \phi/2\pi$, $\phi = 2\pi x/\lambda = 2\pi \times 0.20/0.60 = \mathbf{2.1\ rad}$.
5 a Only longitudinal waves can pass through liquids, so no S waves can pass through the Earth's outer, liquid core.
b $T = 60/10 = 6.0$ s.
$f = 1/T = 1/6.0 = 0.17$ Hz.
$v = s/t = 100/20 = 5.0$ m/s.
$v = f\lambda$, $\lambda = v/f = 5.0/(1/6) = \mathbf{30\ m}$.
6 Path difference (PD) = $(n + (1/2))\lambda$ for destructive interference.
Distance travelled, $s = 2 \times 1.4 \times 10^{-7} = 2.8 \times 10^{-7}$ m.
$s = (1/2)\lambda$, so $\lambda = 2s = 2 \times 2.8 \times 10^{-7} = 5.6 \times 10^{-7}$ m.
$v = f\lambda$, $f = v/\lambda = (3.0 \times 10^8)/(5.6 \times 10^{-7}) = \mathbf{5.4 \times 10^{14}\ Hz}$.
7 The aircraft reflects some of the waves, acting as a second source. This wave has travelled further so it arrives a bit later, producing the ghost image. The flickering occurs because of the constructive or destructive interference produced by the reflected waves combining with direct waves.
If the beam crosses the screen 625 times per second it crosses it once in 1/625 seconds. The time delay between the direct and reflected signal arriving,
$t = (3.0/40) \times (1/625) = 1.2 \times 10^{-4}$ s.
$v = d/t$, $d = vt = (3.0 \times 10^8) \times (1.2 \times 10^{-4}) = \mathbf{36\ km}$.
8 The man acts as a second source, so the variation in loudness occurs because of constructive or destructive interference produced by the two waves arriving together.
Distance walked, $s = (1/2)\lambda$, $\lambda = 2.0/(1/2)$, $\lambda = 4.0$ m.
$v = f\lambda$, $f = v/\lambda = (3.0 \times 10^8)/4.0 = \mathbf{7.5 \times 10^7\ Hz}$.
9 The intensity of the source (I_o) is spread over the surface of a sphere at radius r.
$I = I_o/4\pi r^2$
$I = (2.0 \times 10^{-2})/(4 \times \pi \times 5.0^2) = \mathbf{6.4 \times 10^{-5}\ W/m^2}$.
$I \propto A^2$, $I = kA^2$, where k is a constant.
$k = I/A^2 = (2.0 \times 10^{-2})/(3.0 \times 10^{-10})^2 = 2.2 \times 10^{17}$.
$I = kA^2$, $6.4 \times 10^{-5} = 2.2 \times 10^{17} \times A^2$, $\mathbf{A = 5.4 \times 10^{-12}\ m}$.

UNIT 12

RECALL TEST

1 The change in direction of a wave as it passes from one material to another when there is a change in speed.
2 The bending of a wave as it passes by the edge of an obstacle.
3 It bends towards the normal.
4 It is the angle of incidence at which the refracted wave will no longer be transmitted, but will pass along the boundary between the two materials.
5 When the angle of incidence is greater than the critical angle a wave is reflected off the boundary between two mediums.
6 The waves must be monochromatic (that is, they must have the same wavelength) and coherent (they must have a constant phase difference).
7 It is the same distance behind the reflecting surface as the object is in front of it, on a line perpendicular to the surface.
8 Frequency stays the same. Wavelength, speed, and direction change.
9 A piece of transparent material with lots of lines drawn on it very close together.
10 A progressive wave transmits energy; a standing wave does not.

CONCEPT TEST

1 $d = 1/m$ where m is the number of lines per metre
$d = 1/(5.0 \times 10^5) = 2.0 \times 10^{-6}$ m.
$d \sin\theta = n\lambda$,
$\sin\theta = n\lambda/d = (2 \times 3.86 \times 10^{-7})/(2.0 \times 10^{-6})$, $\boldsymbol{\theta} = \mathbf{23°}$.
Max. n occurs when θ is 90°, $n = (d \sin\theta)/\lambda$
$n = (2.0 \times 10^{-6} \times \sin 90°)/(3.86 \times 10^{-7}) = \mathbf{5}$.
2 Fringe spacing, $y = 20/(4 \times 4)$
$y = 1.25$, $y = 1.3$ mm.
$y = \lambda D/a$, $\lambda = ya/D$
$\lambda = (1.25 \times 10^{-3} \times 1.0 \times 10^{-3})/2$
$\lambda = 630$ nm.

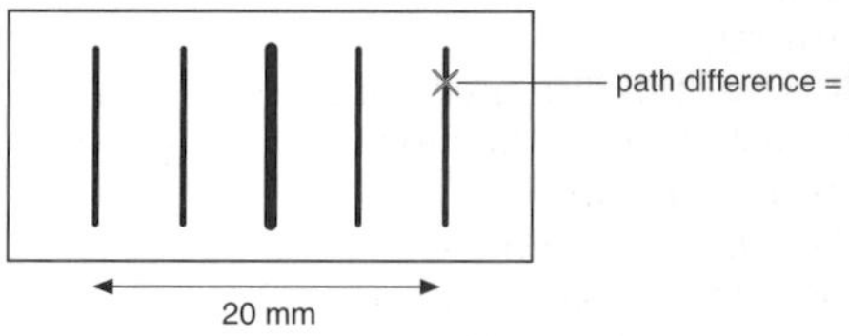

3 For the angler to see the fish, r must be
$r = \tan^{-1}(3.0/1.0) = 72°$.
$n_1 \sin\theta_1 = n_2 \sin\theta_2$, so $1.33 \sin i = 1 \times \sin 72°$, $\boldsymbol{i} = \mathbf{46°}$.
4 When wavetrains from each slit arrive together they produce interference. If the path difference is a whole number of wavelengths they arrive in phase and produce constructive interference. If the path difference is a half number of wavelengths they arrive π out of phase and produce destructive interference. This produces a series of equally spaced light and dark fringes. a = 1–2 mm, D = 1–2 m. From $v = f\lambda$, if f goes up λ goes down because v is constant. From $y = \lambda D/a$, if λ goes down y decreases. If a goes down, y increases.
5

$v = f\lambda = 6.0 \times 4.0 = \mathbf{24\ m/s}$.
The fundamental frequency without the prop would be $f = v/\lambda = 24/6.0 = 4.0$ Hz. It is the third harmonic or second overtone, so $f = 3 \times 4.0 = \mathbf{12\ Hz}$.

6 ${}_1n_2 = c_1/c_2$, $1.4 = 3.0 \times 10^8/c_2$, $\mathbf{c_2 = 2.1 \times 10^8}$ **m/s.**
If it just emerges it must be hitting the second face at the critical angle: $n_1 \sin \theta_1 = n_2 \sin \theta_2$,
so $1.4 \sin i_c = 1 \times \sin 90°$, $i_c = 46°$. $90 - 46 = 44°$.
$\theta = 180 - (60 + 44) = 76°$.
$90 - 76 = 14$,
$n_1 \sin \theta_1 = n_2 \sin \theta_2$, so $1 \times \sin i = 1.4 \times \sin 14°$, $\boldsymbol{i = 20°}$.
A smaller angle of incidence for the light entering the prism will produce total internal reflection where the light tries to leave the prism.

7 The molecules vibrate from side to side with maximum amplitude at X, producing an antinode. At Y they oscillate with minimum or zero amplitude, producing a node. At Z the molecules oscillate π out of phase with those at X.
$\lambda/2 = l$, $\lambda = 2 \times 12 = 24$ m.
$v = f\lambda$, $f = 320/24 = 13$ Hz.
$\lambda = 2 \times 0.010 = 0.020$ m.
$v = f\lambda$, $f = 320/0.020 = 16$ kHz.
Range = 13 to 16 000 Hz.
$\lambda/4 = l$, $\lambda = 4 \times 12 = 48$ m.
$v = f\lambda$, $f = 320/48 = 6.7$ Hz.
$\lambda = 4 \times 0.01 = 0.04$ m.
$v = f\lambda$, $f = 320/0.04 = 8$ kHz.
Range = 6.7 to 8000 Hz.

8 Use $n_1 \sin \theta_1 = n_2 \sin \theta_2$
$1.6 \sin 12° = n_2 \sin 90°$, $n_2 = 0.33$.
${}_1n_2 = c_1/c_2$, $c_2 = (3.0 \times 10^8)/1.6 = \mathbf{1.9 \times 10^8}$ **m/s.**
Take $c_2 = v$, $v = s/t$,
$t = s/v = 500/(1.9 \times 10^8) = \mathbf{2.6 \times 10^{-6}}$ **s.**

UNIT 13

RECALL TEST

1 The emission of electrons from the surface of a metal when electromagnetic radiation is incident upon it.
2 The energy of a photon is directly proportional to its frequency.
3 It is the minimum energy required by an electron to break free of the surface of a metal.
4 It is a device used to detect the presence of charge.
5 The potential difference required to stop a photoelectron being emitted from a surface.
6 It is the minimum frequency that will produce the emission of photoelectrons.
7 A de Broglie wave is the wave associated with a moving particle. Its wavelength is equal to the Planck constant divided by its momentum.
8 More photoelectrons are emitted.
9 The probability of the electron being there.
10 Electron and neutron diffraction.

CONCEPT TEST

1 $\lambda = h/p = (6.6 \times 10^{-34})/(0.05 \times 3) = \mathbf{4.4 \times 10^{-33}}$ **m.**

2 A photon is a small wave 'packet' with a fixed amount of energy which is directly proportional to its frequency.
$c = f\lambda$, $f = (3.0 \times 10^8)/(5.5 \times 10^{-7})$, $f = 5.45 \times 10^{14}$,
$f = 5.5 \times 10^{14}$ Hz.
$E = hf = 6.6 \times 10^{-34} \times 5.45 \times 10^{14} = \mathbf{3.6 \times 10^{19}}$ **J.**
Energy in eV, $E = (3.6 \times 10^{19})/(1.6 \times 10^{-19}) = \mathbf{2.2}$ **eV.**

3 $E = hf = 6.6 \times 10^{-34} \times 5.80 \times 10^{14} = 3.83 \times 10^{-19}$ J.
$E = \phi + KE_{max}$,
$\phi = (3.83 \times 10^{-19}) - (1.2 \times 10^{-19}) = \mathbf{2.63 \times 10^{-19}}$ **J.**
$hf_0 = \phi$, $f_0 = (2.63 \times 10^{-19})/(6.6 \times 10^{-34}) = \mathbf{3.98 \times 10^{14}}$ **Hz.**

4 $eV = \frac{1}{2}mv^2$, $1.6 \times 10^{-19} \times 2000 = \frac{1}{2} \times 9.1 \times 10^{-31} \times v^2$,
$\boldsymbol{v = 2.7 \times 10^7}$ **m/s.**
$\lambda = h/p = (6.6 \times 10^{-34})/(9.1 \times 10^{-31} \times 2.7 \times 10^7)$
$\boldsymbol{\lambda = 2.7 \times 10^{-11}}$ **m.**

5 The y intercept is the work function, ϕ; $\boldsymbol{\varphi = 2.4 \times 10^{-19}}$
The x intercept is the threshold frequency, f_0;
$\boldsymbol{f_0 = 3.6 \times 10^{14}}$ **Hz.**
The gradient equals the Planck constant.

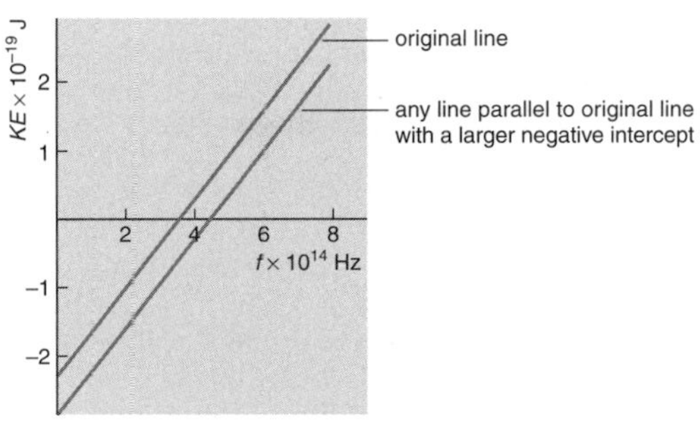

6 $\lambda = h/p = (6.6 \times 10^{-34})/(60 \times 2.0) = \mathbf{5.5 \times 10^{-36}}$ **m.**
Good diffraction occurs if the slit width is just a little bit bigger than the wavelength. Quite clearly it would be impossible for you to walk through a gap 5.5×10^{-36} m wide, so diffraction is impossible.

7 The photoelectric effect provides evidence for the particle-like nature of light, and interference and diffraction provide evidence for its wave-like nature.
$c = f\lambda$, $f = c/\lambda = (3.0 \times 10^8)/(6.0 \times 10^{-7})$, $f_0 = 5.0 \times 10^{14}$ Hz.
$f = c/\lambda = (3.0 \times 10^8)/(1.6 \times 10^{-7}) = 1.9 \times 10^{15}$ Hz.
$hf_0 = \phi$, $hf = \phi + KE_{max}$,
$h \times 1.9 \times 10^{15} = (h \times 5.0 \times 10^{14}) + (8.9 \times 10^{-19})$
$\boldsymbol{h = 6.4 \times 10^{-34}}$ **J s.**

8 Take two points and see if the product of λp is the same.
$0.50 \times 10^{-24} \times 13 \times 10^{-10} = 6.5 \times 10^{-34}$.
$1.5 \times 10^{-24} \times 4.3 \times 10^{-10} = 6.5 \times 10^{-34}$.
$h = 6.5 \times 10^{-34}$ J s.
The amplitude of the de Broglie wave is directly proportional to the probability of the electron being there.

9 $f = c/\lambda = (3.0 \times 10^8)/(1.9 \times 10^{-7})$, $f_0 = 1.6 \times 10^{15}$ Hz.
$hf = \phi + KE_{max}$
$6.6 \times 10^{-34} \times 1.6 \times 10^{15} = (7.2 \times 10^{-19}) + KE_{max}$,
$KE_{max} = 3.2 \times 10^{-19}$ J.
$eV = \frac{1}{2}mv^2$, $1.6 \times 10^{-19} \times V = 3.4 \times 10^{-19}$, $\boldsymbol{V = 2.0}$ **V.**
If the frequency is increased the energy of the photons is greater, so the maximum KE of the emitted electrons is greater, so the stopping potential is greater. If the intensity is increased more photons hit the surface, so more electrons are emitted, so the current is greater.

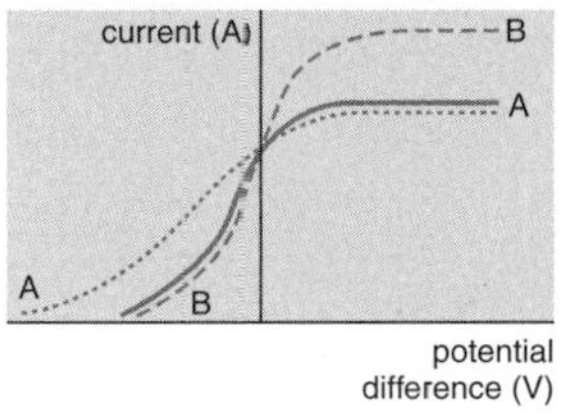

UNIT 14

RECALL TEST

1 The energy given to an electron accelerated across a potential difference of 1 volt.
2 All of an atom's electrons are in the lowest available energy levels.
3 The minimum energy required to raise an electron from one level to another for an atom in its ground state.
4 The minimum energy required to remove an electron from an atom, for an atom in its ground state.
5 It is a normal white-light spectrum with a series of dark lines upon it. The dark lines are produced by photons absorbed during electron transitions.
6 It is a series of different-coloured bright lines on a dark background. The bright lines are produced by photons emitted during electron transitions.
7 Light emitted by distant objects in space appears to have lower frequencies than it actually has because of the Doppler effect.
8 It is the constant in the relationship velocity $\propto$ distance, which shows that distant objects in space are moving faster the further they are away from us.

9 The apparent change in frequency of waves emitted by a wave source moving relative to an observer.

10 The Universe began with a huge explosion occurring at a single point, so now all the galaxies are moving away from each other.

CONCEPT TEST

1 Change in energy, $E_2 - E_1$, $-1.8 - -4.4 = 2.6$ eV.
Convert eV to joules: $2.6 \times 1.6 \times 10^{-19} = 4.2 \times 10^{-19}$ J.
$E_2 - E_1 = hc/\lambda$, $4.2 \times 10^{-19} = (6.6 \times 10^{-34} \times 3 \times 10^8)/\lambda$,
$\boldsymbol{\lambda = 480}$ **nm**.

2 Change in energy, $E_2 - E_1$, $-1.5 - -6.2 = 4.7$ eV.
Convert eV to joules: $4.7 \times 1.6 \times 10^{-19} = 7.52 \times 10^{-19}$ J.
$(1.2 \times 10^{-19}) - (7.52 \times 10^{-19}) = 4.48 \times 10^{-19}$ h,
$\boldsymbol{KE = 4.5 \times 10^{-19}}$ **J**.

3 $f' = f(c + v_o)/(c - v_s) = 500(320/(320 - 10.0)) =$ **516 Hz**.

4 The light is re-emitted in all directions so the intensity in the original direction is reduced. Possible transitions are:
$E_2 - E_1$, $-0.96 - -1.92 = 0.96$ eV.
$0.96 \times 1.6 \times 10^{-19} = 1.5 \times 10^{-19}$ J.
$E_2 - E_1 = hf$,
$f = (1.5 \times 10^{-19})/(6.6 \times 10^{-34}) = \boldsymbol{2.3 \times 10^{14}}$ **Hz**.
$E_2 - E_1$, $-0.96 - -11.6 = 10.6$ eV.
$10.6 \times 1.6 \times 10^{-19} = 1.7 \times 10^{-18}$ J.
$E_2 - E_1 = hf$,
$f = (1.7 \times 10^{-18})/(6.6 \times 10^{-34}) = \boldsymbol{2.6 \times 10^{15}}$ **Hz**.

5 $v = H_0 d = 75 \times 200 = 15\,000$ km/s.
$f' = f(c + v_o)/(c + v_s)$
$f' = 2.8 \times 10^{14}(3.0 \times 10^{-8})/((3.0 \times 10^8) + (1.5 \times 10^7))$
$\boldsymbol{f' = 2.7 \times 10^{14}}$ **Hz**.

6

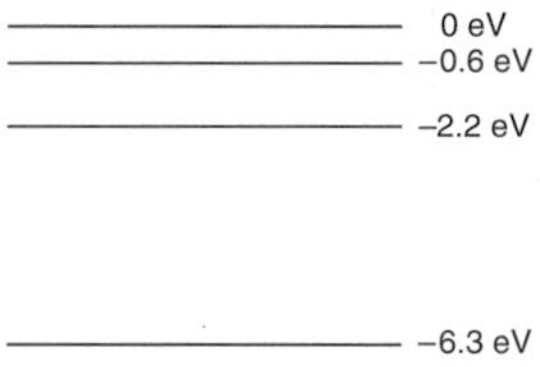

$6.3 - 4.1 = 2.2$.
$6.3 - 5.7 = 0.6$.

$E_2 - E_1$, $-0.6 - -6.3 = 5.7$ eV.
$5.7 \times 1.6 \times 10^{-19} = 9.12 \times 10^{-19}$ J.
$E_2 - E_1 = hf$,
$f = (9.12 \times 10^{-19})/(6.6 \times 10^{-34})$
$f = 1.38 \times 10^{15}$ Hz.
$\lambda = v/f = (3.0 \times 10^8)/(1.38 \times 10^{15})$
$\boldsymbol{\lambda = 2.2 \times 10^{-7}}$ **m**.
$E_2 - E_1$, $-0.6 - -2.2 = 1.6$ eV.
$1.6 \times 1.6 \times 10^{-19} = 2.56 \times 10^{-19}$ J.
$E_2 - E_1 = hf$,
$f = (2.56 \times 10^{-19})/(6.6 \times 10^{-34}) = 3.88 \times 10^{14}$ Hz.
$v = f\lambda$, $\lambda = v/f = (3 \times 10^8)/(3.88 \times 10^{14}) = \boldsymbol{7.7 \times 10^{-7}}$ **m**.
$E_2 - E_1$, $-2.2 - -6.3 = 4.1$ eV.
$4.1 \times 1.6 \times 10^{-19} = 6.56 \times 10^{-19}$ J.
$E_2 - E_1 = hf$,
$f = (6.56 \times 10^{-19})/(6.6 \times 10^{-34}) = 9.9 \times 10^{14}$ Hz.
$v = f\lambda$, $\lambda = v/f = (3 \times 10^8)/(9.9 \times 10^{14}) = \boldsymbol{3.0 \times 10^{-7}}$ **m**.
An electron is taken as having zero potential energy when it has left an atom. The force on the electron due to the nucleus is attractive, so as the electron gets closer its potential energy decreases. If it decreases from zero all values will be negative.

7 Convert to parsecs: $(2.0 \times 10^{10})/3.258 = 6.1 \times 10^3$ Mpc.
$v = H_0 d = 75 \times 6.1 \times 10^3 = \boldsymbol{4.6 \times 10^5}$ **km/s**.
$f' = f(c + v_o)/(c + v_s)$
$= 4.7 \times 10^{14}((3.0 \times 10^8) + (3.0 \times 10^6))/((3.0 \times 10^8) + (4.6 \times 10^8))$
$\boldsymbol{f' = 1.87 \times 10^{14}}$ **Hz**.

8 $c = f\lambda$, $f = (3.0 \times 10^8)/(4.4 \times 10^{-7}) = 6.8 \times 10^{14}$ Hz.
$E = hf = 6.6 \times 10^{-34} \times 6.8 \times 10^{14} = 4.5 \times 10^{-19}$ J.
$P = 60$ W means 60 J per second.
Total number of photons emitted per second $N = P/E$
$N = 60/(4.5 \times 10^{-19}) = 1.3 \times 10^{20}$.
The surface area of a sphere at 0.5 m, $A = 4\pi r^2 = 3.1$ m^2.
The number of sheets of paper required to cover this area, $n = 3.1/0.25 = 12.4$.
So number of photons hitting paper,
$N_0 = (1.3 \times 10^{20})/12.4 = \boldsymbol{1 \times 10^{19}}$.

9 If a pure gas is heated and the emitted light is analysed using a spectrometer, an emission spectrum is produced. If a gas made of molecules is heated, the vibrational and rotational energy of the molecules combines with the electron transitions to produce a spectrum consisting of a series of bands. The zero level corresponds to when an electron has left the atom.
Ionization energy is 10.4 eV.
$c = f\lambda$, $f = (3.0 \times 10^8)/(1.41 \times 10^{-7})$
$f = 2.13 \times 10^{15}$ Hz.
$E = hf = 6.6 \times 10^{-34} \times 2.13 \times 10^{15} = 1.4 \times 10^{-18}$ J.
Convert to eV: $(1.4 \times 10^{-18})/(1.6 \times 10^{-19}) =$ **8.8 eV**.
This corresponds to a jump between the 1.6 eV level and the 10.4 eV level.

UNIT 15

RECALL TEST

1 An atom that has a central positively charged nucleus containing most of the atom's mass.

2 The nucleus is about 10^{-15}, and the atom is about 10^{-10}. This gives a difference of about 10^5.

3 Most of the positively charged alpha particles pass through undeflected, indicating that the mass is concentrated in a small volume (the nucleus). A few particles bounce back in the original direction, indicating that the nucleus is positively charged.

4 It is the time taken for half of the number of nuclei present to decay, or the time taken for the rate of decay to be reduced by half.

5 A helium nucleus.

6 It can have two forms: an electron or a positron.

7 The electrostatic repulsion between positively charged nuclei means that the nuclei must have a lot of kinetic energy in order to get close enough for fusion to take place.

8 A slow-moving neutron interacts with a uranium nucleus, causing it to split into two 'daughter products', some more neutrons and a lot of energy.

9 The energy required to break a nucleus up into its individual protons and neutrons. (The mass of the nucleus is less than the total mass of all the isolated nucleons. This energy is converted into mass.)

10 A chain reaction is a succession of fissions. (Each of the neutrons produced in one fission go on to produce more fissions, and so on.)

CONCEPT TEST

1 $T_{1/2} = 3.6$ min, $\lambda = \ln 2/(3.6 \times 60) = \boldsymbol{3.2 \times 10^{-3}}$ $\mathbf{s^{-1}}$.
$A = -\lambda N$, $N = A/\lambda$
$N = (3.5 \times 10^{3)}/(3.2 \times 10^{-3}) = \boldsymbol{1.1 \times 10^6}$.

2 Mass defect,
$M = ((6 \times 1.008) + (6 \times 1.0087)) - 12.0074$
$\boldsymbol{M = 0.0928}$ **u**.
In kg: $m = 0.0928 \times 1.66 \times 10^{-27} = \boldsymbol{1.54 \times 10^{-28}}$ **kg**.
$E = mc^2 = 1.54 \times 10^{-28} \times (3.0 \times 10^8)^2 = \boldsymbol{1.4 \times 10^{-11}}$ **J**.

3 Number of moles, $n = M/A = (160 \times 10^{-3})/238$
$n = 6.72 \times 10^{-4}$ mol.
Number of nuclei, $N = n \times N_A$
$N = 6.72 \times 10^{-4} \times 6.0 \times 10^{23} = \boldsymbol{4.0 \times 10^{20}}$.
$A = -\lambda N = 2.4 \times 10^{-10} \times 4.0 \times 10^{20} = \boldsymbol{9.6 \times 10^{10}}$ **Bq**.
Power $= A \times E$ (in joules)
$P = 9.6 \times 10^{10} \times 5.4 \times 10^6 \times 1.6 \times 10^{-19} =$ **0.083 W**.

4 $\boldsymbol{A = 4, Z = 2}$.
Energy is released because the total mass of the particles after the reaction is less than before the reaction, so some of the energy has been converted into energy.
Mass defect,
$m = ((3.342\,50 + 5.005\,73) - (6.626\,09 + 1.674\,38)) \times 10^{-27}$

m = **0.047 76 × 10^{-27} kg**.
$E = mc^2 = 0.047\ 76 \times 10^{-27} \times (3.0 \times 10^8)^2 =$ **4.3 × 10^{-12} J**.

5 It is a random process because it is not possible to predict which nuclei present will decay or how long it will take before they decay.
Using d for days:
$A = A_0e^{-\lambda t} = 4.8 \times 10^4 e^{-\ln 2 \times 13.5\ d/2\ d}$
$A = 4.8 \times 10^4 e^{-4.68} = 445.4 =$ **450 Bq**.

6 Mass defect, M
$= (235.044 + 1.0087) - (94.906 + 138.906 + 2(1.0087))$
$M = 0.223$ u.
$m = 0.223 \times 1.66 \times 10^{-27} = 3.7 \times 10^{-28}$ kg.
$E = mc^2 = 3.7 \times 10^{-28} \times (3.0 \times 10^8)^2 = 3.33 \times 10^{-11}$ J.
number of moles, n = total mass/atomic mass
$n = 2.0/235 = 8.5 \times 10^{-3}$ mol.
$N = n \times N_A = 8.5 \times 10^{-3} \times 6.0 \times 10^{23} = 5.11 \times 10^{21}$.
Total energy, $E_T = E \times N = 3.33 \times 10^{-11} \times 5.11 \times 10^{21}$
E_T = **1.7 × 10^{11} J**.

7 8 He atoms represents 1 U decay, so for every 10 U present, 1 has decayed.
$N = N_0e^{-\lambda t}$, $9 = 10e^{-\ln 2 \times t/(\text{half-life of U})}$,
$\ln(10/9) = 0.693 \times t/(4.5 \times 10^9)$, t = **6.8 × 10^8 years**.

8 From graph: $T_{1/2} = 8$ days.
$\lambda = (\ln 2)/T_{1/2} = 0.693/(8 \times 24 \times 60 \times 60) = 1.0 \times 10^{-6}\ s^{-1}$.
$A = -\lambda N$, $2.0 \times 10^4 = 1.0 \times 10^{-6} \times N$, $N = 2.0 \times 10^{10}$.
Total $N_T = N + (N \times 5.0 \times 10^7)$,
$N_T = (2.0 \times 10^{10}) + ((2.0 \times 10^{10}) \times 5.0 \times 10^7)$
N_T = **1.0 × 10^{18}**.
Number of mol,
$n = (1.0 \times 10^{18})/(6.0 \times 10^{23}) = 1.667 \times 10^{-6}$ mol.
Mass = n × mass number
$m = 1.667 \times 10^{-6} \times 127 =$ **2.1 × 10^{-4} g**.

9 $A = A_0e^{-\lambda t} = 100e^{-\ln 2 \times 20\ h/15\ h}$
$A = 100e^{-0.924} = 40$ Bq.
Total volume corresponds to activity $A = 40$ Bq.
1.0 cm^3 corresponds to 0.004 Bq, so by ratios:
$V/1.0 = 40/0.004$, V = **1.0 × 10^4 cm^3**.

UNIT 16

RECALL TEST

1. The study of subatomic particles and phenomena, using the idea that things can only change by discrete or fixed amounts.
2. Electromagnetism, gravity, and strong and weak nuclear.
3. It is a particle that moves between two objects and mediates one of the fundamental interactions.
4. A particle which cannot be broken up or decay into any smaller particles.
5. Electromagnetic, gravitational, and weak nuclear interactions.
6. Leptons = 1, anti-leptons = –1, hadrons = 0, baryons = 0.
7. Fermions have half-integer spin, i.e. 1/2, 3/2, etc. Bosons have whole-integer spin, i.e. 0, 1, 2, etc.
8. It is a particle with zero rest mass produced in beta decay.
9. The total lepton number in any particle interaction must always be the same, or the total of each type of lepton number must remain constant.
10. The range of an interaction (force) is inversely proportional to the mass of the exchange particle. The mass of the gravitational and electromagnetic exchange particles is zero, therefore their range is infinite.

CONCEPT TEST

1 a Gravity. **b** Electromagnetic. **c** Weak nuclear.

2 a Gravity, electromagnetic, and weak nuclear.
b Gravity and weak nuclear.
c Gravity, electromagnetic, strong and weak nuclear.
d Gravity, electromagnetic, and weak nuclear.

3 a A neutron decays into a proton, an electron, and an electron anti-neutrino. **b** A muon (mu minus) decays into an electron, a muon neutrino, and an electron anti-neutrino. **c** A pion (π^+) decays into an anti-muon (mu plus) and muon neutrino.

4 Pair production occurs when a gamma ray changes into an electron–positron pair. Another particle is required so that the conservation of energy and momentum can occur.

5 a The repulsion between two electrons with the exchange of a photon. **b** The decay of a pion (π^+) into an anti-muon (mu plus) and muon neutrino with the exchange of a W$^+$ boson. **c** A proton interacting with an anti-neutrino to produce a positron and a neutron with a W$^-$ boson exchange.

6

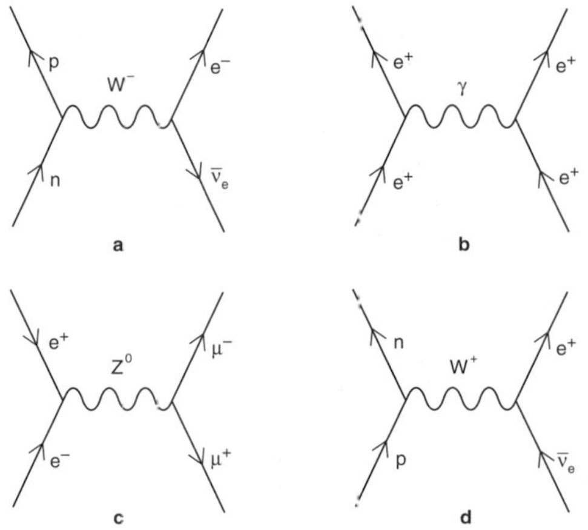

7 a

	μ^-	→	e^-	+	$\bar{\nu}_e$	+	ν_μ	
L_e	0	=	1	–	1	+	0	Y
L_μ	1	=	0	+	0	+	1	Y
L_τ	0	=	0	+	0	+	0	Y
Tot	1	=	1	–	1	+	1	Y

b

	μ^+	→	e^+	+	ν_e	
L_e	0	=	–1	+	1	Y
L_μ	–1	=	0	+	0	N
L_τ	0	=	0	+	0	Y
Tot	–1	=	–1	+	1	N

c

	ν_e	+	n	→	p	+	e^-	
L_e	1	+	0	=	0	+	1	Y
L_μ	0	+	0	=	0	+	0	Y
L_τ	0	+	0	=	0	+	0	Y
Tot	1	+	0	=	0	+	1	Y

a and **c** are allowed, **b** is not.

8 a

	μ^+	→	e^+	+	X	+	ν_e
L_e	0	=	–1	+	?	+	1
L_μ	–1	=	0	+	?	+	0
L_τ	0	=	0	+	?	+	0

So it must have
$L_e = 0$, $L_\mu = -1$, $L_\tau = 0$ and not be μ^+
which is a muon anti-neutrino ($\bar{\nu}_\mu$). So
$\boldsymbol{\mu^+ \rightarrow e^+ + \bar{\nu}_\mu + \nu_e}$

b

	τ^-	+	μ^+	+	p	→	e^+	+	ν_e	+	$\bar{\nu}_\mu$	+	X
L_e	0	+	0	+	0	=	–1	+	1	+	0	+	?
L_μ	0	–	1	+	0	=	0	+	0	–	1	+	?
L_τ	1	+	0	+	0	=	0	+	0	+	0	+	?

So it must have:
$L_e = 0$, $L_\mu = 0$, $L_\tau = 1$ and not be τ^-
which is a tau neutrino (ν_τ). So
$\boldsymbol{\tau^- + \mu^- + p \rightarrow e^+ + \nu_e + \bar{\nu}_\mu + \nu_\tau}$
Charge is also conserved in both particle interactions.

UNIT 17

RECALL TEST

1. Any particle which 'feels' the strong interaction as well as the other three interactions.
2. A baryon is made up of three quarks, and a meson is made up of only two quarks.
3. The gluon.

4 It is the name given to the strange behaviour whereby some particles which are formed by the strong interaction decay by the weak interaction.
5 Hadrons 'feel' the strong interaction, leptons do not. Hadrons are not fundamental particles, leptons are.
6 Baryon numbers are quantum numbers given to represent baryons. (The baryon number of baryons is 1, and of anti-baryons is –1; all other particles have baryon number 0.)
7 A quark is a fundamental particle which makes up hadrons, yet cannot exist in isolation.
8 Deep inelastic scattering of electrons produced a diffraction pattern which was the first evidence of quarks.
9 All hadrons; baryons and mesons.
10 Diffraction of de Broglie waves or collision experiments.

CONCEPT TEST

1 A down quark changes into an up quark. Weak interaction.
2 **a** Electromagnetic. **b** Weak nuclear.
3
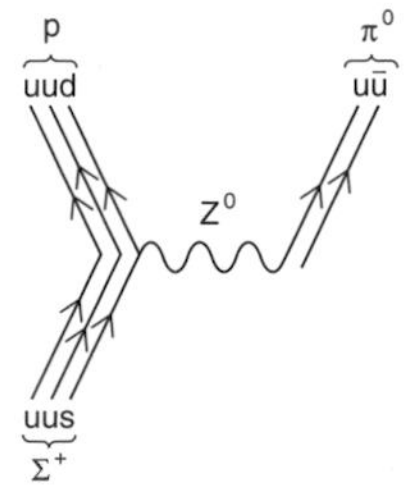

4 Mesons contain two quarks, baryons contain three. All baryons are fermions and all mesons are bosons.

a	Λ	$\rightarrow$	π^-	+	p					
Q	0	=	–1	+	1	Y				
B	1	=	0	+	1	Y				
S	–1	=	0	+	0	N				
b	K^+	+	K^-	$\rightarrow$	π^0					
Q	1	–	1	=	0	Y				
B	0	+	0	=	0	Y				
S	1	–	1	=	0	Y				
c	n	$\rightarrow$	p	+	e^-	+	ν_e			
Q	0	=	1	+	0	+	0	N		
B	1	=	1	+	0	+	0	Y		
S	0	=	0	+	0	+	0	Y		

In **a**, Λ is a 'strange' particle and decays by the weak interaction, so S may change by +1 or –1. It changes by +1 so this interaction is allowed. **b** is allowed. **c** is not.

5 **a**

	π^+	+	n	$\rightarrow$	Λ	+	X	
Q	1	+	0	=	0	+	?	
B	0	+	1	=	1	+	?	
S	0	+	0	=	–1	+	?	

So it must have
$Q = 1$, $B = 0$, $S = 1$
which is a kaon (K^+). So
$\boldsymbol{\pi^+}$ **+ n** $\rightarrow$ $\boldsymbol{\Lambda}$ **+** $\mathbf{K^+}$

b

	p	+	p	$\rightarrow$	p	+	X	+	π^0	
Q	1	+	1	=	1	+	?	+	0	
B	1	+	1	=	1	+	?	+	0	
S	0	+	0	=	0	+	?	+	0	

So it must have
$Q = 1$, $B = 1$, $S = 0$
which is a proton (p). So
p + p $\rightarrow$ **p + p +** $\boldsymbol{\pi^0}$

6 Proton uud, neutron udd.
Baryon number = (1/3) + (1/3) + (1/3) = 1.
Charge = (2/3) + (–1/3) + (–1/3) = 0.
Weak interactions can change quark flavour (type).
a All mesons are made of a quark and an anti-quark. So:
K^0 has $Q = 0$, $B = 0$, $S = 1$. This can only be made up from $\mathbf{d\bar{s}}$.
Q (–1/3) + (1/3) = 0
B (1/3) + (–1/3) = 0
S 0 + 1 = 1
K^- has $Q = –1$, $B = 0$, $S = –1$. This can only be made up from $\mathbf{\bar{u}s}$.
Q (–1/3) + (–2/3) = –1
B (1/3) + (–1/3) = 0
S –1 + 0 = –1
π^- has $Q = –1$, $B = 0$, $S = 0$. This can only be made up from $\mathbf{\bar{u}d}$.
Q (–2/3) + (–1/3) = –1
B (–1/3) + (1/3) = 0
S 0 + 0 = 0
b All baryons have three quarks. So:
Λ has $Q = 0$, $B = 1$, $S = –1$. This can only be made up from dus.
Q (–1/3) + (2/3) + (–1/3) = 0
B (1/3) + (1/3) + (1/3) = 1
S 0 + 0 + (–1) = –1
Ω^- has $Q = –1$, $B = 1$, $S = –3$. This can only be made up from **sss**.
Q (–1/3) + (–1/3) + (–1/3) = –1
B (1/3) + (1/3) + (1/3) = 1
S (–1) + (–1) + (–1) = –3
Ω^+ has $Q = 1$, $B = –1$, $S = 3$. This can only be made up from $\mathbf{\bar{s}\bar{s}\bar{s}}$.
Q (1/3) + (1/3) + (1/3) = 1
B (–1/3) + (–1/3) + (–1/3) = –1
S 1 + 1 + 1 = 3
7 Positively charged particles are curving clockwise, so there must be a magnetic field acting downwards (see unit 22), and therefore negatively charged particles will curve anti-clockwise. At B the particle decays into an uncharged particle and a positively charged particle, and also produces a gamma ray. The gamma ray goes on to produce an electron–positron pair at E. The uncharged particle then decays into a positively charged particle and a negatively charged particle at C. This positively charged particle then stops at D.

UNIT 18

RECALL TEST

1 A region of space where a mass will experience a force.
2 A region of space where a charge will experience a force.
3 The force between two masses is directly proportional to the product of their masses and inversely proportional to the distance between their centres.
4 The force between two charges is directly proportional to the product of their charges and inversely proportional to the distance between their centres.
5 It is the force acting on unit mass (1 kg) at a point in space.
6 It is the force acting on unit positive charge (+1 C) at a point in space.
7 It is the work done in moving unit mass charge from infinity to a point in space.
8 It is the work done in moving unit positive charge from infinity to a point in space.
9 The path taken by unit mass in a gravitational field.
10 A line joining points of equal potential.

CONCEPT TEST

1 $F = -Gm_1m_2/r^2$
$F = -6.67 \times 10^{-11} \times 6.0 \times 10^{24} \times 7.4 \times 10^{22}/(3.8 \times 10^8)^2$
$\boldsymbol{F = -2.1 \times 10^{20}}$ **N.**
The force of the Earth on the Moon acts from the centre of the Moon towards the centre of the Earth, and vice versa.
2 Treat the spheres as if they are point charges.
$F = Q_1Q_2/4\pi\varepsilon_0 r^2$
$F = 0.20 \times 10^{-4} \times 6.0 \times 10^{-3}/(4 \times \pi \times 8.8 \times 10^{-12} \times 0.80^2)$
$\boldsymbol{F = 1700}$ **N.**
3 $E = Q/4\pi\varepsilon_0 r^2$
$E = 10 \times 10^{-3}/(4 \times \pi \times 8.8 \times 10^{-12} \times 0.70^2)$
$\boldsymbol{E = 1.8 \times 10^8}$ **N/C.**
$V = Q/4\pi\varepsilon_0 r$
$V = 10 \times 10^{-3}/(4 \times \pi \times 8.8 \times 10^{-12} \times 0.70)$
$\boldsymbol{V = 1.3 \times 10^8}$ **J/C.**

4 Field lines and lines of equipotential are always perpendicular to each other.

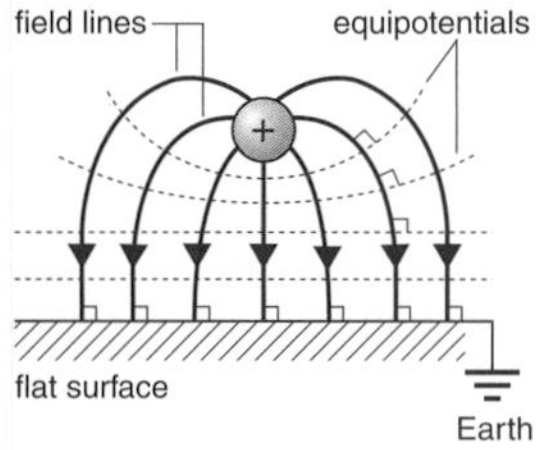

5 $g = GM/r^2$
$16.7 = (6.67 \times 10^{-11} \times M)/(8.20 \times 10^6)^2$, $\boldsymbol{M = 1.68 \times 10^{25}}$ **kg**.
$V_g = -GM/r = -(6.67 \times 10^{-11} \times 1.68 \times 10^{25})/(8.24 \times 10^6)$
$\boldsymbol{V_g = -1.36 \times 10^8}$ **J/kg**.

6 Electric field strength equals the potential gradient. Take a tangent at a point on a potential graph to get the field strength.
$E = -\Delta V/\Delta r$
$E = (-3.6 \times 10^{-2})/0.80 = -0.045$ N/C.

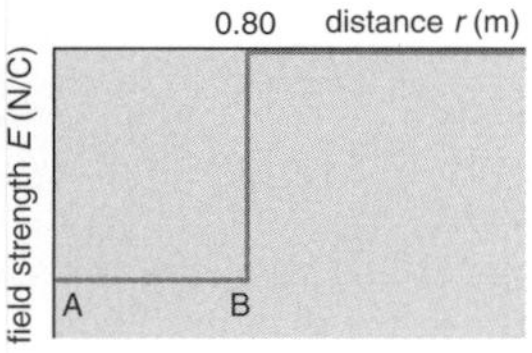

7 $E = Q/4\pi\varepsilon_0 r^2$
$= (92 \times 1.6 \times 10^{-19})/(4 \times \pi \times 8.8 \times 10^{-12} \times (1.0 \times 10^{-8})^2)$
$\boldsymbol{= 1.3 \times 10^9}$ **N/C**.
$V = Q/4\pi\varepsilon_0 r$
$V = (92 \times 1.6 \times 10^{-19})/(4 \times \pi \times 8.8 \times 10^{-12} \times 1.0 \times 10^{-8})$
$\boldsymbol{V = 13}$ **J/C**.

8 Field strength is a vector quantity, potential is a scalar. Work out the field strength due to each mass and add vectorally.
$g_A = -GM/r^2 = -(6.67 \times 10^{-11} \times 2.0 \times 10^{15})/(0.60 \times 10^6)^2$
$g_A = -3.7 \times 10^{-7}$ N/kg.
$g_B = -GM/r^2 = -(6.67 \times 10^{-11} \times 6.2 \times 10^{15})/(0.80 \times 10^6)$
$g_B = -6.5 \times 10^{-7}$ N/kg.
$g_A^2 + g_B^2 = g^2$, $g = \boldsymbol{-7.5 \times 10^{-7}}$ **N/kg**.
Work out potential due to each mass and just add:
$V_{g1} = -GM/r = -(6.67 \times 10^{-11} \times 2.0 \times 10^{15})/(0.60 \times 10^6)$
$V_{g1} = -0.22$ J/kg.
$V_{g2} = -GM/r = -(6.67 \times 10^{-11} \times 6.2 \times 10^{15})/(0.80 \times 10^6)$
$V_{g2} = -0.52$ J/kg.
$V_g = V_{g2} + V_{g1} = \boldsymbol{-0.74}$ **J/kg**.

9 $F_1 = -Gm_{water}m_{Moon}/r^2$
$F_1 = -(6.67 \times 10^{-11} \times 1.0 \times 7.4 \times 10^{22})/(4.0 \times 10^8)^2$
$F_1 = -3.1 \times 10^{-5}$ N.
$F_2 = -Gm_{water}m_{Sun}/r^2$
$F = -(6.67 \times 10^{-11} \times 1.0 \times 2.0 \times 10^{30})/(1.5 \times 10^{11})^2$
$\boldsymbol{F = -5.9 \times 10^{-3}}$ **N**.
The vector sum of these:
$F_1^2 + F_2^2 = F^2$, $F^2 = -(3.1 \times 10^{-5})^2 + (-5.9 \times 10^{-3})^2$,
$F = -5.9 \times 10^{-3}$ N.
The direction is given by $\tan\theta = F_1/F_2$, $\boldsymbol{\theta = 0.3°}$.
The Sun has a much greater effect than the Moon. (This does not take into account centripetal force required for orbit.)

10 Take a tangent to the curve at a distance of 1.0×10^{-10} m and work out its gradient.
$E = -\Delta V/\Delta r = 20/(14.5 \times 10^{-11})$
$\boldsymbol{E = -1.4 \times 10^{11}}$ **N/C**.
Force = field strength × charge
$F = -1.4 \times 10^{11} \times 1.6 \times 10^{-19} = \boldsymbol{-2.2 \times 10^{-8}}$ **N**.

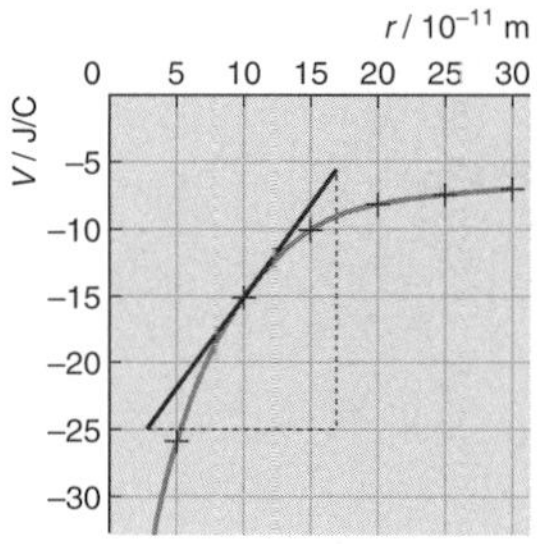

UNIT 19

RECALL TEST

1. Multiply its mass by its gravitational field potential.
2. The velocity with which an object must be projected from the surface of a planet in order for it to completely leave the gravitational pull of that planet.
3. The gravitational force of the Earth's mass acting on the mass of the satellite.
4. Charge density is the charge per unit volume and surface charge density is the charge per unit area.
5. Charge can take two forms, positive and negative, while mass only has one form.
6. It is a field where the field strength is the same at all points.
7. G is the same throughout space, whereas ε varies depending upon the material in which the charges lie.
8. Charge concentrates at points on surfaces.
9. g is directly proportional to r.
10. Field strength equals the potential gradient, so if the gradient is zero the field strength is zero.

CONCEPT TEST

1 $g = -GM/r^2$
$g = -(6.7 \times 10^{-11} \times 6.0 \times 10^{24})/(6.7 \times 10^6)^2 = \boldsymbol{-9.0}$ **m/s²**.
$T^2 = 4\pi^2 r^3/GM$
$T^2 = 4\pi^2 \times (6.7 \times 10^6)^3/(6.7 \times 10^{-11} \times 6.0 \times 10^{24})$
$\boldsymbol{T = 5.4 \times 10^3}$ **s**, or $\boldsymbol{T = 1.5}$ **h**.

2 $V_E = Q/4\pi\varepsilon_0 r$
$V_E = (1.6 \times 10^{-19})/(4 \times \pi \times 8.8 \times 10^{-12} \times 1.0 \times 10^{-10})$
$V_E = 14.5$ J/C.
Energy = $V_E \times Q$
$E = 14.5 \times 1.6 \times 10^{-19} = \boldsymbol{2.3 \times 10^{-18}}$ **J**.

3 $v^2 = 2GM/r$,
$v^2 = (2 \times 6.7 \times 10^{-19} \times 6.0 \times 10^{26})/(5.0 \times 10^7)$
$\boldsymbol{v = 40}$ **km/s**.
$KE = mv^2/2$, $= 0.5 \times 2.0 \times (4.0 \times 10^5)^2 = \boldsymbol{1.6 \times 10^{11}}$ **J**.

4 $T^2 = Kr^3$, $K = T^2/r^3 = (1.00)^2/(1.50 \times 10^{11})^3$
$K = 2.96 \times 10^{-34}$.
$T^2 = 2.96 \times 10^{-34} \times (8.00 \times 10^{11})^3$, $\boldsymbol{T = 12.3}$ **Earth years**.

5 $g = 4G\rho r/3$,
$g_1 = 4 \times G \times 0.50 \times \rho \times 0.75 \times r/3 = 0.375(4G\rho r/3)$
$g_1 = 0.375g$.
Assume initial velocity is the same on both planets:
On Earth, $v^2 = u^2 + 2as$, $0^2 = u^2 - (2 \times g \times 1.2)$
$u = (2.4g)^{0.5}$.
On planet, $v^2 = u^2 + 2as$, $0^2 = (2.4g) - (2 \times 0.375gs)$
$2.4g = (2 \times 0.375)gs$, $\boldsymbol{s = 3.2}$ **m**.

6 21 squares of $1.6 \times 10^6 \times 1.25$.
$\Delta V_g = 21 \times 2.0 \times 10^6$
$\Delta V_g = 4.2 \times 10^7$ J/kg.
energy = $\Delta V_g \times m$
energy = $4.0 \times 10^7 \times 600 = \boldsymbol{2.5 \times 10^{10}}$ **J**.

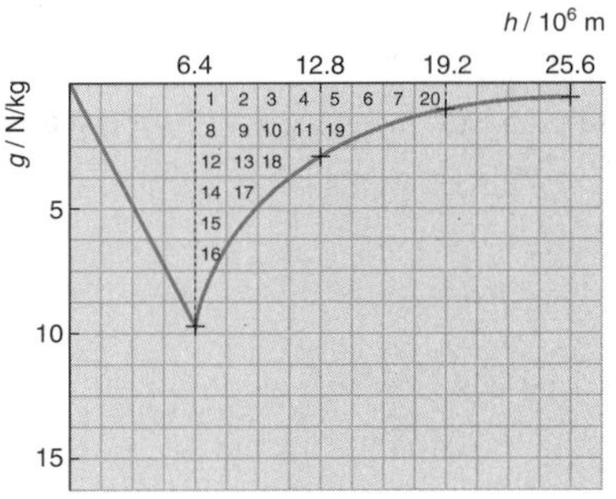

7 Potential at point, $V_E = Q/4\pi\varepsilon_0 r$.
$V_E = (79 \times 1.6 \times 10^{-19})/(4 \times \pi \times 8.8 \times 10^{-12} \times r)$.
$PE = V_E \times Q$, Q for alpha particle $= 2 \times 1.6 \times 10^{-19}$:
$PE = (79 \times 1.6 \times 10^{-19} \times 2 \times 1.6 \times 10^{-19})/(4 \times \pi \times 8.8 \times 10^{-12} \times r)$
$PE = (4.0 \times 10^{-36})/(1.1 \times 10^{-10} \times r)$,
KE lost = PE gained
$2.4 \times 10^{-13} = (4.0 \times 10^{-36})/(1.1 \times 10^{-10} \times r)$,
$\mathbf{r = 1.5 \times 10^{-13}}$ **m.**

8 The field strength is the same at all points, which is usually indicated by the presence of parallel field lines.
$F_E = W$, $F_E = QE$, $E = V/d$, $W = mg$, $QV/d = mg$, $Q = mgd/V$
$Q = (4.0 \times 10^{-15} \times 9.8 \times 0.0050)/400 = 4.9 \times 10^{-19}$ C.
$n = (4.9 \times 10^{-19})/(1.6 \times 10^{-19}) =$ **3**.

9 $F = -Gm_1m_2/r^2$
$= -(6.7 \times 10^{-11} \times 2.0 \times 10^{32} \times 2.0 \times 10^{32})/(1.0 \times 10^{12})^2$
$\mathbf{F = -2.7 \times 10^{30}}$ **N.**
The common centre of mass lies midway between them. Centripetal force is provided by the force between them.
$mr\omega^2 = F$,
$\omega^2 = (2.7 \times 10^{30})/(2.0 \times 10^{32} \times 0.50 \times 10^{12})$
$\omega = 1.6 \times 10^{-7}$ rad/s.
$T = 2\pi/\omega = (2 \times \pi)/(1.6 \times 10^{-7}) = \mathbf{3.9 \times 10^7}$ **s.**

UNIT 20

RECALL TEST

1 It is the ability to store charge, and is equal to the charge stored per unit voltage.
2 The ratio of the permittivity of a material to the permittivity of a vacuum.
3 The charge stored on each capacitor is the same.
4 An exponential variation is one in which quantity changes by the same factor over equal changes in the other quantity e.g. the current in a *CR* circuit always halves over a fixed period of time.
5 The time taken for a value (*I*, *Q*, or *V*)to decrease to 1/e of its original value.
6 The time taken for a value (*I*, *Q*, or *V*) to decrease to 1/2 of its original value.
7 $1/C_T = 1/C_1 + 1/C_2$.
8 $C_T = C_1 + C_2$.
9 Work is done against the repulsive force produced between the charge already present on the capacitor and more charge moving onto it.
10 The separation of the plates, the area of plate overlap, and the material between the plates.

CONCEPT TEST

1 $C = \varepsilon A/d = (8.8 \times 10^{-12} \times 0.4)/0.0040 = 8.8 \times 10^{-10}$ *F*,
$\mathbf{C = 9 \times 10^{-10}}$ ***F.***
$Q = CV = 8.8 \times 10^{-10} \times 4.0 = 3.5 \times 10^{-9}$, $\mathbf{Q = 4 \times 10^{-9}}$ **C.**

2 $C = C_1 + C_2 = (2 \times 10^{-6}) + (4 \times 10^{-6}) = 6 \times 10^{-6}$ F, $C = 6$ µF.
$1/C_T = 1/C_1 + 1/C_2 = (1/(6 \times 10^{-6})) + (1/(6 \times 10^{-6}))$, $\mathbf{C = 3}$ **µF.**
$Q = CV = (3 \times 10^{-6}) \times 12 = 36 \times 10^{-6} =$ **36 µC.**
p.d. across 6 F capacitor, $V = Q/C$
$V = (36 \times 10^{-6})/(6 \times 10^{-6}) = 6$ V.
Therefore p.d. across capacitor P: 12 – 6 = 6 V.
$Q = CV = (4 \times 10^{-6}) \times 6 = 24 \times 10^{-6} = 20$ µC.
$E = \frac{1}{2}CV^2 = \frac{1}{2} \times 4 \times 10^{-6} \times 6^2 = \mathbf{7 \times 10^{-5}}$ **J.**

3 $V = Q/4\pi\varepsilon_0 r$
$V = (4.8 \times 10^{-9})/(4 \times \pi \times 8.8 \times 10^{-12} \times 0.040)$
$\mathbf{V = 1.1 \times 10^3}$ **V.**
$C = Q/V = (4.8 \times 10^{-9})/(1.1 \times 10^3) = 4.4 \times 10^{-12} =$ **4.4 pF.**

4 $C = \varepsilon A/d = (8.8 \times 10^{-12} \times 10 \times 10^6)/500$
$C = 1.76 \times 10^{-7} = \mathbf{1.8 \times 10^{-7}}$ **F.**
$E = \frac{1}{2}CV^2 = 0.5 \times 1.76 \times 10^{-7} \times (1.0 \times 10^4)^2 =$ **8.8 J.**
$Q = CV = 1.76 \times 10^{-7} \times 1.0 \times 10^4 = 1.76 \times 10^{-3}$ C.
$C = \varepsilon A/d = (8.8 \times 10^{-12} \times 10 \times 10^6/1000)$
$C = 8.8 \times 10^{-8}$ F.
$E = \frac{1}{2}Q^2/C = 0.5 \times (1.76 \times 10^{-3})^2/8.8 \times 10^{-8} =$ **18 J.**
Energy increases because as the cloud is moved up, work is done against the attractive force between the cloud and the ground.

5 $Q = CV = (1.2 \times 10^{-6}) \times 12 =$ **14 µC.**
Total capacitance, $C_T = 1.2$ µF + 4 µF = 5.2 µF.
$V = Q/C = (14 \times 10^{-6})/(5.2 \times 10^{-6}) =$ **2.7 V.**
$E = \frac{1}{2}Q^2/C = 0.5 \times (14 \times 10^{-6})^2/(1.2 \times 10^{-6})$
$E = 8.2 \times 10^{-5}$ J.
$E = \frac{1}{2}Q^2/C = 0.5 \times (14 \times 10^{-6})^2/(5.2 \times 10^{-6})$
$E = 1.9 \times 10^{-5}$ J.
Energy lost, $E_L = (8.2 - 1.9) \times 10^{-5} =$ **63 µJ.**

6 Time before switching off $t = 0.5$ s.
$V = V_0 e^{-t/CR} = 12e^{-0.5/4.0\,\mathrm{n} \times 200\,\mathrm{M}} = 12e^{-0.63} =$ **6.4 V.**
(n = nano = 10^{-9}, M = mega = 10^6)

7 Area under graph equals charge stored. 1 square is worth $2.0 \times 10^{-4} \times 1.0 = 2.0 \times 10^{-4}$ C.
There are approximately 5 squares:
$Q = 5.0 \times 2.0 \times 10^{-4} =$ **1.0 mC.**
$C = Q/V$
$C = 1.0 \times 10^{-3}/6.0 = \mathbf{1.7 \times 10^{-4}}$ **F.**

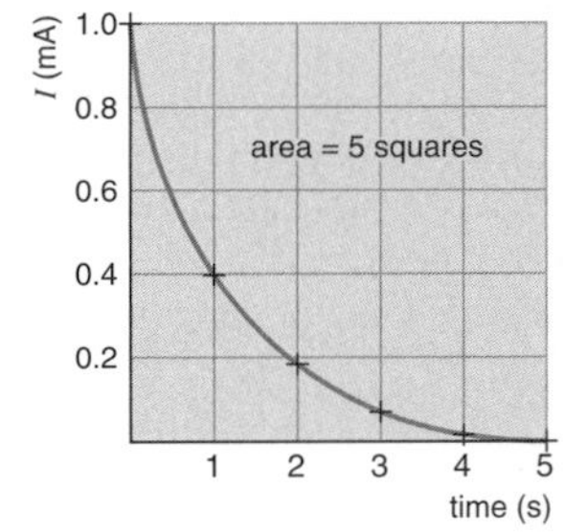

8 Maximum charge that could be stored, $Q_0 = CV$
$Q_0 = 6.0 \times 10^{-6} \times 12 = 7.2 \times 10^{-5}$ C.
$Q = Q_0(1 - e^{-t/CR}) = 7.2 \times 10^{-5}(1 - e^{-0.020/(6.0\,\mu \times 4.0\,\mathrm{M})})$
$Q = 7.2 \times 10^{-5}(1 - e^{-833\,\mu}) =$ **60 nC.**
$I = Q/t = (60 \times 10^{-9} \times 50)/1 = \mathbf{3.0 \times 10^{-6}}$ **A.**
(µ = micro = 10^{-6})

9 The combined capacitance of X and Y is C_T:
$1/C_T = 1/C_1 + 1/C_2$
$1/C_T = 1/2\,\mu + 1/4\,\mu$, $C_T = 1.3$ µF.
$Q = CV = 1.3 \times 10^{-6} \times 12 = 1.6 \times 10^{-5}$ C.
$V_X = Q/C = (1.6 \times 10^{-5})/(2 \times 10^{-6}) =$ **8 V.** So $\mathbf{V_Y = 4}$ **V.**
This could be done by ratios.
When the switches are closed the resistors (A and B) fix the p.d.s across the capacitors.
$R = R_A + R_B = 6$ kΩ.
$I = V/R = 12/(6 \times 10^3) = 2 \times 10^{-3}$ A.
$V_B = IR$, $V = 2 \times 10^{-3} \times 4 \times 10^3$, $V_B = 8$ V. $V_X = 4$ V.
Before the switch is closed the potentials would be 8 V at P and 4 V at T, so current would flow from P to T.

UNIT 21

RECALL TEST

1 A region where all the magnetic fields produced by individual atoms point in the same direction.
2 The product of magnetic flux density and area, or how much magnetic field is passing through an area perpendicular to the direction of the field.
3 The force per unit length per unit current acting upon a wire lying in a magnetic field, or how much magnetic field is passing through unit area (1 m^2) perpendicular to the direction of the field.

4 A region of space where a magnetic material will experience a force.
5 It is a line showing the direction of the magnetic flux, or the path that an isolated north pole would take in the field.
6 Iron filings, and a plotting compass.
7 Use the right-hand grip rule.
8 A coil of wire that produces a magnetic field like a bar magnet.
9 It is a measure of how easily a magnetic field passes through a material.
10 Look at the ends of the solenoid. If the current is flowing clockwise it is a south pole; if it is flowing anticlockwise it is a north pole.

CONCEPT TEST

1 $B = (\mu_r\mu_0 I)/(2\pi a)$
$B = (2.5 \times 4\pi \times 10^{-7} \times 2.0)/(2 \times \pi \times 0.80) = \mathbf{1.3 \times 10^{-6}\ T}$.
2 $B = (\mu_0 NI)/l$
$B = (4\pi \times 10^{-7} \times 1000 \times 4.0)/0.16 = \mathbf{3.1 \times 10^{-2}\ T}$.
At the end: $B_E = B/2 = \mathbf{1.6 \times 10^{-2}\ T}$.
$A = \pi r^2 = \pi \times 0.020^2 = 1.3 \times 10^{-3}\ m^2$.
$\phi = BA = 3.1 \times 10^{-2} \times 1.3 \times 10^{-3} = \mathbf{4.0 \times 10^{-5}\ Wb}$.
3 $B = (\mu_0 NI)/(2r)$
$2.0 \times 10^{-3} = (4\pi \times 10^{-7} \times 200 \times 12)/(2 \times r)$, $\boldsymbol{r} = \mathbf{0.75\ m}$.
4 Vertical component of field $B_V = B \cos 60°$
$B_V = 6.0 \times 10^{-4} \cos 60° = 3.0 \times 10^{-4}$ T.
$B = \mu_0 nI$, $3.0 \times 10^{-4} = 4\pi \times 10^{-7} \times 4000 \times I$, $\boldsymbol{I} = \mathbf{60\ mA}$.
5 $B = \mu_0 NI/l$, $B_H = (4\pi \times 10^{-7} \times 500 \times 4.0)/0.010 = 0.25$ T.
$B = \mu_0 NI/l$, $B_V = (4\pi \times 10^{-7} \times 1000 \times 3.0)/0.010 = 0.38$ T.
Pythagoras's theorem gives:
$B_R = (0.38^2 + 0.25^2)^{0.5} = \mathbf{0.45\ T}$.
Angle to horizontal θ: $\tan \theta = 0.38/0.25$, $\boldsymbol{\theta} = \mathbf{57°}$.
6 Take the wire:
at X, B (from the right-hand grip rule) acts vertically upwards:
$B = (\mu_0 I)/(2\pi a)$
$B_V = (4\pi \times 10^{-7} \times 10)/(2 \times \pi \times 0.015)$, $= 1.3 \times 10^{-4}$ T.
Take the solenoid: At X, B acts horizontally:
$B = \mu_0 nI$
$B_H = 4\pi \times 10^{-7} \times 200 \times 0.6 = 1.5 \times 10^{-4}$ T.
Pythagoras's theorem gives
$B_R = ((1.3 \times 10^{-4})^2 + (1.5 \times 10^{-4})^2)^{0.5}$
$\mathbf{B_R = 2.0 \times 10^{-4}\ T}$.
Angle to horizontal θ: $\tan \theta = 1.3/1.5$, $\boldsymbol{\theta} = \mathbf{41°}$.
7 $B = (8\mu_0 NI)/(125^{0.5} r)$
$B = (8 \times 4\pi \times 10^{-7} \times 1000 \times 6.0)/(125^{0.5} \times 0.050)$
$\mathbf{B = 0.11\ T}$.
$A = \pi r^2 = \pi \times 0.050^2 = 7.9 \times 10^{-3}\ m^2$.
$\phi = BA = 0.11 \times 7.9 \times 10^{-3} = \mathbf{8.7 \times 10^{-4}\ Wb}$.
8 $B = (\mu_0 NIa^2)/(2(a^2 + x^2)^{1.5})$
$B = (4\pi \times 10^{-7} \times 50 \times 6.0 \times 0.20^2)/(2(0.20^2 + 0.50^2)^{1.5})$
$B = (1.5 \times 10^{-5})/0.31 = \mathbf{4.8 \times 10^{-5}\ T}$.
For the graph work out a series of values for B.
When $x = 0$,
$B = (4\pi \times 10^{-7} \times 50 \times 6.0 \times 0.20^2)/(2(0.20^2)^{1.5}$
$B = (1.5 \times 10^{-5})/0.016 = 9.4 \times 10^{-4}$ T.
When $x = 0.25$,
$B =(4\pi \times 10^{-7} \times 50 \times 6.0 \times 0.20^2)/(2(0.20^2 + 0.25^2)^{1.5})$
$B = (1.5 \times 10^{-5})/0.066 = 2.3 \times 10^{-4}$ T.
When $x = 1.0$,
$B = (4\pi \times 10^{-7} \times 50 \times 6.0 \times 0.20^2)/(2(0.20^2 \times 1.0^2)^{1.5})$
$B = (1.5 \times 10^{-5})/2.1 = 7.1 \times 10^{-6}$ T.

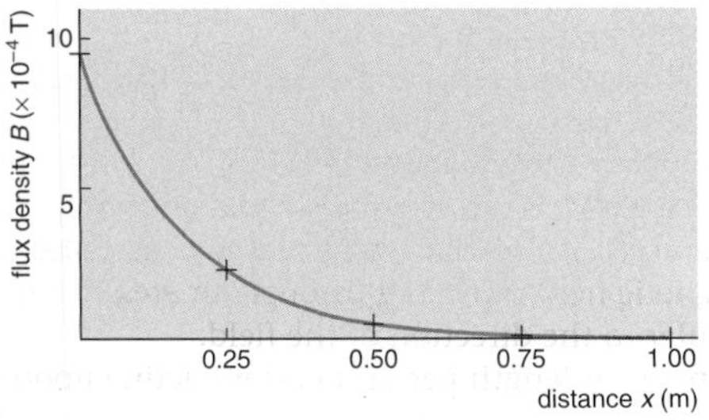

UNIT 22

RECALL TEST

1 1 tesla is the magnetic flux density that will produce a force of 1 N on a wire 1 m in length, carrying a current of 1 A, and perpendicular to the direction of the field.
2 It is used for determining the direction of the force acting on a current-carrying conductor lying in a magnetic field.
3 One ampere is the current flowing in each of two infinitely long parallel wires of negligible cross-sectional area, 1 metre apart in a vacuum, that produces a force of 2.0×10^{-7} N per metre of length.
4 The length, the size of the current, the flux density, and the angle between the conductor and the field.
5 The force is attractive.
6 A split-ring commutator is used in a d.c. motor, and a slip-ring commutator is used in an a.c. motor.
7 By using either a slip-ring or a split-ring commutator.
8 It moves into a circular path.
9 The flux density, the size of the current, the number of turns of coil, and the area of the coil.
10 It is a device that only allows charged particles travelling at a particular velocity to pass through it.

CONCEPT TEST

1 $F = BIl = 1.2 \times 10^{-4} \times 2.0 \times 3.0 = 7.2 \times 10^{-4}$ N.
$F = BIl \sin \theta = 1.2 \times 10^{-4} \times 2.0 \times 3.0 \times \sin 30°$
$\mathbf{F = 3.6 \times 10^{-4}\ N}$.
2 $F = BQv = 2.0 \times 10^{-2} \times 1.6 \times 10^{-19} \times 1.0 \times 10^2$
$\mathbf{F = 3.2 \times 10^{-19}\ N}$.
Initially it is perpendicular to the original direction of motion, and the direction is given by Fleming's left-hand rule, where the direction of current is 'conventional', i.e the opposite direction to electron motion.
3 $B_1 = (\mu_0 I)/(2\pi a) = (4\pi \times 10^{-7} \times 4.0)/(2 \times \pi \times 0.25)$
$B_1 = 3.2 \times 10^{-6}$ T.
$B_2 = (\mu_0 I)/(2\pi a) = (4\pi \times 10^{-7} \times 7.0)/(2 \times \pi \times 0.25)$
$B_2 = 5.6 \times 10^{-6}$ T.
The fields act in opposite directions, so
$B_R = (5.6 \times 10^{-6}) - (3.2 \times 10^{-6}) = \mathbf{2.4 \times 10^{-6}\ T}$.
The wires move towards each other.
4 $F/l = (\mu_0 I_1 I_2)/(2\pi a)$
$F/l = (4\pi \times 10^{-7} \times 6.0 \times 6.0)/(2 \times \pi \times 3.0 \times 10^{-3})$,
$\mathbf{F = 2.4 \times 10^{-3}\ N/m}$.
The force is repulsive, because the currents are flowing in opposite directions.

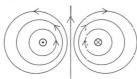

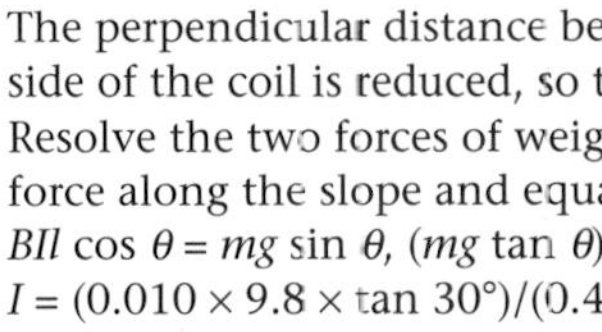

5 $\boldsymbol{T} = BIAN \cos \theta = 4.0 \times 10^{-2} \times 6.0 \times 0.30 \times 200 \times 1.0$
$\mathbf{T = 14\ N\ m}$.
$\boldsymbol{T} = BIAN \cos \theta = 4.0 \times 10^{-2} \times 6.0 \times 0.30 \times 200 \times \cos 30°$
$\mathbf{T = 12\ N\ m}$.
The perpendicular distance between the forces on either side of the coil is reduced, so the torque is reduced.
6 Resolve the two forces of weight and the magnetic field force along the slope and equate them.
$BIl \cos \theta = mg \sin \theta$, $(mg \tan \theta)/Bl = I$,
$I = (0.010 \times 9.8 \times \tan 30°)/(0.40 \times 3.0) = \mathbf{47\ mA}$.
7 The force always acts perpendicular to the direction of motion; this causes the particle to change its direction of motion and move into a circular path.
$F = BQv = 4.0 \times 10^{-3} \times 2 \times 1.6 \times 10^{-19} \times 20 = 2.6 \times 10^{-20}$ N.
$T = 2\pi m/BQ$
$= (2 \times \pi \times 4 \times 1.7 \times 10^{-27})/(4.0 \times 10^{-3} \times 2 \times 1.6 \times 10^{-19})$
$\mathbf{T = 3.3 \times 10^{-5}\ s}$.

8 A velocity selector is a device that only allows charged particles travelling at a particular speed to pass through undeflected. Their value of charge does not matter.
$E = V/d = 10\,000/0.040 = 2.5 \times 10^5$ V/m.
$v = E/B = (2.5 \times 10^5)/(3.0 \times 10^{-3}) =$ **8.3×10^7 m/s**.

UNIT 23

RECALL TEST

1 It is the production of electricity by moving a conductor through a magnetic field or by changing the magnetic field passing through a conductor.
2 The size of an induced e.m.f. is directly proportional to the rate of change of flux linkage that produced it.
3 The direction of an induced e.m.f. is such that it opposes the change that produced it.
4 It is the product of the flux passing through a coil and its number of turns. $\Phi = N\phi$.
5 It is used to work out the direction of induced e.m.f. or current.
6 It is a device used for changing the voltage of an a.c. supply from higher to lower or lower to higher.
7 A small circular current induced in the soft iron core of a transformer.
8 If the voltage is high ,the current will be low, so there will be less power dissipated because of the resistance of the cables. ($P = I^2R$).
9 An alternator is a device that produces a.c. electricity.
10 When a conductor has a changing current passing through it and induces an opposing e.m.f. in itself.

CONCEPT TEST

1 $E = \Delta(BAN)/\Delta t = (6.0 \times 10^{-3} \times 0.080 \times 50)/0.50$
$E = 4.8 \times 10^{-2}$ V.
$V = IR$, $I = V/R = (4.8 \times 10^{-2})/20 =$ **2.4 mA**.
2 $E = -Blv = -4.0 \times 10^{-4} \times 0.10 \times 0.050 =$ **-2.0×10^{-6} V**.
From the right-hand dynamo rule, current flows from X to Y, so X is positive.
3 $E = -L\Delta I/\Delta t = -(10 \times 10^{-3} \times 3.0)/0.020 = -1.5$ V.
p.d. = 2.0 – 1.5 = **0.5 V**.
4 $V_p/V_s = N_p/N_s$
$240/12 = 2000/N_s$, **$N_s = 100$**.
$P = IV = 0.050 \times 240 = 12$ W.
P out $= 12 \times 0.88 = 10.56$ W.
$P = IV$, $10.56 = I \times 12$, **$I = 0.88$ A**.
5 Total resistance of cable, $R = 100 \times 50 = 5000\ \Omega$.
Current in cables I:
$P_{in} = IV$, $2.0 \times 10^6 = I \times 4.0 \times 10^5$, $I = 5.0$ A.
$P = I^2R = 5.0^2 \times 5000 = 1.3 \times 10^5$ W.
Power output,
$P_{out} = (2.0 \times 10^6) - (1.3 \times 10^5) =$ **1.9×10^6 W**.
To lower the current and so reduce power loss. ($P = I^2R$.)
6 The initial current is zero so back e.m.f. equals the supply.
$V_B = 12$ V.
$E = -L\Delta I/\Delta t$, $12 = 2.5 \times 10^{-3}(\Delta I/\Delta t)$,
$\Delta I/\Delta t = 4800$ A/s.
$I = V/R = 12/200 =$ **6.0×10^{-2} A**.
$E = -L\Delta I/\Delta t = 2.5 \times 10^{-3} \times 3000 = 7.5$ V.
$V = 12 - 7.5 =$ **4.5 V**.
7 As the motor coil rotates it cuts through the magnetic field around it and induces an e.m.f. in itself, which opposes the rotation. This 'back' e.m.f. increases as it speeds up until it is just less than the supply e.m.f., at which point the motor moves at a constant speed. (The difference is equal to the potential difference across the resistance of the wire.) If it is jammed there is no back e.m.f., so the current is greater and the energy dissipated by the resistance of the wires is a lot greater ($P = I^2R$), so it gets very hot.
8 a **Anticlockwise**. When the switch S is closed current flows around loop B in a clockwise direction. This produces a magnetic field going into the paper in the centre of the loop. (Use the right-hand grip rule to show this.) So there is a change from no field to a field going into the paper. A current is induced in A that opposes the change that produced it, so it must produce a field coming out of the paper in the centre of the loop B. To do this the current must flow **anticlockwise in A** for a short time until the field is no longer changing.
b **Clockwise**. Before the switch is opened there is a steady clockwise current flowing, producing a steady magnetic field acting into the paper in the centre of the coil. When the switch is opened this field disappears, so a current is induced in A which tries to oppose the change and maintain the field. It must therefore flow in a **clockwise direction**.
c **Clockwise**. There is a steady field in the centre of B acting into the paper. This is like a thin solenoid with a south pole on top and a north pole on the bottom. When A is moved upwards a current is induced in it that tries to oppose the change that produced it, so it must have a south pole on top and a north pole on the bottom. Therefore the current must flow in a **clockwise direction in A**.
9 $E = -Blv = -2.0 \times 10^{-4} \times (\cos 30°) \times 10 \times 200 =$ **-0.35 V**.
No, because any wire connecting the two wing tips will also have an identical e.m.f. induced in it, which will cancel out the e.m.f. between the wing tips.
Max e.m.f. is given by
$E = -Blv = -2.0 \times 10^{-4} \times 10 \times 200 =$ **-0.40 V**.
This occurs when the aircraft is at 150°, and at 330°.
Minimum e.m.f. of 0 V occurs when the aircraft is flying parallel to the field at 60° and at 240°.

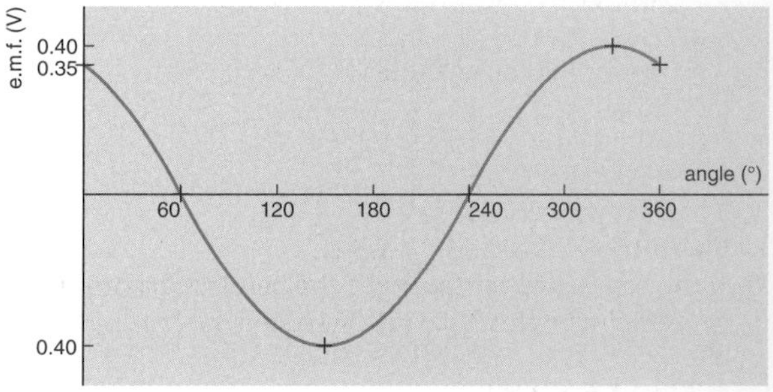

UNIT 24

RECALL TEST

1 SHM occurs when an object is oscillating so that its acceleration is directly proportional to its displacement from a fixed point and is always directed towards that fixed point.
2 Where displacement from the fixed point is greatest.
3 It is greatest at the centre of the oscillation.
4 It is the frequency with which it will oscillate at greatest amplitude.
5 When the frequency of something which is causing an object to oscillate corresponds to the natural frequency of that object.
6 When displacement is a maximum.
7 When displacement is zero, at the centre of the oscillation where velocity is greatest.
8 It must be able to have KE and PE.
9 Measuring time.
10 When critical damping occurs a displaced object returns to its equilibrium position and stops, without oscillating.

CONCEPT TEST

1 $\omega = 2\pi/T = 2\pi/2.0 = \pi$.
$v_{max} = \omega A = \pi \times 0.030 =$ **0.094 m/s.**
$a_{max} = -\omega^2 A = -\pi^2 \times 0.030 =$ **−0.30 m/s².**
At 1 cm from centre:
$a = -\omega^2 x = -\pi^2 \times 0.010 =$ **−0.099 m/s².**

2 $T = 2\pi(l/g)^{0.5} = 2\pi(2.0/9.8)^{0.5} =$ **2.8 s.**
$\omega = 2\pi/T = 2\pi/2.8 = 2.24$ rad/s.
$v = \omega A = 2.2 \times 0.20 = 0.45$ m/s.
$KE = \frac{1}{2}mv^2 = \frac{1}{2} \times 0.20 \times 0.45^2$, **$E$ = 0.020 J.**

3 $T = 2\pi(m/k)^{0.5} = 2\pi(0.50/12)^{0.5} = 1.3$ s.
$f = 1/T =$ **0.78 Hz.**
$\omega = 2\pi/T = 2\pi/1.3 = 4.8$ rad/s.
$a = -\omega^2 A = -4.8^2 \times 0.12 =$ **−2.8 m/s².**

4 $T = 2\pi(m/k)^{0.5} = 2\pi(2.0/25)^{0.5} = 1.8$ s.
$\omega = 2\pi/T = 2\pi/1.8 = 3.5$ rad/s.
$v = \omega A = 3.5 \times 0.020 =$ **0.070 m/s.**
$PE = \frac{1}{2}m\omega^2 x^2 = \frac{1}{2} \times 2.0 \times 3.5^2 \times 0.020^2$
PE = 4.9 × 10⁻³ J.

5 $T = 2\pi(m/k)^{0.5} = 2\pi(2.0/(4.0 \times 10^5))^{0.5} = 0.014$ s.
Rev/s = f, $f = 1/T$, rev/s = **71 Hz.**
$\omega = 2\pi/T = 2\pi/0.014 = 450$.
$KE = \frac{1}{2}m\omega^2 x_0^2 = \frac{1}{2} \times 2.0 \times 450^2 \times 0.050^2$
KE = 510 J.

6 **0 m/s²**, because it is at the centre.
Or $a = -\omega^2 x = -\omega^2 \times 0 = 0$ **m/s².**
$T = 2\pi(l/g)^{0.5} = 2\pi(2.0/9.8)^{0.5} = 2.8$ s.
$\omega = 2\pi/T = 2\pi/2.8 = 2.2$ rad/s.
$v = \omega A = 2.2 \times 0.10 =$ **0.22 m/s.**
Time from A to B is 1/4 of T, which is 0.70 s. From B to C use $x = A \sin \omega t$, and work in radians.
$\sin^{-1}(0.05/0.10) = 2.2t$, $t = 0.24$ s.
$t_{AC} = 0.24 + 0.70 =$ **0.94 s.**

7 $F = kx$, $k = (0.15 \times 9.8)/0.60 =$ **2.5 N/kg.**
$T = 2\pi(m/k)^{0.5} = 2\pi(0.15/2.5)^{0.5} =$ **1.5 s.**
$\omega = 2\pi/T = 2\pi/1.5 = 4.2$ rad/s.
$v = \omega A = 4.2 \times 0.60 =$ **2.5 m/s.**
$a = -\omega^2 x = -4.2^2 \times 0.020 = -0.35$ m/s².
Resultant upward force is given by
$F_R = ma = 0.15 \times 0.35 = 0.053$ N.
$T - W = F$, $T = F + W$
$T = (0.15 \times 9.8) + 0.053 =$ **1.5 N.**

8 If the bags leave contact with the belt the maximum acceleration acting upwards must be equal to gravity.
$a = -\omega^2 A$, $-9.8 = -\omega^2 \times 0.040$, $\omega = 15.6 = 16$.
$\omega = 2\pi f$, **f = 2.5 Hz.**
$t = 1/f = 0.40$.
$v = s/t = 0.20/0.40 =$ **0.50 m/s.**

UNIT 25

RECALL TEST

1 It is a measure of how hot or cold something is.
2 It the energy that will flow between two objects at different temperatures.
3 The energy required to raise the temperature of one kilogram of a material by 1 °C.
4 The energy required to change the state of one kilogram of a material from liquid to gas.
5 The pressure of a gas is inversely proportional to its volume if the temperature is constant. pV = constant or $p_1V_1 = p_2V_2$.
6 The volume of a gas is directly proportional to its absolute temperature if the pressure is constant. V/T = constant or $V_1/T_1 = V_2/T_2$.
7 The pressure of a gas is directly proportional to its absolute temperature if the volume is constant. P/T = constant or $P_1/T_1 = P_2/T_2$.
8 At low pressures and high temperatures.
9 Brownian motion and diffusion.
10 The amount of substance that contains as many units (atoms, molecules, or ions) as there are atoms in 0.012 kg of carbon-12.

CONCEPT TEST

1 $E = mc\Delta\theta = 4.0 \times 380 \times 55 =$ **8.4 × 10⁴ J.**

2 $IVt = mc\Delta\theta$, $\Delta\theta = (2.0 \times 12 \times 12 \times 60)/(3.0 \times 500)$
$\Delta\theta = 11.5 = 12$ °C.
Final temp., $\theta = 20 + 12 = 32 =$ **32 °C.**

3 A car has four wheels and four brakes.
KE lost = heat gained
$\frac{1}{2}Mv^2 = mc\Delta\theta$, $0.5 \times 1400 \times 30^2 = 4 \times 26 \times 600 \times \Delta\theta$
$\Delta\theta$ = 10 °C.
Assume that all the kinetic energy is converted into heat energy in the brakes.

4 PE lost = heat gained, $mgh = mc\Delta\theta$
$gh = c\Delta\theta$, $\Delta\theta = (9.8 \times 400)/4200 =$ **0.93 °C.**
This is too small a difference to detect, relative to other factors such as weather conditions and sunlight.

5 Heat lost = heat gained, heat lost by water + heat lost by container = heat used to melt ice + heat used to increase temperature of water from melted ice:
$m_w c_w \Delta\theta_1 + m_c c_c \Delta\theta_1 = m_i L_i + m_i c_w \Delta\theta_2$
$(0.30 \times 4200 \times (21 - \theta)) + (0.10 \times 400 \times (21 - \theta))$
$= (0.20 \times 3.3 \times 10^5) + (0.20 \times 4200 \times (\theta - 0.0))$
$1260(21 - \theta) + 40(21 - \theta) = 660 + 840(\theta - 0.0)$
$1300(21 - \theta) = 660 + 840\theta$
$27\,300 - 1300\theta = 660 + 840\theta$
$2140\theta = 26\,640$, $\theta = 12.44 =$ **12 °C.**

6 $p_1V_1/T_1 = p_2V_2/T_2$
$(4.5 \times 10^3 \times 4.0)/298 = (p_2 \times 1.0)/329$,
p_2 = 2.0 × 10⁴ Pa.

7 $pV = nRT$
$2.4 \times 10^4 \times 10 = n \times 8.3 \times 300$, **$n$ = 96 mol.**
$N = N_A \times n = 6.0 \times 10^{23} \times 96 =$ **5.8 × 10²⁵ atoms.**
Mass, $M = n \times$ molar mass
$M = 96 \times 0.029 =$ **2.8 kg.**

8 $(c^2)^{0.5} = (3p/\rho)^{0.5}$, $(c^2)^{0.5} = ((3 \times 2.0 \times 10^3)/0.067)^{0.5}$
$(c^2)^{0.5} =$ **300 m/s.**

9 Rate of heat loss by body = rate of heat gain by evaporating water.
$Mc\Delta\theta/\Delta t = \Delta mL/\Delta t$
$60 \times 4200 \times 0.75 = 2.4 \times 10^6 \Delta m/\Delta t$, **$m/\Delta t$ = 80 g/s.**

10 If the change is isothermal, pV = constant:
$2.0 \times 10^2 \times 2.0 = 4.0 \times 10^5$,
$1.0 \times 10^5 \times 4.0 = 4.0 \times 10^5$,
$0.5 \times 10^5 \times 8.0 = 4.0 \times 10^5$, etc.
Therefore the change is isothermal.
At A, $pV = nRT$
$5.0 \times 10^5 \times 2.0 \times 10^{-2} = 4 \times 8.3 \times T$, **$T$ = 300 K.**
At B, $pV = nRT$
$3.0 \times 10^5 \times 2.0 \times 10^{-2} = 4 \times 8.3 \times T$, **$T$ = 180 K.**
At C, $pV = nRT$
$2.0 \times 10^5 \times 8.0 \times 10^{-2} = 4 \times 8.3 \times T$, **$T$ = 480 K.**
At D, $pV = nRT$
$1.0 \times 10^5 \times 8.0 \times 10^{-2} = 4 \times 8.3 \times T$, **$T$ = 240 K.**
The maximum temperature occurs at C and is 480 K.

UNIT 26

RECALL TEST

1 If object A is in thermal equilibrium with object B and object B is in thermal equilibrium with object C, then object A is in thermal equilibrium with object C.
2 The change in internal energy of a system is equal to the sum of the changes of heat energy and work done on or by the system. $\Delta U = \Delta Q + \Delta W$.
3 A heat engine can never be 100% efficient.
4 The sum of a system's atoms' kinetic and potential energies.
5 A change at constant temperature.
6 A change during which no heat energy enters or leaves the system.
7 A change at constant pressure.
8 The heat energy required to raise the temperature of one mol of gas by 1 K at constant volume.
9 A device that converts heat energy into mechanical energy.

10 When a gas is heated at constant pressure it expands, therefore some extra energy is required to do work against the surroundings.

CONCEPT TEST

1 $\theta = 100 \times (X_1 - X_0)/(X_{100} - X_0)$
$\theta = 100 \times (42 - 20)/(100 - 20) =$ **28 °C**.

2 $\theta = 100 \times (X_1 - X_0)/(X_{100} - X_0)$
$\theta = 100 \times ((4.2 \times 10^6) - (1.0 \times 10^5))/((2.4 \times 10^6) - (1.0 \times 10^5))$
$\boldsymbol{\theta} =$ **180 °C**.

3 If $\Delta W = 0$, $\Delta U = \Delta Q$, $\Delta Q = nC_V\Delta K$
$\Delta Q = 2.00 \times 20.8 \times 161 =$ **6700 J**.
No work is done because the gas does not expand; volume is constant.

4 a Heat the gas or compress it.
b ΔQ is +ve because heat energy goes in.
ΔU is +ve because the temperature increases, so the atoms' KE increases.
ΔW is –ve because the gas expands.
c ΔQ is +ve because heat energy goes in.
ΔU is +ve because the temperature increases so the atoms' KE increases.
ΔW is 0 because the gas does not expand.

5 $\Delta Q = nC_p\Delta K = 4.5 \times 29.1 \times 330 = 43$ kJ.
$\Delta W = p\Delta V = 1.6 \times 10^4 \times 1.5 = 24$ kJ.
ΔW is negative because the gas expands.
$\Delta U = \Delta Q + \Delta W = 43\,000 - 24\,000 =$ **19 kJ**.

6 A–B is an isobaric compression.
B–C is an isothermal expansion.
C–A is an adiabatic compression.

7 Adiabatic change, so $\Delta Q = 0$, so $\Delta U = \Delta W$:
$\Delta U = nC_p\Delta T = 3.5 \times 29.1 \times 367 = 37.4$ kJ.
$\Delta W = p\Delta V$, $\Delta V = 37400/(2.4 \times 10^3)$, $\boldsymbol{\Delta V = 16\ \mathrm{m}^3}$.

8 The fridge as a system:
ΔQ is –ve because heat energy goes out.
ΔU is 0, because the temperature is constant so the atoms' KE is constant.
ΔW is +ve because work is being done on the system.
The room as a system:
ΔQ is +ve because heat energy is going in.
ΔU is +ve because the temperature is increasing.
ΔW is 0 because no work is being done on the system, as the volume of the room is constant.

9 Efficiency $= 100 \times (T_1 - T_2)/T_1$
Efficiency $= 100 \times (473 - 323)/473$
Efficiency = **32%**.

UNIT 27a

RECALL TEST

1 The extension of a wire is directly proportional to the force applied to it.
2 Stress is the force per unit cross-sectional area applied to an object (stress = force/area).
3 Strain is the fractional increase in length of an object being stretched (strain = extension/length).
4 The Young modulus is equal to the ratio of stress to strain. The Young modulus = stress/strain.
5 If an object shows elastic deformation it will return to its original shape when the applied force is removed.

CONCEPT TEST

1 $F = k\Delta l$, $k = 9.8 \times 0.200/0.04 =$ **49 N/m**.
Energy, $E = \frac{1}{2} \times F \times \Delta l = \frac{1}{2} \times 9.8 \times 0.20 \times 0.04$
$\boldsymbol{E} =$ **0.039 J**.

2 Stress, $\sigma = F/A = (9.8 \times 40)/(3.0 \times 10^{-6}) = 1.31 \times 10^8$.
Young modulus, E = stress/strain
$2.0 \times 10^{11} = 1.31 \times 10^8/\varepsilon$,
strain, $\varepsilon = 6.53 \times 10^{-4}$.
$\varepsilon = \Delta l/l$, $\Delta l = 2.0 \times 6.53 \times 10^{-4} =$ **1.3 mm**.

3 $E = (F/A)/(\Delta l/l) = (Fl)/(\Delta lA)$, $F = (E\Delta lA)/l$.
The force is the same in each wire so
$(E\Delta lA/l)_s = (E\Delta lA/l)_b$
$(2.0 \times 10^{11} \times \Delta l_s)/2.2 = (9.2 \times 10^{10} \times \Delta l_b)/1.5$
$1.48\Delta l_s = \Delta l_b$
$\Delta l_T = \Delta l_s + \Delta l_b$, $0.05 = \Delta l_s + 1.48\Delta l_s$, $0.05 = 2.48\Delta l_s$
$\boldsymbol{\Delta l_s}$ **= 0.02 m**, $\boldsymbol{\Delta l_b}$ **= 0.03 m**.

4 One square equals $0.2 \times 0.1 = 0.02$ J and there are 15.5 squares:
$0.02 \times 15.5 =$ **0.31 J**.

5
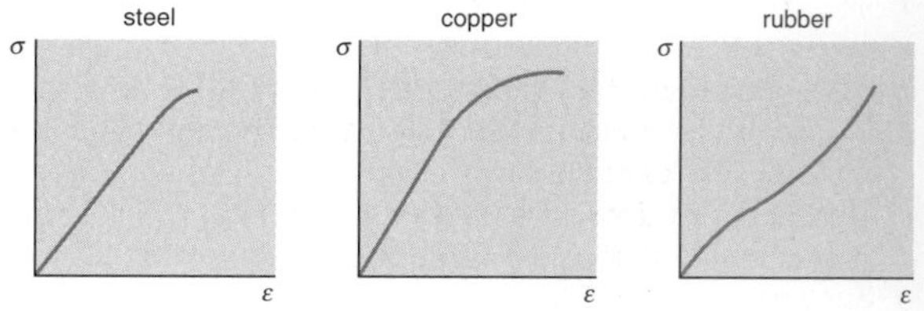

UNIT 27b

RECALL TEST

1 It is one of the fundamental units from which other units may be derived.
2 Dimensions represent quantities which have one of the fundamental base units.
3 Each term must have the same base units or dimensions.
4 Some constants have base units in order to make the equations that contain them homogenous.

CONCEPT TEST

1 a Power = work done/time: use $\frac{1}{2}mv^2/t$, which gives kg m^2 s^{-2}/s, **kgm^2 s^{-3}**.
b p.d. = work done/charge: use $\frac{1}{2}mv^2/It$, which gives kg m^2 s^{-2}/(A s), **kg A^{-1} m^2 s^{-3}**.
c Resistance = p.d./current, kg A^{-1} m^2 s^{-3}/A, **kg A^{-2} m^2 s^{-3}**.
d $E = F/Q$, $E = ma/It$, kg m s^{-2}/(A s), **kg m s^{-3} A^{-1}**.
e $F = BIl$, $B = F/(Il)$, kg m s^{-2}/(A m), **kg A^{-1} s^{-2}**.
f $C = Q/V$, $C = It/V$, A s/(kg A^{-1} m^2 s^{-3}), **kg^{-1} A^2 m^{-2} s^4**.

2 a $A = -\lambda N$, $\lambda = A/N$, **s^{-1}**.
b $E = hf$, $h = E/f$, kg m^2 s^{-2}/s^{-1}, **kg m^2 s^{-1}**.
c $\Delta Q = mc\Delta\theta$, $c = \Delta Q/m\Delta\theta$, kg m^2 s^{-2}/kg K, **m^2 s^{-2} K^{-1}**.
d $pV = nRT$, $R = pV/nT = (F/A)V/(nT)$,
(kg m s^{-2}/m^2) m^3/mol K, **kg m^2 s^{-2} mol^{-1} K^{-1}**.
e $F = Q_1Q_2/4\pi\varepsilon r^2$, $\varepsilon = Q_1Q_2/(4\pi Fr^2)$,
A^2 s^2/(kg m s^{-2} m^2), **A^2 s^4 kg^{-1} s^2 m^{-3}**.
f $B = \mu_0 I/2\pi a$, $\mu_0 = B2\pi a/I$, kg A^{-1} s^{-2} m/A, **kg A^{-2} s^{-2} m**.

3 a $v^2 = u^2 + 2as$, (m s^{-1})^2 = (m s^{-1})^2 + (m s^{-2})m,
m^2 s^{-2} = m^2 s^{-2} + m^2 s^{-2}.
b $T = 2\pi(l/g)^{0.5}$, s = (m/m s^{-2})^2, s = s.
c $v = f\lambda$, m s^{-1} = m s^{-1}.
d $F = (mv - mu)/t$, kg m s^{-2} = (kg m s^{-1} + kg m s^{-1})/s,
kg m s^{-2} = kg m s^{-2} + kg m s^{-2}.
e $I = nAve$, A = m^{-3} m^2 m s^{-1} A s, A = A.
f $W = \frac{1}{2}CV^2$, kg m^2 s^{-2} = kg^{-1} A^2 m^{-2} s^4 (kg A^{-1} m^2 s^{-3})^2,
kg m^2 s^{-2} = kg m^2 s^{-2}.

UNIT 28

RECALL TEST

1 Scientific method is a process where a hypothesis is formulated, then tested experimentally, and then changed if necessary in the light of the results.
2 To find the values of constants, to verify laws, and to find the relationship between variables.
3 It is a variable which is affected or changes when another variable is changed by the experimenter.
4 It is a variable which is not affected by the change of another variable, or is one of the variables changed by the experimenter.
5 A 'fair test' is the name given to an experiment in which all the independent variables are kept constant except one, which is changed by the experimenter.

6 A 'control' is the term used to describe an experimental set-up which is identical to the main experiment. No variables are changed in the control, but the two experiments are compared. The effects on variables (due solely to any changes in the environment or surroundings) which are seen in the control are subtracted from changes in the actual experiment.

7 Any form of error due to the measuring devices or equipment.

8 An error due to the experimenter taking readings incorrectly, or random fluctuations in the environment affecting the experiment.

9 There should always be the same number of significant figures in the derived results as there are in the experimental readings.

10 A line of best fit is a line drawn through a series of points which represents the general trend of the data. The sum of perpendicular distances of the points to the line on one side of the line should equal the sum of those on the other side.

11 Gradient and intercept can be determined, which can give the values of constants; or the relationship between two variables may be determined.

12 Logs are taken to change curves into straight lines and thereby enable the values of constants or powers (indices) to be determined.

13 Datalogging is a system in which electronic sensors and computers are used to gather and process data.

14 A constant systematic error will mean that all values of data will be out by the same amount and will affect values derived from y intercepts, but not gradients.

15 The effect of a random error is reduced by repeating readings, taking averages, and plotting graphs.

INDEX

A

absolute zero 99, 102
absorption line spectra 55
acceleration–time graph 6
acceleration 3
adiabatic change 102
alpha decay 58
alternating current 87, 90, 91
ampere 87
angular displacement 30
angular velocity 30
anti-particles 62
area under graph 6
atomic number 58

B

band spectra 55
baryon number 66
baryons 63, 66
base units 107
beta decay 58, 62
Big Bang 55
binding energy 59
binding energy per nucleon 59
Bohr model of the atom 51, 54, 58
bosons 63
Boyle's law 98
brittle 106
Brownian motion 99
bubble chamber 67

C

capacitance 78
capacitor half-life 79
capacitors in series and in parallel 78
centre of gravity 14
centre of mass 14
centripetal acceleration 30
centripetal force 30
chain reaction 59
characteristic curves 39
charge 34
charge and discharge of a capacitor 79
charge density 75
Charles's law 99
chemical potential energy 19
circular motion 26
cloud chamber 67
coefficient of restitution 27
coherent 46
compression 14
conservation of charge 38
conservation of energy 18, 38, 90, 102
conservation of momentum 26
constructive interference 43
control 110
Coulomb's law 70
couple 11
critical angle 46
critical mass 59
current 34

D

damping 95
datalogging 111
de Broglie waves 51
decay constant 58
deep inelastic scattering 67
density 2
destructive interference 43
diffraction 47
diffraction grating 47
diffraction of electron de Broglie waves 67
diffusion 99
dimensions 107
direct current 87
displacement 2
displacement–time graph 6
distance 2
Doppler effect 55
drag force 15
drift velocity 34
ductile 106
dynamos and alternators 90

E

eddy currents 90
efficiency 19
elastic behaviour in solids 106
elastic collisions 26
elastic energy 19
elastic limit 106
elastic potential energy 31
electric circuit 38
electric field potential 71
electric field strength 70
electric motors 88
electrical energy 19, 35
electrial power 35
electricity 34
electromagnetic force 15, 62
electromagnetic induction 90
electromagnetic waves 42
electromotive force (e.m.f.) 34
electron–positron pair 62
electron transitions 54
electronvolt 54
electrostatic field 70
electrostatics 70
emission line spectra 55
energy 18
energy in SHM 95
energy levels in atoms 54
energy stored in a capacitor 78
equating forces 27
equations of motion 3
equipotential 71
escape velocity 74
exchange particles 62
excitation energy 55

F

Faraday's law 90

fermions 63
Feynman diagrams 62
field line 71
fields 102
fields, similarities and differences between 75
fixed points 102
flavours 66
Fleming's left-hand rule 86
force 10, 11, 22
forced oscillation 95
free-body force diagram 15
free oscillation 95
frequency 42
friction 14
fundamental forces 15, 62
fundamental particles 63

G
gamma rays 58, 62
gauge bosons 62
gluon 66
gold-leaf electroscope 50
gradient of graph 6
graphs 6, 111
gravitational field 70
gravitational field potential 71
gravitational field strength 70
gravitational potential energy 31
gravity 70
gravity, variation inside and outside the Earth 75
ground state 55

H
hadrons 63, 66
half-life 58
heat 98
heat energy 19, 98
heat engine 103
Helmholtz coils 83
homogeneity in equations 107
Hooke's law 106
Hubble constant 55

I
ideal gas equation 99
impulse 27
inelastic collisions 26
inertia 2
interactions 62
internal energy 102
internal resistance 39
ionization energy 55
isobaric change 103
isothermal change 103
isotope 58

K
Kelvin scale 99, 102
kinetic energy 19, 26
kinetic theory of gases 99
Kirchhoff's first law 38
Kirchhoff's second law 38

L
latent heat 98
laws of thermodynamics 102
Lenz's law 90
lepton number 63
leptons 63
light energy 19, 54
limit of proportionality in solid deformation 106
longitudinal wave 42
lost volts 39

M
magnetic field 82
magnetic field forces 87
magnetic field line 82
magnetic flux 82
magentic flux density 82
mass defect 59
mass 2
mass number 58
mechanical waves 42
mesons 63, 66
mol 99
molar heat capacities 103
moment 11, 31
momentum 22, 26
monochromatic 49

N
natural frequency 95
neutrinos and anti-neutrinos 62
neutrons 58, 66
newton, definition of 19
Newton's law of gravitation 70
Newton's laws of motion 22
nodes and antinodes 47
normal contact or normal reaction 14
nuclear energy 19, 59
nuclear fission 59
nuclear fusion 59
nuclear radius 58
nucleon number 58

O
Ohm's law 35
orbital speed 74
orbital time period 74

P
P waves 42
permeability 83
permittivity 70, 78
phase angle 43
photoelectric effect 50
photons 50, 54
Planck constant 50
Planck's hypothesis 50
plane polarization 43
plastic behaviour in solids 106
potential difference 34
potential divider 39
potential energy 19, 102

potential energy in fields 74
potential gradient 71
power 30
pressure 10
pressure law 98
principle of moments 11, 31
progressive waves 42
projectile motion 3
protons 58, 66

Q
quantization of energy 54
quantum mechanics 62
quark numbers 67
quarks 63, 66

R
radioactivity 58
random error 110
rate of change of momentum 18
redshift 55
reflection 46
refraction 46
relative permittivity 78
relative velocity 2
resistance 35
resistivity 35
resistors in series and parallel 38
resonance 95
resultant force 18
right-hand grip rule 83
Rutherford's gold-leaf experiment 58

S
S waves 42
scalar 2
self-inductance 90
semiconductor 34
significant figures 7, 111
simple harmonic motion (SHM) 94
solenoid 83
sound energy 19
specific heat capacity 98
speed 3
spin 63
static equilibrium 31
standing or stationary wave 47
statics 27
stopping potential 51
strain 106
strangeness 66
stress 106
strong nuclear force 15, 62, 67
superconductors 35
systematic error 110

T
temperature scales 102
temperature 98, 102
tension 14
terminal velocity 10
tesla 86
the principle of superposition 43
thermometric property 102
threshold frequency 50
thrust force 15
time constant 79
time period 30, 42
torque 11
total internal reflection 46
transformers 90
transverse wave 42

U
uniform fields 75
universal gravitational constant 70
upthrust force 15

V
variables 110
vector 2
vector component 11
vector resultant 10
velocity 3
velocity selector 86
velocity–time graph 6

W
wave–particle duality 50
wave displacement 42
wavefront 47
wavelength 43
weak nuclear force 15, 62, 67
weight 14
work done 18, 90, 103
work function 50

Y
yield point 106
Young modulus of elasticity 106
Young's slits experiment 46